The Struggle for Accountability

The Struggle for Accountability
The World Bank, NGOs, and Grassroots Movements

edited by Jonathan A. Fox and L. David Brown

The MIT Press
Cambridge, Massachusetts and London, England

This book was set in Sabon on the Monotype "Prism Plus" PostScript Imagesetter by Asco Trade Typesetting Ltd., Hong Kong.

Printed and bound in the United States of America.

Library of Congress Cataloging-in-Publication Data

The struggle for accountability : the World Bank, NGOs, and grassroots
  movements / edited by Jonathan A. Fox and L. David Brown.
    p.   cm. — (Global environmental accord)
  Includes bibliographical references and index.
  ISBN 0-262-06199-6 (hc : alk. paper). — ISBN 0-262-56117-4
  (pbk. : alk. paper)
    1. Economic development—Environmental aspects—Developing
  countries. 2. Political participation. 3. World Bank. 4. Nongovernmental organizations. 5. Social movements—Developing
  countries. I. Fox, Jonathan A. II. Brown, L. David (L. Dave)
  III. Series: Global environmental accords.
  HD75.6.S77   1998
  332.1′532—dc21   98-10305
                                                                 CIP

Contents

Series Foreword

A new recognition of profound interconnections between social and natural systems is challenging conventional intellectual constructs as well as the policy predispositions informed by them. Our current intellectual challenge is to develop the analytical and theoretical underpinnings crucial to our understanding of the relationships between the two systems. Our policy challenge is to identify and implement effective decision-making approaches to managing the global environment.

The Series on Global Environmental Accords adopts an integrated perspective on national, international, cross-border, and cross-jurisdictional problems, priorities, and purposes. It examines the sources and consequences of social transactions as these relate to environmental conditions and concerns. Our goal is to make a contribution to both the intellectual and the policy endeavors.

Preface and Acknowledgments

This book is the product of ongoing exchanges between scholars, environmental activists, and development practitioners. The chapter authors include researchers with activist experience and activists with research experience. The editors combine experience as university-based researchers, as collaborators with diverse nongovernmental organizations (NGOs) in both North and South, and as consultants to international development agencies, including the World Bank.

Balancing the exchanges among scholarly and activist perspectives is often a very delicate process. Our efforts to develop a balanced assessment have been greatly aided by the advice of Chad Dobson and Kay Treakle of the Bank Information Center, a Washington-based public interest group concerned with Bank-NGO relations, and by the insights of Aubrey Williams and John Clark, successive directors of the NGO Unit of the World Bank. Where we have failed to achieve that balance, of course, the responsibility lies with ourselves rather than those who have sought to advise us.

The book was made possible by support from many sources. Initial studies of project campaigns and a first conference of case writers were supported by grants from the Ford Foundation and the Social Science Research Council. Subsequent studies of Bank policies and a second case-writer conference were made possible by a grant from the Environment Program of the Charles Stewart Mott Foundation. The Ford and C.S. Mott Foundations have also contributed directly to many of the initiatives whose work is documented in these pages.

Professor Fox's early work in this project was also supported by the Massachusetts Institute of Technology's Center for International Studies, which hosted the first research workshop. Professors Nazli Choucri and Hayward Alker of the MIT Political Science Department provided critical intellectual encouragement early on. Professor Fox's later editorial work and research for the concluding chapters were supported in part by a research grant from the John D. and Catherine T. MacArthur Foundation's Program on Peace and International Cooperation, as well as a Council on Foreign Relations International Affairs Fellowship, which permitted him to spend a year as a visiting researcher at the Bank Information Center. Professor Brown has been supported by the Department of Organizational Behavior and the School of Management at Boston University and by the Institute for Development Research (IDR).

The Institute for Development Research has provided invaluable institutional and logistical support throughout the project. The project's initial conception and its evolving directions owe much to the advice and support of Jane Covey, IDR's executive director. A project of this complexity—with authors in many countries and chapters that have evolved as the events and institutions on which they are based have changed—makes great administrative and editing demands. Maria Hernandez, Marian Doub, Sheara Cohen, Johanna Martinez, and Dane Machado have made heroic efforts to cope with a myriad of chapter drafts, software discrepancies, and complexities created by the intensive travel schedules of editors on two coasts.

The studies were greatly improved by insightful feedback from diverse readers, most of whom are acknowledged in individual chapters. Five anonymous external reviewers for MIT Press provided thoughtful and provocative feedback that influenced subsequent drafts of the manuscript. The editors are especially grateful to John Clark of the World Bank for his incisive comments on earlier drafts, Jonathan Schlefer and Jennifer Smith for their support in developing the chapters on Indonesia and the Philippines, and IDR Research Associate John Gershman for his insightful analytical and editorial input throughout the project.

This volume examines efforts to build linkages among actors with very different perspectives and interests—social activists, nongovernmental organizations, national governments, and the World Bank—to influence large-scale development programs over decades. The project on which this book is based has itself involved a five-year process, a process that has imposed substantial burdens on our families. We want to acknowledge and express our gratitude for the support of our spouses, Helen and Jane, and the tolerance of our children—Benjamin, and Rachel, and Nathan—during this venture.

Finally, we want to recognize the courage and tenacity of many grassroots groups, the sustained advocacy work of many nongovernmental organizations, and the quiet persistence of those World Bank staff committed to serious reform that together made possible the changes documented here. Our belief in the possibilities of just and sustainable social change has been refreshed by your examples.

Contributors

L. David Brown
President
Institute for Development Research
Boston

Jane G. Covey
Executive Director
Institute for Development Research
Boston

Jonathan A. Fox
Associate Professor of Social Sciences
Latin American and Latino Studies
University of California, Santa Cruz

Andrew Gray
Researcher
International Work Group on
Indigenous Affairs
Oxford, United Kingdom

Margaret E. Keck
Associate Professor of Political Science
Johns Hopkins University
Baltimore

Deborah Moore
Senior Scientist
Environmental Defense Fund
Oakland

Antoinette G. Royo
Environmental Lawyer
Legal Rights and Natural Resources
Center
Manila, Philippines

Augustinus Rumansara
Executive Director
YPPWI (Foundation for Enterprise
Initiative Development)
Jakarta, Indonesia

Leonard Sklar
Geology Department
University of California, Berkeley
Formerly Research Director
International Rivers Network

Kay Treakle
Co-Director
The Bank Information Center
Washington, D.C.

Lori Udall
Environmental Lawyer
Formerly Washington Director
International Rivers Network
Washington, D.C.

David A. Wirth
Associate Professor of Law
Washington and Lee University
Lexington, Virginia

1

Introduction

Jonathan A. Fox and L. David Brown

The World Bank has long been accused of "bankrolling disasters"—funding economically questionable megaprojects with devastating social and environmental costs.[1] Yet, as early as 1972 the World Bank was the first major international aid agency to declare that developmental and environmental goals were compatible.[2] Since 1987, its presidents have admitted past mistakes, and by the mid-1990s the World Bank claimed to be a leading force for "environmentally sustainable development." As the world's most influential international aid agency, which side is it really on?

Many environmental critics charge that the "greening" of the World Bank is largely a facade, and that its projects have changed little on the ground.[3] Social critics add that the Bank's commitment to technocratic, export-led growth models favors the rich and blocks more equitable and environmentally sustainable development alternatives. It is a multilateral *bank*, after all, formally accountable only to its member governments. World Bank defenders recognize that many of its projects have problems, but attribute them primarily to the borrower governments that carry them out. Bank officials often add that even "problem projects" would have been much worse without their involvement.

What began in the early 1980s as an international debate over rainforest road building and dam evictions became by the early 1990s a much broader effort to hold the World Bank more publicly accountable to civil societies in both donor and borrowing countries. For more than a decade, nongovernmental environmental and development organizations (NGOs) have formed diverse transnational advocacy coalitions, both within the Northern industrial countries and across the developing

South, including varying degrees of participation by grassroots movements of people directly affected by internationally funded projects.[4]
These campaigns have had an impact, most notably by pressuring donor
governments to encourage the World Bank to adopt more rigorous
environmental and social policies. Controversial projects that the World
Bank would probably have funded a decade ago are today much more
likely to be vetoed or modified in the design phase. Although its practice
continues to lag behind policy promises, the World Bank's reform policies are important because they create benchmark standards that public
interest groups can use to hold the institution accountable (standards
that the Bank's growing private sector investments escape, however).
Public debate continues over appropriate standards and institutions for
accountability, but the World Bank's own "sustainable development"
reforms make commitments that are ambitious compared to its own
practice, past and present.[5]

This book analyzes the origins of these policy reforms and the subsequent conflicts over how and whether to follow them in practice, an
international struggle for accountability that involves the World Bank,
donor and borrowing governments, nongovernmental public interest
organizations, and grassroots movements. It asks two specific questions.
First, how has the Bank responded to the NGO/grassroots environmental
and social critique? In other words, to what degree has protest had a
tangible impact on changing Bank and borrowing government behavior?
Some frame the question in dichotomous terms: has the World Bank
either "learned its lesson" or "failed to reform"? The debate is heated
and the stakes are high. Ironically, those critics who claim that no
meaningful change has occurred are in the position of devaluing their
own influence and creativity. If the Bank has not changed at all, then
NGO campaigns mounted over the last fifteen years have failed. This
volume's diverse collection of case studies of project and policy conflicts
attempts to assess *degrees* of change, based on the assumption that even
relatively small changes in the behavior of large, powerful institutions
matter a great deal to directly affected populations.[6] The conclusions
suggest that the Bank has to a small and uneven but significant degree
become more publicly accountable as the result of protest, ongoing
public scrutiny, and the empowering effect on insider reformists.

Second, to what degree have these advocacy campaigns, often led by NGOs, represented the organizations of people most directly affected by Bank projects? Most NGOs are intermediary organizations, not direct representatives of grassroots groups, so how have they bridged the vast differences in culture and power between Washington lobbyists and distant villagers? To what degree have NGO critics been accountable to their own coalition partners? Some—though not all—of the advocacy campaigns included a significant grassroots voice, especially those focused on specific projects. The more general policy reform campaigns, in contrast, tended to be dominated by international NGOs. At the same time, these broader policy reform efforts depended on the public concern and legitimacy generated by the more locally driven project-specific campaigns. Some NGOs learned how to buffer the vast cultural and power gaps between coalition partners by building "bridging organizations." The studies collected here found that, over time, transnational NGO advocacy networks have become more accountable to their local coalition partners—partly because of more vocal and autonomous grassroots movements, and partly in response to the Bank's challenge to the legitimacy of international NGO critics.

These uneven processes of change are illustrated by one of the most intense project controversies of the 1990s. Just before it was to be approved, the World Bank's new president canceled a proposed loan for Nepal's Arun III Hydroelectric Dam, shocking both the project's critics and supporters. As the case discussion in the concluding chapter argues, this unusually clear-cut victory showed that an emerging transnational NGO media and lobbying alarm system had gained a growing capacity to block questionable development projects *before* they were built. This partial veto power contrasts with most previous NGO and grassroots protests, which had led to only partial mitigation of project impact.

The cancellation of the Arun III Dam also opened a window on the broader political conflict between the World Bank, civil societies, and national governments over how to allocate resources in the name of development. The debate over the project was not just an exchange between the Bank and its NGO critics, but was mediated by the fragile and uneven process of democratization within Nepal, as well as by new World Bank institutions and policies created in response to previous protest and lobbying campaigns. Specifically, the decision to cancel Arun III cannot be explained without taking into account the Bank's newly bolstered reform policies regarding environmental impact assessment, public information access, involuntary resettlement, and indigenous peoples. In 1994, these policy commitments were reinforced by the Bank's creation of the Inspection Panel, a relatively autonomous official appeals channel designed to investigate claims made by project-affected people that World Bank reform policies have been violated.[7]

The Arun III case also revealed the importance of the *reciprocal interaction* between external critics and internal Bank dissidents. Neither insider advocates of environmental and social concerns within the Bank, nor external criticism alone were sufficient to defeat the Arun project; each reinforced the other, with the external critique tipping the balance in an internally divided Bank.[8] In contrast, when World Bank ranks have remained more unified—for example, in their belief that structural adjustment programs (SAPs) are the only legitimate solution to macroeconomic problems—external critics appear to have had much less tangible impact. The World Bank now does lend more to cushion the social impact of adjustment; it also claims that since the late 1980s its macroeconomic conditionalities have made "maintaining or increasing social expenditures a condition of adjustment loans."[9] Nevertheless, most SAP critics respond that the core of the model remains largely unchanged.[10] This may be due to their lack of influential insider allies. This link between insider-outsider interaction and advocacy *impact* dates back to the very beginnings of NGO efforts to reform the multilateral development banks (MDBs).

Stepping Back: The Multilateral Development Bank Campaign(s)

What came to be known as "the MDB campaign" began in a coordinated way among Washington-based environmental activists in 1983. It was predated, however, by two main streams of protest. Human rights activists had highlighted the World Bank's Cold War political biases. It opposed loans to Chile's radical reform government elected in 1970, and then quickly renewed lending to the military regime that overthrew the reform government. The World Bank had also faced local protest movements on the ground, such as the Cordillera peoples' successful resistance to the Philippines' Chico River Dam in the late 1970s and tribal protest against India's social forestry programs.

By the early 1980s, growing awareness of rainforest destruction and violations of indigenous rights had created a widely accessible political and visual shorthand that allowed public interest groups to raise much broader questions about the environmental and social impact of World Bank projects. The U.S.-based environmental NGOs began with congressional hearings in the early 1980s, showcasing expert testimony from Brazilian and Indian experts, as well as World Bank consultants and staffers who had seen their internal warnings of environmental and social damage go unheeded. Brazil's Polonoroeste Amazon road project became the "paradigm case."[11] Although only a small part of the World Bank's portfolio directly destroyed the rainforest, the NGOs' "case study" strategy highlighted the Bank's decision-making processes, which allowed such projects to be funded. This case study approach was not a social science concept; it was a media and lobbying strategy to highlight egregious projects that resonated with widely accepted "frames" for understanding environmental problems—such as burning rainforests—in order to underscore more general institutional problems at the World Bank.

U.S. NGO critics kept up the pressure in Congress, contributing to the Reagan administration's choice of a former congressman—Barber Conable—to head the Bank in 1986. The U.S. Treasury Department's concern with defusing the debt crisis led some conservatives to seek to allay public criticism of the World Bank, while a few pragmatic environmentalists sought legislative clout through coalitions with Republican aid critics. Partly in response, the Bank created a high-profile Environment Department as part of its broader reorganization in 1987. Growing international media criticism and grassroots protest in borrowing countries such as Brazil and India also put the World Bank on the defensive. These pressures bolstered U.S. environmental NGOs, whose influence in Congress and the U.S. Treasury created leverage over a strategic institutional pressure point: donor funding. The political threat to always uncertain U.S. foreign aid appropriations for the World Bank directly led its management to start listening more closely to a small group of insider environmental professionals that had been quietly developing reform policy proposals, according to Robert Wade's recent study, which is based on full access to the World Bank's internal files.[12] The result was the precedent-setting establishment of an environmental assessment policy in 1989 and its subsequent strengthening in 1991.

The U.S. NGO effort to reform the multilateral development banks was originally led by a small group of individuals, strategically located in the international departments of large membership environmental organizations, such as the National Wildlife Federation, the Sierra Club, and the Environmental Defense Fund.[13] Their millions of dues-paying members gave them congressional clout and media credibility, while the small number of full-time activists in Washington facilitated relatively high levels of coordination of what they called "the MDB campaign." Though this term implies that there was one single campaign, diverse campaigns against the multilateral development banks multiplied around the world. Though the growing number of active groups focused on shared targets, they often differed in basic analysis, priorities, strategies, and tactics. By the late 1980s, advocacy groups in developing countries began playing more leading roles, and more aggressive NGO campaigns spread to Europe, Canada, and the Pacific. An unprecedented mass street protest at the World Bank's 1988 annual meetings in Berlin began to change the tone of the campaign, broadening it beyond small lobbyist circles and encouraging some U.S. groups to become more radical. At the same time, many of the European NGOs began to experiment with the U.S. advocacy strategy of "following the money" and began to target government appropriations. By the late 1980s, rainforest destruction and large dams brought environmentalists, development NGOs, human rights activists, and indigenous activists together into a broad effort to question multilateral economic development decision making more generally. As the social, geographic, and institutional base of the campaigns broadened, anti-poverty concerns and structural adjustment became growing priorities on the MDB campaign agenda.

During the 1980s, the MDB campaign pursued a "sandwich" strategy: from below, Southern groups provided local project information and political legitimacy, while from above Northern advocacy groups lobbied donor governments to push for reform, through their representation on the Bank's board of directors. Bank reforms, in turn, were intended to encourage borrowing governments to become responsive to social and environmental concerns. Recall that the World Bank lends primarily to governments, and some Northern NGO advocacy groups hoped that its famous leverage over borrowing governments could be used for reformist ends, a hope that led to sovereignty concerns among Southern groups. In the process, different participants in these North-South NGO coalitions had different priorities. Northern groups tended to be more focused on changing the Bank, whereas Southern groups were often more concerned with creating the political space needed to challenge their own governments' development strategies.

The Bank's need for regular donor government contributions to the International Development Association (IDA), its low-interest lending window for the poorest countries, created a three-year cycle of political opportunities for the NGO campaigns, and most major Bank policy reforms were directly associated with the need to respond to donor government debates over IDA contributions. NGOs differed, however, over whether to call for cuts in IDA contributions or to take the less confrontational approach of using the appropriations debates to encourage reform. This issue divided both Northern and Southern advocacy groups into more moderate and more radical wings.[14] Pressure from radical environmental and social critics to cut IDA funds encouraged the World Bank to engage in dialogue with more moderate NGOs, especially those from the South. The Bank began to accept the principle of direct NGO and grassroots participation in projects. The 1996 panel discussion that followed World Bank president James Wolfensohn's official blessing of the new *Participation Sourcebook* revealed one of the key political impulses behind the growing internal legitimacy of the "pro-participation" current within the World Bank. As one senior official reported to the group,

I was in charge of trying to ... help ... to raise money for IDA and you all know it's gotten very difficult to do so. One thing I learned very quickly is that we need the support of NGOs in the North in order to do that. It also became very clear very quickly that the NGOs in the North are very closely related to their work and experience with the NGOs in the South. And so it became very quickly clear that we had to build better bridges to the NGOs in the South. *We had to let them into our tent. We had to work with them, not only because it does better projects, but because it's very important for us to be able to convince the[ir] partners in the North that we're doing a reasonable job.* And I can assure you the line of communication between the NGOs in the South and the NGOs in the North is extremely short, meaning quick (emphasis added).[15]

The Bank's wave of reforms in the early 1990s not only strengthened earlier involuntary resettlement, indigenous peoples, and environmental assessment policies, but also included more funding for poverty-targeted and "green" projects—the recognition that the social costs of macro-economic structural adjustment should be taken more into account. Bank discourse was changing, but its "unreformed" practice caught up with it at the same time, as the unprecedented grassroots/international NGO campaign against India's Narmada (also known as the Sardar Sarovar Dam) Dam severely damaged the Bank's international credibility. Sharp conflicts between official descriptions of the social and environmental impact and field reports coming from grassroots and NGO sources led the Bank's board of directors, dominated by donor governments, to commission a full-scale independent review. The 1992 Morse Commission report concluded that the Bank's environmental and resettlement policies had been flouted, and took the dramatic step of recommending that the Bank "step back" from the project.[16] The militancy of grassroots opposition to the Narmada Dam echoed throughout international advocacy networks, as thousands of villagers vowed to drown rather than leave their homes.

By 1993, bolstered by the Morse Commission's official legitimation of key elements of their critique, international NGO coalitions used their leverage over donor government IDA funding to push for broader policy changes at the Bank—including more public access to project information, the creation of an appeals mechanism for project-affected people (to become the Inspection Panel) and more accountability for the actual implementation of the resettlement policy. The 1993 information policy reform and the establishment of the Inspection Panel were directly negotiated with the U.S. congressional leaders who controlled IDA funding (see chapter 11 in this volume). The Inspection Panel was especially important because it was created to verify actual implementation of the rest of the Bank reform policy package.

Seeing limited change on the ground, NGOs differed over whether the Bank should ultimately be reformed or abolished, but most agreed on the immediate goal of pushing for institutional reforms. By the World Bank's

1994 anniversary year, many groups—especially in the United States—agreed on a compromise strategy, under the deliberately ambiguous slogan "50 Years is Enough." The premise was that the Bank should be reformed, and if it could not, then its funds should be reallocated to more socially and environmentally responsible development agencies. The campaign had a high degree of success in reaching the media, which was a key channel for influencing the then-Democratic U.S. Congress.[17] As Lewis Preston, president of the World Bank at the time, put it,

I think that the mistake the bank has paid the highest price for was not recognizing the importance of the environment. Initially they perceived that there was a conflict between development and the environment. That mistake, I think, has in terms of the criticisms of the bank, eroded some of the support the bank is entitled to.... [But this criticism is] very much out of date.[18]

As this volume shows, World Bank practice still falls far short of its promises of reform. According to the World Bank's director of environmental economics, at a basic conceptual level changes have been mainly limited to "grafting environmental concerns onto business as usual," at least until recently.[19] These changes have involved some attention to environmental mitigation measures; for most projects, however, social impact, participation, and environmental sustainability criteria have not been built into their original conceptualization, a situation analogous to handling pollution problems with "end-of-the-pipe" remedies rather than source reduction. As even the politically moderate World Wildlife Fund notes, "The Bank's 'end-of-the-pipeline' approach to incorporating environmental concerns has demonstrated that environmental sustainability cannot be added on the 'business-as-usual' approach to development."[20] NGO critics of structural adjustment make similar argument about the Bank's growing portfolio of targeted antipoverty projects: until the macroeconomic models stop increasing poverty and promote a more broad-based pattern of growth, poverty-reduction investments will at best deal with the symptoms rather than the underlying causes of poverty, such as inequitable access to productive resources and employment.

Frame One: Explaining Institutional Change

The question of the degree to which the World Bank is changing raises the more general analytical issue of how and why large, relatively autonomous institutions change.[21] The literature on organizations explains the dynamics that make large hierarchies guard their autonomy and therefore resist power sharing with other actors, and it distinguishes between mere adaptation and learning. Organizational adaptation involves changes in behavior in response to new pressures or incentives, but without any adjustments in the organization's underlying goals, priorities, or decision-making processes. Learning, in contrast, involves disseminating new conceptual frameworks and institutional changes throughout an organization, thus leading to qualitatively new goals and priorities, as well as changes in behavior.[22]

Although there is widespread acceptance of the notion of organizational learning and its importance to institutional performance, it turns out that the literature has not developed any widely accepted theory to explain it. Moreover, many approaches conflate adaptation and learning, a problem because changes in organizational behavior may occur with or without learning (defined as changes in the ways problems are perceived and explained). Moreover, some staff may learn without necessarily changing organizational priorities and internal incentives, leading to little change in institutional behavior.[23]

The distinction between organizational adaptation and learning translates into the difference between assessing change at the World Bank as the result of external political pressure and changing institutional incentives, versus attributing its reforms to increased internal acceptance of new ideas, bolstered by their technical credibility, expertise, and intellectual acceptability.[24] Many World Bank staff would argue that the institutional greening process is driven primarily by internal learning, changes in patterns of recruitment, research and staff education, and technical progress in finding "win-win" approaches to environment-development trade-offs.[25] Others would attribute less weight to the power of expertise, suasion, and intellectual prestige, and would stress the importance of changing institutional incentives and power relations, so that those staff who have "learned" to be more socially and environmentally responsible are actually *heeded* by those who make the funding decisions. These choices are not dichotomous, however, because ideas and interests in an organization clearly interact and influence one another.

In the Bank context, external political pressure can encourage learning as well as adaptation (by politically bolstering those insider reformists willing to learn). Institutional rethinking and redesign can help to avoid problem projects in the first place. However, most observers tend to conclude that the Bank does much more adapting than learning. One study of a Bank effort to change states that "they continue to fit the task to the organization rather than the organization to the task."[26] As Fox's chapter (chapter 9) on the Bank's resettlement policy shows, for example, that although project managers can "learn" that tens and hundreds of thousands of poor people are evicted and immiserated by large infrastructure projects, this "learning" does not necessarily lead them to respect the people's rights. This chapter's analysis of internal learning shows that greater institutional accountability is driven less by conceptual insights than by changes in the balance of political power between pro-reform forces both inside and outside the Bank on the one hand, and old guard, anti-reform forces entrenched in the operational apparatus and borrowing governments on the other.

Organizational learning that threatens dominant paradigms—such as the hegemony of neoclassical economics at the World Bank—is likely to provoke resistance.[27] The core analytical models and country strategies that guide key lending decisions still tend to treat social and environmental costs as externalities and therefore as secondary considerations.[28]

As Ernst B. Haas's comparison of international organizations concludes, "adaptive behavior is common, whereas true learning is rare. The very nature of institutions is such that the dice are loaded in favor of the less demanding behavior associated with adaptation."[29] Over time, significant organizational learning is therefore likely to involve long periods of rising tension with existing paradigms, followed by periods of transformational disruption (a "punctuated equilibrium"), which in turn are followed by periods of incremental change and consolidation.[30] In this view, if there were to be "true" organizational learning at the Bank, it would follow a long period of gradually building pressures for change—such as the NGO advocacy campaigns from the mid-1980s through the early 1990s and the resulting increase in insider environmentalist influence. As of the mid-1990s, however, these gradual changes had not yet been translated into dramatic changes in the Bank's primary activity:

development lending. Whether or not the further greening of the Bank in the mid-1990s will lead to the kind of concentrated period of qualitative transformation stressed by the literature on organizational change remains an open empirical question because the actual impact of many key changes has yet to be felt on the ground. As seen throughout this volume, Bank discourse and actions are both *moving targets*, which greatly complicates assessments of the consistency between them.[31]

Frame Two: The Difficult Construction of Accountability

Accountability refers to the process of holding actors responsible for actions. Operationalizing such an open-ended concept is fraught with complications, starting with the politically and technically contested issue of assessing performance. Even if the measurement problem were solved, the factors explaining the process have received remarkably little research attention. For example, although political science has sought broad generalizations to explain wars, treaties, military coups, legislation, electoral behavior, and transitions to democracy, it has not produced empirically grounded conceptual frameworks that can explain how public accountability is constructed across diverse institutions.[32]

At the most general level, the World Bank is one of many large bureaucracies that claim to operate in the public interest, but which have only discretionary mechanisms for civil society input. When reflecting about how such relatively impermeable institutions respond to calls for public accountability, one could venture the following series of general propositions. First, most hierarchical organizations resist power sharing and attempt to control their environment. Second, when under-represented groups manage to overcome the usual obstacles to collective action and attempt to hold public institutions accountable, the resulting conflict will lead to, at best, partial change. Concessions can be won by broad advocacy coalitions that manage to target institutional pressure points and ideological inconsistencies (such as challenging funding sources and producing delegitimizing "facts" that show violation of the public interest). Third, through cycles of conflict and reform, concessions may cumulate. If some of those concessions take the form of changes in decision-making processes, such as increased public access to information or the strengthening of insider reformists, then the rules of engagement between public interest advocacy groups and the bureaucracy shift, thus facilitating future changes. Such institutional changes can also be rolled back or hollowed out by bureaucratic backlash.[33]

Officially, the multilateral development banks, including the regional banks as well as the World Bank, are accountable only to their member nation-states, whose votes are weighted according to their respective financial contributions. But the relative autonomy of Bank managers and staff vis-à-vis the agency's formal owners (the member-states) is widely recognized. The most powerful evidence is that the Bank's board has never rejected a loan proposal from management. Even if the Bank were effectively accountable to its member governments, most critics would reject this formal accountability criterion because of the board's "one dollar–one vote" distribution of power. Once one leaves this formal institutional domain, however, producing a consistent, measurable, widely accepted definition of accountability becomes problematic.

This study operationalizes the concept of accountability by focusing on only one of many relevant indicators: compliance with the World Bank's own social and environmental reform goals as benchmarks of institutional change. The indicators stressed here are the Bank's own policies that are supposed to guide staff behavior and resource allocation. The focus here is on actual Bank compliance in practice, rather than on changes in official discourse and affirmations of intent. This methodological choice is not meant to suggest that the editors (or much less the chapter authors) are taking a position regarding whether these official reform policies are "sufficient" from a normative standpoint. The choice is based on the view that most civil society protest produces at best partial institutional change and therefore that indicators of *degrees* of change are needed to assess protest impact.

Using Bank compliance with its own social and environmental policies as an accountability indicator also helps to disentangle the issues of World Bank accountability and advocacy group representation. World Bank officials note that many other stakeholders are involved in development projects, and some would question the degree to which advocacy groups and mass movements represent broader public interests. However, because the Bank itself accepts that it should abide by its own reform policies, when NGOs and grassroots movements manage to hold the Bank accountable *for its own promises*, they then have unambiguous pro-accountability impact on the institution—regardless of who they represent.

Conceptualizing the role of civil society critics in holding the World Bank accountable requires more than a simple two-actor model of conflict. Nation-states play determinative roles as well.[34] In donor countries, NGO advocacy groups lobby governments to try to influence the World Bank through their country's executive directors. Donor country NGO influence on the World Bank is therefore mediated by the NGOs' capacity to forge coalitions in the executive and legislative branches of their own governments, bolstered by the political legitimacy of taxpayers' calls for effective and transparent public spending.[35]

In developing countries, the influence of NGOs and grassroots groups on Bank-funded projects and policies is mediated by the degree of democratization of their own national political institutions. Even in borrowing countries with elected governments, however, World Bank funding decisions largely bypass the democratic process. Borrowing country state managers are crucial in determining the actual impact of Bank projects because they are the ones who end up allocating Bank funds in practice. The World Bank's capacity to force compliance with its famous loan conditionalities is actually quite uneven and has been more often assumed than demonstrated. Many borrowing government state managers are also quite adept at creating merely the appearance of compliance.[36] Conclusions based on the subordination of small, very poor countries do not necessarily hold for large countries with nationalist traditions and strong ruling classes—such as Brazil, India, or Mexico. Moreover, many assume a two-actor, zero-sum negotiation process between the Bank and borrowing governments, but more often Bank and national finance ministry officials agree on what needs to be done and then use their combined leverage to influence the rest of the state apparatus.[37]

The issue of Bank accountability is inherently mediated by the actions of the borrowing governments, especially when it is seen from the receiving end of socially and environmentally destructive projects. Borrowing government responsibility for project implementation has long provided the Bank with a very convenient way of avoiding responsibility for "problem projects." Because the bargaining processes between national governments and the Bank are not public, it is difficult to ascertain in any given case whether Bank officials actually try to use their political capital in an effort to prevent or mitigate social and environmental damage. Their "it would have been worse without us" argument is undermined by the internal Bank histories of the Brazil's Amazon road and India's Narmada Dam projects. In both cases Bank project staff and managers were fully aware of the high social and environmental risks *before* deciding to extend the loans, but they failed to use their leverage to reduce those risks until crises erupted.[38]

Most NGO advocacy campaigners are fully aware that national governments share responsibility with the World Bank for problem projects, but they have targeted the Bank for three main reasons. First, the project may well not have proceeded at all without the Bank's official seal of approval, which provides legitimacy and leverages significant counterpart funding from other multilateral and bilateral sources. Second, the dynamics of the transnational advocacy process itself impels campaigns to focus on available pressure points—for example, in the case of U.S. environmental NGOs lobbying Congress to pressure the Bank.[39] Indeed, because the World Bank depends in part on donor government contributions, constituencies of those governments can claim that such transfers require the Bank to be accountable to them, as taxpayers. This raises important sovereignty issues because the corollary is that donor governments are asking the Bank to impose social and green conditionalities on borrowing governments (such as environmental assessments). Third, many of the borrowing governments themselves are less than democratic (which weakens the sovereignty counterargument). Although authoritarian borrowing governments are often the direct perpetrators of socially and environmentally costly projects, advocacy groups would argue that the World Bank—as "intellectual author"—has added responsibility in such cases precisely because citizens lack channels for holding their own states accountable.[40]

The prominent role of authoritarian regimes in the World Bank's portfolio is also crucial for grappling with the question of how well transnational NGO advocacy coalitions represent those most directly affected by Bank projects. Because many of the most controversial Bank-funded projects are carried out by authoritarian regimes, it is difficult to expect the groups most affected to have access to public information about the project or the ability to organize freely to defend their interests. Indeed, projects with high social and environmental costs are often most likely to proceed where those most affected are denied basic rights.

Authoritarian political contexts also influence the potential civil society response in terms of the repertoires of resistance and coalition building available. Indeed, the classic early Bank project protests involved few voices from the directly affected. In the early campaigns against Brazil's Amazon roads and Indonesia's transmigration projects, for example, NGOs spoke out on behalf of those affected. Over time, grassroots organizations increased their direct representation within the transnational Bank campaigns, though their influence is still often uneven.[41] Even recently, in Nepal's Arun III case, the campaign was led by national NGOs, but local communities in the valley did not organize mass protests. The government had promised villagers that the project would bring prosperity, they lacked access to information about project risks and alternatives, and Nepal's nascent democratic process had yet to reach into the more remote rural areas. Project critics did begin organizing in the Arun Valley, but faced at least one death threat.[42] After the successful NGO campaign to block the project, Nepalese project critics felt a strong responsibility to the residents of the valley (who had been promised rapid "development") to continue to pressure for more sustainable development alternatives.[43]

The chapter authors in this volume are all concerned with drawing broader lessons from the diverse range of conflicts between the Bank, governments, NGOs, and grassroots groups over accountability, but they do not share a unified conceptual framework. Some analyze project and policy cases that have changed more than others, leading to different assessments of the degree of Bank reform. One of the most important general lessons for those interested in building on these cases to create a more comprehensive analytical framework is that the key "actors," such as the World Bank, civil society, or the state, are actually contested arenas. Notably, since beginning to institutionalize its environmental reforms, the Bank is internally divided over whether and how to promote sustainable development. Borrowing governments are also often divided over whether and how to respond to sometimes contradictory Bank pressures as well as to grassroots and NGO concerns. Civil societies split over Bank-funded projects, with some influential stakeholders standing to gain economically. NGOs and local groups sometimes take different positions regarding controversial Bank-funded projects: some choose to work within the system, whereas others criticize from the outside. External critics can be split into those who push for mitigation or compensation measures and those who attempt to block a project entirely. In any controversial project, there is no reason to assume that either NGOs or grassroots groups are united. Representation, like accountability, is relational and a matter of degree.

Research Questions and Methods

As stated earlier, this book focuses on two specific questions: first, to what degree did NGO/grassroots advocacy campaigns influence specific projects or policies at issue? Second, to what degree were directly affected grassroots communities represented in the MDB campaign process? Different chapters focus on each question to varying degrees.

The question of advocacy coalition impact on the World Bank actually subsumes two quite distinct issues: what was the degree of social and environmental policy change, and what relative weight did NGOs and grassroots movements have in driving such change? The answer to the first question is largely empirical; at the level of specific projects, development investments have winners and losers that are tangible enough to be measurable, at least in principle. Such assessments are still methodologically complicated by the persistent problem of distinguishing between where the Bank role ends and that of national governments begins. Even subnational governments can play critical roles in federal states—as in India, Brazil, and Mexico.

Assessing the specific role of external pressure and scrutiny poses even greater methodological problems. How does one take stock of the impact of transnational advocacy campaigns on an institution as large, opaque, and slow-moving as the World Bank? The first problem involves the question of the counterfactual. Without external scrutiny and protest, would World Bank projects have been even *more* socially and environmentally costly? Would a *larger number* of them have been destructive?

Second, where the Bank does appear to have responded to public pressure, how does one enter the black box of official decision making to disentangle the relative weights of the various different factors—internal and external, ideological and material—that come into play? Projects that seem to have been blocked because of protest could actually have collapsed because unrelated economic changes undermined their predicted rate of return or because of internal bureaucratic conflicts.

Third, there is the risk of conflating normative and analytical criteria for assessing change. Protest could well have significant damage control impact, but such changes could easily be undervalued if they are deemed to fall short of minimum demands. This undervaluation is common in cases of projects that are widely seen as illegitimate because compensatory or mitigating measures can facilitate project implementation. This particular issue is part of the broader problem of the counterfactual within social movement analysis. How does one know whether mitigating measures were the most that could have been won or were somehow accepted "instead of" hypothetically larger victories that could have been won had the movement not been divided by partial victories?

The fourth methodological dilemma confronting assessments of protest impact appears when there is no consistent relationship between the breadth or intensity of advocacy campaign mobilization and the institutional response. The diverse range of advocacy coalitions discussed in the cases recounted here include some where relatively little mobilization had a large impact on a project, as well as some where intense mobilization had little direct impact on the project. Most often, their main impact has been *indirect,* on broader Bank policies designed to prevent such problem projects in the future. This finding is the main reason why analyses of Bank policies as well as project cases have been included here. A focus on project campaigns alone would miss one of their main outcomes—policy reform; yet a focus on general policies alone would in turn miss the highly contested processes that determine their degree of implementation—the projects themselves.

Bank and NGO leaders have tended to differ sharply over the volume's second broad concern—the accountability of advocacy coalitions themselves. Many World Bank operational staff assume that advocacy groups usually do not represent the ostensible project beneficiaries. Campaigners often assume the opposite—that advocacy groups necessarily represent those on the ground.[44] It is much safer to start from a different assumption: given the sharp differences in power, culture, and goals between grassroots groups, national NGOs, and international NGOs, there will *necessarily* be tensions around the process of representation, and their resolution will be a matter of degree.

Public interest advocacy strategies in North and South often differ greatly, as do elite and grassroots styles within any one society, especially because national societies themselves are highly diverse.[45] For example, some U.S.-based groups are accustomed to confrontational approaches, whereas NGOs in countries that lack consolidated democratic regimes may prefer subtler ways of creating political space in which to be heard (as in Indonesia, for example). In India, however, "struggle-oriented" groups have found their Washington-based counterparts to be much more moderate. This diversity poses a challenge for efforts to explain why some transnational advocacy campaigns come closer to the widely discussed ideal of building a transnational civil society than others do. This volume's chapters address, to varying degrees, how transnational advocacy campaign strategies shape relations among alliance partners, how the distribution of power among them waxes and wanes in the course of a campaign, and how that process relates to the campaign's broader impact.[46]

It is useful to recall that the subject of *transnational civil society* has generated a great deal of discussion, but not much of it, so far, is presented in terms of detailed case studies. Although growing, scholarly analysis of these new emerging nongovernmental actors is still largely based on limited secondary sources. In order to help to fill this empirical gap, the volume editors invited the participation of a diverse range of campaign strategists and engaged scholars, all distinguished by their longstanding direct experience as "relays" who bridge the local, national, and transnational advocacy arenas.[47]

This volume was designed to generate hypotheses. The goal was to ask similar questions of cases that were sufficiently diverse and representative of Bank–civil society conflict as to suggest both general patterns and explanations. The first challenge was to fill an empirical gap; all the case studies are based on primary research and firsthand experience, and most document their campaigns for the first time. Case analysts were also asked to explain the key turning points in terms of both campaign impact and decision making. Though these participants were encouraged to address similar variables, the general approach to the collection as a whole was inductive rather than deductive. The two concluding chapters synthesize the cases' general patterns and causal propositions, which now lend themselves to more systematic hypothesis testing in future research.

The campaigns under study were limited to the World Bank itself, rather than its associated agencies (the International Finance Corporation [IFC], Multilateral Investment Guarantee Agency [MIGA], or the Global Environmental Facility [GEF]) or regional multilateral development banks.[48] In retrospect, given the IFC and MIGA's rapidly growing importance, the decision to exclude them was problematic. Several criteria informed case selection. First, project and policy cases were chosen from among second generation World Bank conflicts of the late 1980s and early 1990s, rather than the paradigm cases of the early and mid-1980s that launched the MDB campaign(s), because more recent conflicts could offer more evidence of change both within the Bank and among NGO/grassroots coalitions. Second, each campaign chosen had to involve relatively sustained North-South collaboration over time. Third, campaigns were selected from geographic regions and sectors that broadly represented the set of civil society–World Bank environmental conflicts (mainly large infrastructure projects in Asia and Latin America). Fourth, each project campaign case involved *conflict over whether to implement already established World Bank social and environmental reform policies* (including the environmental assessment, indigenous peoples, and involuntary resettlement policies). This last factor is perhaps most important for understanding the broader pattern that emerges from this volume as a whole, a pattern in which sustained external pressure turns out to be necessary to encourage compliance with Bank policies.[49]

The study's original focus on projects was based on the premise that the most revealing indicator of institutional change is where the money really goes. The early findings suggested, however, that this focus on the World Bank's actions would be strengthened by a complementary analysis of the policies that ostensibly shape resource allocation—even if they were not implemented consistently in practice. As it happened, however, the analysis of policy change came around full circle because the impact of new policies depended on the degree to which they in turn actually influenced projects.

Organization of the Book

To frame this book's specific campaign cases, the first thematic section provides background on diverse NGO coalitions. *Nongovernmental organizations* are defined here as nonprofit civil society organizations that do grassroots support and advocacy work.[50] They are intermediary organizations, in contrast to membership groups with relatively defined social constituencies. They therefore do not directly represent the grassroots constituencies they attempt to serve.[51] Unlike private firms, they are not directly accountable to market forces, though they are subject to the ebbs and flows of the more intangible and opaque market for private foundation, religious, and public sector funding. NGOs also make important contributions as bridging organizations, catalyzing collaboration across public and private boundaries or between elite and grassroots actors.[52] Because NGOs' credibility and legitimacy depends heavily on their capacity to provide technical, organizational, and political services to underrepresented groups, they are highly vulnerable to charges that their practice falls short of their discourse. As a result, although they usually lack formal institutional accountability mechanisms, their dependence on maintaining at least the appearance of consistency between theory and practice creates informal, inconsistent, but often powerful accountability pressures.[53]

As with large bureaucracies, accountability within grassroots organizations is most usefully understood as a matter of degree. In contrast to bureaucracies, however, the accountability relations within such organizations are supposed to flow from the bottom up rather than from the top down. The tendency for leaders to hold on to power—the "Iron Law of Oligarchy"—is a powerful force that undermines leadership accountability within large membership organizations. In practice, however, the balance of power between leadership and rank-and-file often ebbs and flows over time. There is no basis for assuming that grassroots organizations necessarily represent their members, but most leaders do depend on their claim of representation to sustain their organizational power over time. Even leaders who become brokers are pressured to represent at least some of their members, to some degree, some of the time. In spite of the importance of internal power relations for understanding the struggle for accountability, this volume does not attempt to disentangle these relations *within* NGOs or grassroots movements. The task of exploring power relations *between* them proved more than sufficiently challenging.[54]

Part I: Mapping NGOs

It is in this context that the volume's first two chapters analyze different kinds of NGO/grassroots advocacy coalitions. David Wirth's chapter examines the history of U.S. environmental NGO efforts to reform the World Bank. With its combination of legislative lobbying, media campaigns, legal challenge, and grassroots protest, the domestic environmental reform movement of the 1970s created the institutional levers and political culture from which the MDB campaign first emerged in the United States. The kinds of procedural innovations proposed for the World Bank echoed previous domestic U.S. reforms, including greater public access to information, the right to public hearings and consultation in the context of environmental impact assessments, as well as the eventual creation of an appeals channel. Following the environmental legislative legacy of 1970s, Wirth also shows how the U.S. legislative process provided some of the key NGO leverage over the World Bank. This legal-lobbying culture left its imprint on NGO coalition dynamics as environmental groups in the United States and in developing countries created what he calls *"partnership advocacy."*

In the third chapter, Jane Covey places the NGO strategies described by Wirth within a broader international context. Not all NGOs that deal with the World Bank follow the partnership advocacy approach. Some seek what is officially called *operational collaboration* to participate as partners in project implementation. Collaborative NGO influence over the *design* or evaluation of World Bank projects is much rarer, though reportedly growing.[55] This collaborative process involves an increasingly dense network of NGO ties, including international and grassroots-oriented NGOs that are more politically moderate than those involved in more public and sometimes more confrontational partnership advocacy.

Political differences among NGOs cannot be attributed solely to North-South differences because many civil societies include both moderate and radical approaches to the World Bank. Covey's chapter shows how NGO and grassroots approaches to the World Bank range from confrontational protest to operational collaboration, but an ambiguous new gray area of *"critical cooperation"* is emerging in between the two extremes. She examines the history of the World Bank–NGO Working Group—the oldest formal NGO liaison group—to assess the nature and impact of critical cooperation in policy dialogue, as well as efforts to use operational cooperation as an approach to making Bank practice more accountable to grassroots groups. The limited results from NGO efforts to use the official channels available help to explain why some advocacy groups turned to more confrontational approaches by the end of the 1980s.

Part II: Project Campaigns

The second part includes four case studies of conflictive projects that provoked the emergence of sustained transnational advocacy coalitions. Each chapter traces the campaign history and attempts to assess both its impact and internal dynamics, presented in rough chronological order. Chapter 4, by Augustinus Rumansara, focuses on resistance to displacement by Indonesia's Kedung Ombo Dam in the late 1980s. The dam was designed for irrigation, urban water, and hydro power. Rather than questioning the dam per se, the villagers and NGOs fought to be able to resettle near their original homes and for better terms of compensation for those evicted. From the point of view of this volume's broader concern with the relationship between policy and practice, Kedung Ombo was a clear case where coordinated international and local protest was necessary to highlight prolonged noncompliance with an official Bank reform policy. Indeed, the Bank's own project completion report recognizes both that the project involved especially blatant violations of its official resettlement policy and that advocacy groups played a positive role in drawing attention to resettlement problems.[56]

The critique of the World Bank's funding for Indonesia's transmigration project was well known internationally because the project necessitated the organized, "voluntary" relocation of landless farmers from densely populated central islands to rainforests occupied by ethnic minority tribal groups.[57] Indeed, the transmigration project was such a high priority for the Indonesian regime that limited political space kept most local resistance to a minimum, with the eventual exception of some of the displaced indigenous groups. Rumansara chose to analyze the Kedung Ombo case, however, because it was the first Bank-funded project in Indonesia to provoke sustained grassroots mobilization.[58] The Kedung Ombo Dam project was linked to transmigration, however, because thousands of families that were to be displaced were originally supposed to be "transmigrated" to the outer islands. In practice, few wanted to move, but the government built the dam, closed its gates, and the waters began to rise before many villagers were able to find alternative settlements. In the end, most found ways of remaining in the area, whereas others continued to utilize legal strategies to gain better compensation. The economic conditions for most worsened, but at least they managed to avoid forced transmigration. In addition to showing the difficulty of achieving even partial damage control late in a project cycle, the case also highlights the complex problems involved in building balanced bridges between actors within international campaigns, especially under authoritarian conditions. Kedung Ombo involved delicate and sometimes tense relations not only between Northern and Southern NGOs, but also between national NGOs, local NGOs, and local communities pursuing diverse political and survival strategies.

The fifth chapter, by Antoinette Royo, focuses on one of the first cases in which a combination of grassroots pressure and national and international lobbying empowered World Bank reformists to veto funding for a project embedded *within* an already approved World Bank loan. In the late 1980s the Philippine government asked the Bank to allocate funds from an already approved energy sector loan to build the Mt. Apo geothermal plant. This was an early test case of the Bank's environmental assessment policy, which requires it to investigate and then either eliminate, redesign, or mitigate high-impact projects. As with many of the more balanced transnational coalitions, international environmental and indigenous rights activists worked directly with both national NGOs and local indigenous leaders to defend ancestral lands. The coalition faced the challenge of balancing both environmental and indigenous rights concerns because the former were easier for the project to mitigate than the latter. Although the coalition effectively vetoed international funding for the project, the Philippine government continued to invest in the drill site.

The Mt. Apo campaign set at least two important precedents. First, environmental and social impact standards were upheld at the level of a "subproject" *within* an already approved sector loan—a point in the project cycle where Bank reformists often have little leverage. Second, social concerns became effectively joined to environmental issues because ancestral land claims were recognized as legitimate grounds for denying project funding.

The sixth chapter, by Margaret Keck, analyzes one of the few World Bank projects designed to rectify mistakes made in a prior project failure in the same area. Brazil's Planafloro project was supposed to be "done right" from the beginning in terms of Bank policies regarding environmental impact assessment, mitigation measures, public consultations, and indigenous rights. Its predecessor project in the early 1980s, Polonoroeste, funded the Amazon road that had become the first paradigm case for Bank critics. After criticism in the U.S. Congress had led the Bank for the first time to suspend loan disbursements for environmental reasons, the Bank began planning the follow-up project that would try to do a better job. This project became known as the Rondônia Natural Resources Management Project, or Planafloro. It included influential input and criticism from local and international NGOs, and was supposed to encourage more sustainable agro-ecological development and to demarcate indigenous and rubber-tapper lands.

Two years into Planafloro's implementation, it turned out to reproduce many of the problems associated with the original road-building project, so local NGOs filed an official complaint with the Bank's new Inspection Panel. In contrast to other Bank campaigns, however, the goal of the Rondônia NGO Forum was not to block the project, but to get it back on track and to deliver the sustainable development benefits it promised. Not only did the project turn into an important test of the environmental assessment policy and the bank's capacity to do "green" projects right, but it was also the second major case brought before the Bank's new appeals channel. Although the Bank's board of directors decided that it did not want to offend such a large borrowing government by officially approving an Inspection Panel inquiry, the process of public scrutiny reportedly improved project implementation on the ground, at least in the short term.[59]

Kay Treakle's chapter documents a grassroots-NGO campaign that linked sector-specific policy reforms to broader macroeconomic conditionalities. In contrast to most NGO lobbying against macroeconomic structural adjustment conditionalities, this effort was rooted in a broad, nationwide mass protest movement. Ecuador's national indigenous movements mobilized against pro-business land tenure and oil "reforms" that turned out to be secretly linked to broader macroeconomic structural adjustment policies. Indigenous rights as well as developmental and environmental concerns were fully joined, drawing from past North-South coalition work against private sector oil pollution in the Ecuadorian Amazon. Unlike the other cases, this study analyzes bargaining with a regional multilateral development bank as well as the World Bank because the same civil society network saw the two targets as linked and so coordinated the two campaigns. In the process, leaders of mass indigenous organizations peoples' managed to win greater recognition from both multilateral development banks than they had from their own government. Beginning with the Ecuadorian indigenous movement's second national civic strike in 1994, their triangular negotiations were eventually able to win a place at the bargaining table. The World Bank even negotiated the design of a new development loan directly with indigenous leaders. The Ecuadorian experience offered a test of the commitment of the Bank's indigenous peoples' policy to "informed participation" in projects, as broad-based grassroots groups began to "follow the money" and to become legitimate participants in national economic policy decisions.

Part III: Policy Campaigns

Part III steps back from specific projects and highlights the politics of five bankwide policies regarding indigenous peoples, involuntary resettlement, water resources, and its twin pro-accountability procedural reforms—the information disclosure policy and the creation of the Inspection Panel. Analysis of social and environmental reform policies is crucial for understanding the broader struggle for accountability for two reasons. First, they were created as an ostensible accountability response to external criticism in order to prevent future problem projects. Reform policies are therefore one of the main ways in which project campaigns have impact on the institution. Second, once a reform policy is created, the consistency of its implementation depends in turn on sustained external monitoring and advocacy. Each policy case documents the varying degrees to which external pressure accounts for both policy formulation and, to some degree, implementation.

The need to uphold indigenous peoples' rights brought together the environmental and human rights critiques of the World Bank. Indigenous rights advocates brought moral authority to the table, and environmentalists added their broad base, media access, and legislative clout in donor countries. The Bank's indigenous peoples policy dates back to 1982, two years after the first social/environmental policy directive, and was a direct response to several high-profile conflicts around the world in the late 1970s and early 1980s. Andrew Gray's chapter explores the causes and the consequences of the Bank's indigenous peoples policy, focusing on the different approaches and relationships both among involved NGOs and between the NGOs and the World Bank. The chapter shows how successive campaigns have broadened and deepened significantly since the early 1980s, as well as the pattern of nonimplementation of official Bank policy requiring informed participation of affected indigenous groups.[60]

Involuntary resettlement is another controversial issue area where the Bank has had a longstanding reform policy on the books. The policy was first formulated in 1980, also in response to a series of high-profile conflictive projects. The Bank led other international development agencies in its development of an official policy on involuntary resettlement, which commits borrowing governments to restore lost homes and livelihoods, but its actual practice of the policy has lagged far behind. In contrast to the Bank's usual denial of its noncompliance with other reform policies, its own resettlement specialists have documented how operational staff and governments systematically ignored the policy's key provisions. Between 1986 and 1993, 15 percent of total lending went to projects that would eventually displace at least two and a half million people by the official count.[61]

In chapter 9, Jonathan Fox explains how, by 1992, external political pressures empowered an internal World Bank team committed to improving Bank compliance with its resettlement policy. International environmental and human rights campaigns, united against forced dam evictions, gained sufficient political momentum to lead top Bank officials to commission an internal review of resettlement policy. This task force managed to gain both autonomy from the operational staff and authority from management, producing a remarkably frank assessment of the lack of consistency between policy and practice. The review process itself contributed to greater policy compliance, at least at the policy design phase. The unprecedented public release of the review, which recognized institutional problems and responsibility, became a first step toward the construction of accountability. In contrast to most of the studies in this volume, this chapter focuses on "unpacking" the Bank to assess the interaction between external pressure and the internal conflict among reformists and the old guard at the Bank.

The tenth chapter analyzes the unprecedented experiment in NGO–World Bank policy dialogue over the revision of the water policy. This policy issue cuts across many different Bank sectors, ranging from hydrodams and irrigation to the Bank's "brown" portfolio of sewage and potable water projects. The sustained dialogue between the World Bank and a transnational advocacy coalition offers a nuanced example of the limits and possibilities for the kind of critical cooperation Covey describes in chapter 3. Deborah Moore and Leonard Sklar detail the exchanges of views, the varying degrees of North-South participation, and the diversity of approaches among NGOs and within the Bank. By the end of the dialogue process, fault lines shifted, with Bank engineers finding that they often had more in common with NGO experts than with Bank economists because of their shared practical experience with the resource. The policy that emerged shows a small but significant degree of NGO impact, and the chapter concludes with an assessment of the uneven degree to which the policy influenced actual lending patterns during its first two years of implementation.

Lori Udall's chapter, the last one in the policy reform section, examines procedural rather than sectoral reforms. In 1994, the World Bank instituted two interlocking accountability reforms: increased public access to information about Bank projects and the creation of an ombudsman's office to investigate claims made by project-affected people of the violation of Bank policies. More than most, these two procedural reforms were the direct result of negotiations between high-level Bank officials and reform advocates in the U.S. Congress, with IDA contributions hanging in the balance.

Although the U.S. environmental NGO/government/World Bank bargaining process was the proximate cause of the twin accountability reforms, the underlying driving force was the Narmada Action Committee's combination of broad local base and transnational civil society alliances. Some NGOs may see the official scope of these two changes as limited, but in practice it appears that even relatively small changes in the Bank's decision-making processes can have significant and open-ended consequences. The twin information and inspection reforms are especially important because they potentially reinforce all the *other* social and environmental policy reforms by increasing the political costs associated with their violation. The information policy, if fully implemented, facilitates civil society input into project design for the first time—precisely the phase in the project cycle when potential external influence is at its maximum. Violations of the information policy provide grounds for an official inspection, as in the Arun Dam case. By permitting increased external scrutiny, these two reforms indirectly change the Bank's internal incentive structure, which has long been dominated by the pressure to lend at the expense of respect for the reform policies. Udall's chapter explains the roots and initial implementation of both policies. Their origins flow directly from the unprecedented international leverage gained by the Narmada Dam protest movement, whereas the Arun Dam campaign was a key test of their first-round implementation.

Part IV: Broad Patterns and Concluding Propositions
The final part returns to our two overarching research questions and outlines the broader patterns that emerge from the project and policy cases. Chapter 12 examines mutual influence and accountability among

diverse NGO/grassroots coalition partners. It frames the comparison of different transnational coalitions in terms of the broader issue of inter-organizational relations and then examines the evolution of mutual influence and accountability within alliances in project and policy campaigns. The conclusion identifies factors associated with intracoalition accountability and considers the implications of these alliances for the development of social capital, social learning, and transnational civil society. Transnational trust and cross-cultural relays turn out to be especially important in a context where great distances and power imbalances make the mutual supervision associated with accountability difficult. The cases suggest broader general patterns of both progress and limits in the efforts to build mutually accountable transnational civil society partnerships. Project campaigns, with their immediate threats and tangible goals, turn out to involve much more direct grassroots participation and control over the agenda than policy advocacy efforts, which bring NGOs closer to the terrain of the Bank itself and into the world of future rather than current Bank action.

These conclusions do not find that Northern-Southern advocacy coalitions are becoming robust transnational social movements. Most are politically contingent, tactical coalitions, not long-term strategic alliances. Many of these relationships are limited to fragile fax-and-cyberspace skeletons, and the strong ones are often based more on key cross-cultural individuals than on dense institutional bonds. Internet access in the civil societies of developing countries remains largely limited to elites, and connected NGOs may or may not be organically linked to social movements. The hegemony of the English language within both the Bank and international NGO networks also reinforces distance with the grassroots. The broadest-based protest movements remain those based on local, previously existing capillary social networks—many of which lack telephones, not to mention computers. As Sidney Tarrow points out, most of what are often presented as transnational social movements

are actually not cases of single movements with national branches, but of political *exchange* between allied actors whose contacts have been facilitated by global economic integration and communication. These are not, strictly speaking, transnational movements, but contingent political alliances linking pre-existing domestic communities with actors from other countries.... [Rather] these networks are linked to one another through transnational "relays."[62]

Many of our cases examine the empirical texture behind these broad processes and focus on how the transnational "relays" manage to link and balance very different coalition partners.

The last chapter takes stock of the impact of transnational advocacy coalitions on the World Bank, highlighting three analytical dilemmas. First, analysts will differ in terms of the criteria for assessing whether change has occurred. The second challenge is to determine the relative weight of a particular set of external pressures from public interest advocacy groups as distinct from other factors. The third dilemma is how to assess the ambiguous relationship between the World Bank's official policies and its actual lending practices. Cutting across all of these is the problem of the counterfactual: how would the sustainable development impact of Bank activity have been different in the absence of criticism and scrutiny?

Against this backdrop, the conclusion assesses the question of impact in terms of three dimensions. First, it addresses the question of how one assesses whether change has occurred by asking how the relative weights of different kinds of lending have changed over time. This conceptual exercise divides resources not by sectors, but in terms of whether their sustainable development impact is on balance "good, bad, or ugly." Second, the chapter turns to the empirical question of placing the project cases analyzed thus far in the context of the broader set of World Bank–funded projects that appear to have been tangibly influenced by public debate and protest. This review synthesizes the available evidence of NGO/grassroots impact and explores the nature of protest impact, which includes the blocking of projects, mitigating their effects, and contributing to momentum for subsequent policy reform. Because policy reform is one of the most important kinds of impact, the study then reviews the available record of Bank compliance with its own policies, drawing from both official and NGO sources, mainly in the "grey" literature of unpublished documents. In addition to the resettlement, indigenous peoples, and water policies covered in previous chapters, this discussion also reviews policies regarding energy, forestry, gender, agricultural pest managment, and environmental assessments. These "portfolio reviews" turn out to be of very uneven quality because few are based on independent, field-based assessments. Some find areas of progress, but even though most of the evidence is limited to official sources, overall Bank compliance with its sustainable-development reforms has been limited and uneven.

The chapter then reflects on the process of institutional change at the World Bank, analyzing the internal trip wires that are supposed to prevent problems before they start, insider-outsider backchannels that can strengthen reform forces, and in-house watchdogs that attempt to institutionalize some degree of accountability. The study finds that damage control is a much more common outcome than positive sustainable development outcomes. Indeed, the factors that contribute to blocking or mitigating costly projects may well be different from the forces that can drive what one might optimistically call "pro–sustainable development lending." One could argue that as the World Bank allocates increased funding in the name of sustainable development, *the center of gravity of civil society's struggle for accountable development funding has shifted closer to the ground*, to the grassroots groups, national advocacy NGOs, and reformist government policymakers who are trying make sure that these projects actually work as intended.

For future research, the final discussion presents a hypothesis about the conditions under which actual implementation of sustainable development projects and policies might be possible. The outcome depends on the contested balance of forces within each of the three key arenas where resource allocation decisions are made. First, project design and supervision must be under the control of pro-reform forces within the Bank itself. Second, projects must be designed to reach pro–sustainable development actors, whether they be in government or civil society. Third, the project must involve a sector or region where civil society actors are well organized and broadly representative of project-affected people. The proposition is that success will be rare because all three actors must be empowered within their respective arenas; otherwise, a missing link leaves the funding flow politically vulnerable to diversion by anti–sustainable development actors.

In response to the multiple processes of globalization, some analysts see the emergence of transnational civil society, a vision in which democratic actors are building bridges toward more equitable and sustainable futures. Others point out that this transnational civil society is still very thin on the ground. Some would go further and suggest that tactical transnational NGO coalitions do not always produce the kind of balanced, long-term partnerships that would really merit being called transnational civil society. The idealistic discourse of intellectuals and NGO leaders sometimes runs ahead of real-life, alternative institution building on the ground. But if transnational civil society ever did start to emerge, what would it look like? It would begin with building blocks like some of the networks under study here: coalitions that specialize in linking diverse social actors who have deep roots in their national societies with communities of potential partners around the world.

Acknowledgments

For generous comments on earlier versions, thanks very much to John Clark, John Gershman, Michael Goldman, Margaret Keck, Sanjeev Khagram, Smitu Kothari, Juliette Majot, Jane Pratt, Bruce Rich, Helen Shapiro, Kay Treakle, Warren Van Wicklin, and Robert Wade.

Notes

1. For the most comprehensive early statements of transnational advocacy campaign strategy, see Bruce Rich, "The Multilateral Development Banks, Environmental Policy and the United States," *Ecology Law Quarterly* 12, no. 4 (1985), and Stephan Schwartzman, *Bankrolling Disasters: International Development Banks and the Global Environment* (San Francisco: Sierra Club Books, 1986).

2. For a comprehensive study of the early years of environmental reform at the World Bank, see Philippe Le Prestre, *The World Bank and the Environmental Challenge* (Selinsgrove: Susquehanna University, 1989). See also his "Environmental Learning at the World Bank," in Robert V. Bartlett, Priya Kurian, and Madhu Malik, eds., *International Organizations and Environmental Policy* (Westport, Conn.: Greenwood, 1995). For a very well informed and insightful recent overview, see Robert Wade, "Greening the Bank: The Struggle over the Environment, 1970–1995," in Devesh Kapur, John P. Lewis, and Richard Webb, eds., *The World Bank: Its First Half-Century* (Washington, Brookings Institution, 1997).

3. See, for example, John Thibodeau, "The World Bank's Persisting Failure to Reform," (Toronto: Probe International, May 1995). A wave of critical publications appeared with the recent fiftieth anniversary of the Bretton Woods institutions. See, among other, Catherine Caufield, *Masters of Illusion: The World Bank and the Poverty of Nations* (New York: Henry Holt, 1996); John Cavanagh, Daphne Wysham, and Marcos Arruda, eds., *Beyond Bretton Woods:*

Alternatives to the Global Economic Order (London: Pluto, 1994); Kevin Danaher, ed., *Fifty Years is Enough* (Boston: South End Press, 1994); Susan George and Fabrizio Sabelli, *Faith and Credit: The World Bank's Secular Empire* (Boulder, Colo.: Westview, 1994); Paul Nelson, *The World Bank and Non-Governmental Organizations: The Limits of Apolitical Development* (New York: St. Martin's Press, 1995); and Bruce Rich, *Mortgaging the Earth* (Boston: Beacon, 1994). See also the regular coverage in the journals *Bankcheck Quarterly, The Ecologist, Lokayan, News and Notices,* and *World Rivers Review*. From the political establishment, see John Lewis et al., *The World Bank: Its First Half-Century,* 2 vols. (Washington: Brookings Institution, 1997); Henry Owen, "The World Bank: Is 50 Years Enough?" *Foreign Affairs* 73, no. 5 (1994); George Cabot Lodge, "The World Bank: Mission Uncertain," *Harvard Business School Case* N9-792-100) (1992); and the comprehensive *Bretton Woods: Looking to the Future* (Washington, D.C.: Bretton Woods Commission, July 1994). The most comprehensive conservative critique is Doug Bandow and Ian Vásquez, eds., *Perpetuating Poverty* (Washington: Cato Institute, 1994). Classic studies also include Walden Bello, David Kinsley, and Elaine Elinson, *Development Debacles: The World Bank in the Philippines* (San Francisco: Institute for Food and Development Policy, 1982); Robin Broad, *Unequal Alliance: The World Bank, the IMF, and the Philippines* (Berkeley: University of California Press, 1988); Teresa Hayter and Catherine Watson, *Aid: Rhetoric and Reality* (London: Pluto, 1985); Raymond Mikesell and Lawrence Williams, *International Banks and the Environment* (San Francisco: Sierra Club, 1992); Cheryl Payer, *The World Bank: A Critical Analysis* (New York: Monthly Review, 1982); and Graham Searle, *Major World Bank Projects* (Cornwall: Wadebridge Ecological Centre, 1987), among others.

4. These campaigns are often led by "transnational advocacy networks" of actors "working internationally on an issue, who are bound together by shared values, a common discourse and dense exchanges of information and services" as defined in Margaret Keck and Kathryn Sikkink's path-breaking analysis, *Activists Beyond Borders: Transnational Advocacy Networks in International Politics* (Ithaca: Cornell University, 1998), p. 2. Because not all World Bank campaigns are characterized by such unity and density, this study often uses the term "coalition" to underscore the diversity of civil society participants and the contingency of their joint campaigns. On national and subnational advocacy coalitions in the United States, see Paul Sabatier and Hank Jenkins-Smith, eds., *Policy Change and Learning: An Advocacy Coalition Approach* (Boulder, Colo.: Westview, 1993). In contrast to the vast literature on the Bank, relatively little published research has focused on the advocacy campaigns themselves. See Marcos Arruda, "Building Strategies on the International Financial Institutions," unpublished (Rio de Janeiro: IAP, 1995); Lisa Jordan and Peter van Tuijl, "Democratizing Global Power Relations," unpublished (The Hague: INGI, July 1993); Paul Nelson, "Internationalising Economic and Environmental Policy: Transnational NGO Networks and the World Bank's Expanding Influence,"

Millenium 25, no. 3 (1996); David Wirth, *Environmental Reform of the Multi-al Development Banks* (Flint, Mich.: Charles Stewart Mott Foundation, April 1992). For a critical discussion, see Seamus Cleary, "In Whose Interest? NGO Advocacy Campaigns and the Poorest: An Exploration of Two Indonesian Examples," *International Relations* 12, no. 5 (August 1995).

5. The World Bank claims that poverty reduction and environmental sustainability are top priorities, in contrast to the International Monetary Fund's (IMF) more narrow focus on pro-business macroeconomic stabilization. Although the two institutions are often seen as very similar, the Bank's different goals and standards have created opportunities for civil society leverage, whereas the IMF has remained largely immune to public criticism.

6. Some critics contend that even full implementation of the World Bank's reform promises would not fundamentally change its development model, whose core assumptions are widely questioned in developing countries. Nevertheless, reform implementation would limit the World Bank's impunity and could significantly mitigate the social and environmental damage incurred. For an analogy, the U.S. government's interventionist impulse did not change qualitatively after losing the Vietnam War and conflicts in Laos and Cambodia, but its freedom to use unbounded military force against civilian populations has never been the same.

7. See chapters by Udall and Fox/Brown, this volume and "Interview with David Hunter," *Bankcheck Quarterly* (May 1996). The World Bank is in the process of converting its policies into a hierarchy, ranging from ostensibly obligatory Operational Policies (Ops) to Bank Procedures (Bps) and Good Practices (Gps, once known as "Best Practices"). Most deal with economic, financial, and administrative issues, but the main social and environmental policies include involuntary resettlement, indigenous peoples, environmental assessment, water resources, pesticide management, cultural property, poverty reduction, gender, forestry, energy efficiency, and conservation. See *World Bank Operational Manual*, (Washington, D.C.: World Bank, September, 1995). Many critics are concerned that this process will weaken reform standards.

8. See chapters 9 and 13, this volume. For one of the few studies that stresses the mutually reinforcing interaction between insider and external advocacy strategies, see Nuket Kardam, "Development Approaches and the Role of Policy Advocacy: The Case of the World Bank," *World Development* 21, 11 (1993). See also Deborah Brautigam's "The World Bank, Policy Change and Political Restructuring in Developing Countries," paper presented at the American Political Science Association Meeting, New York, September 1994, which highlights the influence of international social policy experts inside and outside the institution; and Eva Thorne's "The Politics of Policy Compliance: The World Bank and the Social Dimensions of Development," Ph.D. diss., MIT Political Science Dept., 1998. For a conceptual discussion of an "interactive" approach to state-society relations, see Jonathan Fox, *The Politics of Food in Mexico: State Power and Social Mobilization* (Ithaca, N.Y.: Cornell University Press, 1992).

9. From World Bank, *Implementing the World Bank's Gender Policies,* progress report no. 1 (Washington, D.C.: World Bank, March 1996), p. 5. Among the vast literature on structural adjustment, in addition to the references in note 4 and the bank's own publications and the literature on the IMF, see Samir Amin, "Fifty Years Is Enough," *Monthly Review,* 46, no. 11 (April 1995); Development Gap, *Structural Adjustment and the Spreading Crisis in Latin America* (Washington, D.C.: Development Gap, 1995); John Mihevc, *The Market Tells Them So: The World Bank and Economic Fundamentalism in Africa* (London and Penang: Zed, Third World Network, 1995); David Moore and Gerald Schmitz, eds., *Debating Development Discourse: Institutional and Popular Perspectives* (London: Mac-Millan, 1995); Paul Mosley, Turan Subasat, and John Weeks, "Assessing Adjustment in Africa," *World Development* 23, no. 9 (1995); David Reed, ed., *Structural Adjustment and the Environment* (London: Earthscan, 1992); David Reed, ed., *Structural Adjustment, the Environment and Sustainable Development* (London: Earthscan, WWF, 1996); Pamela Starr, ed., *Mortgaging Women's Lives: Feminist Critiques of Structural Adjustment* (London: Zed, 1994); Witness for Peace, *A High Price to Pay: Structural Adjustment and Women in Nicaragua* (Washington: Witness for Peace, 1996). To begin to address the fundamental differences in perception of the social impact of SAPs, in 1996 the Bank's president Wolfensohn agreed to launch a joint research project between the Bank and some of the most critical advocacy NGOs in both North and South to assess how to improve the social impact of adjustment in the future. In the process, he accepted for the first time the principle that macroeconomic adjustment measures should be designed through a process of open debate with civil society. See web site www.dgap.saprin for more information.

10. The roots of the strength of the World Bank's internal neoclassical economic consensus remain debated. Some emphasize hegemony of Anglo-American-trained economists, some stress institutional interests, and others see the Bank as a direct instrument of the United States and other industrial country governments. Yet even radical international political economists differ in their assessment of the balance of power between the international financial institutions and the United States. For example, whereas Samir Amin asserts that "the policies instituted by the international institutions in obedience to strategies adopted by the G7 [rich countries] are the cause of the brutal and massive impoverishment of popular majorities" ("Fifty Years is Enough," p. 10), Cheryl Payer argues in the same journal that "The Fund and the Bank coopted the Reagan conservatives (who in 1981 were making serious noises about defunding the multilateral agencies) into signing on to their already established enterprise, not vice versa" (in her "Bureaucrats and Autocrats," *Monthly Review* 46, no. 11 [April 1995], pp. 52–3).

11. See Keck's chapter in this volume, as well as Pat Aufderheide and Bruce Rich, "Debacle in the Amazon," *Defenders* (March–April 1985); David Price, *Before the Bulldozer: The Nambiquara Indians and the World Bank* (Cabin John, MD: Seven Locks Press, 1989); Rich, *Mortgaging the Earth;* Operations Evaluation Department, World Bank, *World Bank Approaches to the Environment in Brazil: A Review of Selected Projects,* report no. 10039 (Washington, D.C.: World Bank,

April 1992); Ans Lolk, *Forests in International Environmental Politics: International Organisations, NGOs and the Brazilian Amazon* (Utrecht, Nethe International, 1996); and Maria Guadalupe Rodrigues, "Environmental Protection Issue Networks and the Prospects for Sustainable Development," Ph.D. diss., Boston University Political Science Department, 1996.

12. See Wade, "Greening the Bank." For overviews of the U.S. side of the Bank reform campaigns, see Ian Bowles and Cyril Kormos, "Environmental Reform at the World Bank: The Role of the U.S. Congress," *Virginia Journal of International Law* 35, no. 4 (summer 1995); Barbara Bramble and Gareth Porter, "NGO Influence on United States Environmental Politics Abroad," in Andrew Hurrell and Benedict Kingsbury, eds., *The International Politics of the Environment* (New York: Oxford University Press, 1992); Mikesell and Williams, *International Banks*; Rich, "Multilateral Banks" and *Mortgaging the Earth*; Jonathan Sanford, "U.S. Policy Towards the MDBs: The Role of Congress," *George Washington Journal of International Law and Economics* 1 (1988); David Wirth, "Legitimacy, Accountability, and Partnership: A Model for Advocacy on Third World Environmental Issues," *The Yale Law Journal* 100, no. 8 (June 1991); his *Environmental Reform*, and his chapter in this volume. For the most detailed, year-by-year account of the role of the U.S. Congress, see Ian Bowles and C. F. Kormos, "Environmental Reform at the World Bank: The Role of the U.S. Congress," unpublished ms. (Washington, D.C.: Conservation International, 1995). For a congressional assessment of the bank's compliance with its mandate, see *Multilateral Development: Status of World Bank Reforms* (Washington, D.C.: General Accounting Office, GAO/NSIAD-94-190BR, June 1994). The environment is of course only one of many issues on agenda of U.S.–World Bank relations. For overviews, see Bartram Brown, *The United States and the Politicization of the World Bank* (London: Kegan Paul, 1992); Catherine Gwin, *U.S. Relations with the World Bank, 1945–1992* (Washington, D.C.: Brookings Institution, 1994); and Lars Schoultz, "Politics, Economics and U.S. Participation in Multilateral Development Banks," *International Organization* 36, no. 3 (summer 1982); and Bretton Woods Commission, *Bretton Woods*.

13. The founding U.S. organizations included the Natural Resources Defense Council, National Wildlife Federation, and Environmental Policy Institute (later merged with Friends of the Earth), as well as the Environmental Defense Fund and the Sierra Club.

14. See chapter by Covey, this volume; Nelson, *The World Bank* and "Internationalising"; and Wirth, *Environmental Reform*.

15. Johannes Linn, cited in "The World Bank Participation Sourcebook Launch," panel discussion, 26 February 1996, official transcript, pp. 66–7, emphasis added.

16. See Bradford Morse and Thomas Berger, *Sardar Sarovar: The Report of the Independent Review* (Ottawa: Resources Futures International, 1992), discussed in chapters by Udall and Fox, this volume.

17. Priya Kurian shows this clearly in "The U.S. Congress and the World Bank: Impact of News Media on International Environmental Policy," in Bartlett, Kurian, and Malik, *International Organizations.* For examples of the "50 Years" campaign media strategies, see the polished paid announcements on the *New York Times* editorial page produced by the Rainforest Action Network, one of the more militant members of the coalition. The headline on September 26, 1994, was "See the Forest? The World Bank Can't Even See the Trees," and the article began "For fifty years it has been business as usual." On October 4, 1994 the *New York Times* opinion page ad was headlined, "The Quickest Way to Fix the World Bank?" and pictured a hammer coming down on a piggy bank.

18. Cited in Thomas Friedman, "World Bank, at 50, Vows to Do Better," *New York Times*, 24 July 1995. See Danaher, *Fifty Years*, and the U.S. campaign platform summarized in *WHY* (World Hunger Year), no. 16, summer (1994). For a fourteen-page, single-spaced response, see "World Bank Response to U.S. '50 Years in Enough Platform,'" unpublished (1994). Among developing country approaches to the "50 Years" campaign, see *Third World Resurgence* 49 (September 1994). For a more moderate NGO assessment, see Hilary French, "The World Bank: New Fifty, But How Fit? *Worldwatch*, 7, no. 4 (1994). See also, among others, Clay Chandler, "The Growing Urge to Break the Bank," *Washington Post*, 19 June 1994 (subtitled "A Chorus of Critics Say the World Bank Has Hurt the Poor and the Environment"); and Peter Norman, "Worlds Apart on How to Change the World," *Financial Times*, 30 September 1994, who wrote "The '50 Years Is Enough' campaign, a Washington-based lobby group, has mounted a slick, well-oiled crusade to cut the World Bank and the International Monetary Fund down in size.... It wants the International Development Association, the World Bank's soft loan agency for helping the poorest developing nations, to be removed from the Bank's control."

19. Mohan Munasignhe, public lecture at Center for International Studies, Massachusetts Institute of Technology, Boston, 9 March 1993. For his approach, see "Environmental Issues and Economic Decisions in Development Countries," *World Development* 21, no. 11 (November 1993).

20. Frances Seymour and Kathryn Fuller, "Converging Learning: The World Wildlife Fund, World Bank and the Challenge of Sustainable Development," *The Brown Journal of World Affairs* 3, no. 2 (summer/fall, 1996).

21. For useful and complementary applications of organization theory to the issue of World Bank collaboration with NGOs, see Nelson, *The World Bank,* and Michelle Miller-Adams, "The World Bank in the 1990s: Understanding Institutional Change," paper presented to the American Political Science Association, San Francisco, 1996. See also Miller-Adams' Ph.D. dissertation of the same title, Columbia University Political Science April, 1997.

22. Organizational learning is more than just individual learning by actors in positions to shape immediate organizational responses; it involves organizational changes that endure in spite of individual turnover. Organizational learning at the level of correcting errors in minor performance routines goes on as a matter

of course in many organizations, but learning at the level of basic values, goals, and principles is much more difficult and more disruptive when it occurs. See C. Argyris and D. Schon, *Organizational Learning: A Theory of Action Perspective* (Reading, Mass.: Addison-Wesley, 1978); P. M. Senge, *The Fifth Discipline: The Art and Practice of the Learning Organization* (New York: Doubleday, 1990); and J. Swieringa and A. Wierdsma, *Becoming a Learning Organization: Beyond the Learning Curve* (Reading, Mass.: Addison-Wesley, 1992).

23. See C. Margaret Fiol and Marjorie A. Lyles, "Organizational Learning," *Academy of Management Review* 10, no. 4 (1985), p. 806.

24. Many studies of organizational change suggest that both internal and external factors are involved (see D. A. Nodler and M. L. Tushman, "Organizational Frame Bending," *Academy of Management Executive* 3, no. 3, 1989). For examples of the greening of the Bank's official discourse, see especially the research publications of the Bank's vice presidency for Environmentally Sustainable Development. Although these publications are an important vehicle for expressing the views of insider reformists, their impact on Bank lending operations remains an open empirical question. See in particular the Bank's landmark 1992 *World Development Report* (Washington, D.C.: World Bank), which was dedicated to the idea that environment and (market-led) development goals are quite compatible. For a heterodox economic analysis, see Lance Taylor, "The World Bank and the Environment: The *World Development Report 1992*," *World Development* 21, no. 5 (May 1993). Many analysts have noted, however, that World Bank–sponsored research has at most an indirect impact on actual lending operations. For example, Nathaniel Leff's study of the Bank's intellectual investment in developing social cost-benefit analysis in the late 1970s found that this technique had no influence on lending decisions. See his "Policy Research for Improved Organizational Performance," *Journal of Economic Behavior and Organization* 9 (1988).

25. Peter Haas documents the role of "epistemic communities," which are "networks of knowledge-based experts," in the process of changing the behavior of international environmental agencies. See Haas's *Saving the Mediterranean: The Politics of International Environmental Cooperation* (New York: Columbia University Press, 1990) and his edited thematic issue of *International Organization* 46, no. 1 (winter, 1992), where he argues that "actors can learn new patterns of reasoning and consequently begin to pursue new ... interests," and that "outcomes may be shaped by the distribution of information as well as by the distribution of power capabilities" (pp. 2, 5). This approach stresses the importance of how the range of acceptable policy alternatives is framed: "epistemic communities are channels through which new ideas circulate from societies to governments as well as from country to country. However ... [t]he ideas would be sterile without carriers, who function more or less as cognitive baggage handlers as well as gatekeepers governing the entry of new ideas into institutions (1992, p. 27). Like Keck and Sikkink's advocacy networks, epistemic communities are defined by shared belief systems, which contribute to their persistence and solidarity. In context of the World Bank, Deborah Brautigam uses

the concept of "policy networks" to explore the institution's increasing acceptance of the importance of targeted poverty reduction measures ("The World Bank"). For overviews of NGOs and international environmental policy, see, among others, Pratap Chatterjee and Matthias Finger, *The Earth Brokers* (London: Routledge, 1994); Janet Fisher, *The Road From Rio* (Westport, Conn.: Praeger, 1993); Sheldon Kameiniecki, ed., *Environmental Politics in the International Arena* (Albany: State University of New York, 1993); Keck and Sikkink, *Activists Beyond Borders;* Ronnie Lipschutz and Ken Conca, eds., *The State and Social Power in Global Environmental Politics* (New York: Columbia University Press, 1993); Ronnie Lipschutz, *Global Civil Society and Global Environmental Governance* (Albany: SUNY Press, 1996); Thomas Princen and Matthias Finger, *Environmental NGOs in World Politics* (London: Routledge, 1994); and Paul Wapner, *Environmental Activism and World Civic Politics* (Albany: SUNY Press, 1996).

26. Barbara B. Crane and Jason L. Finkle, "Organizational Impediments to Development Assistance: The World Bank's Population Program," *World Politics* 33, no. 4 (July 1981), p. 518. On the organizational obstacles to change at the Bank, see especially Nelson, *The World Bank,* and Rich, *Mortgaging the Earth*, and his "Memorandum: The World Bank—Institutional Problems and Possible Reforms," unpublished ms. (Environmental Defense Fund, March 1995). Note that the analytical issue here is not *whether* the Bank changes—it certainly does, especially in response to donor government priorities and shifting trends in the economics profession. The focus here is on that subset of changes that have emerged at least partially in response to the transnational "public interest" critique.

27. For a classic framing of this problem, see Aaron Wildavsky, "The Self-Evaluating Organization," *Public Administration Review* 32, no. 5 (September/October 1972). For a masterful analysis of a World Bank "analytical" effort to reconcile East Asian economic success with anti-statist neoclassical assumptions, see Robert Wade, "The World Bank and the Art of Paradigm Maintenance: The *East Asian Miracle* as a Response to Japan's Challenge to the Development Consensus," *New Left Review* 217 (1996).

28. This even holds in the case of ozone-depleting chemicals. World Bank economists continue to insist that this key international environmental cost be left out of development investment calcutions, in spite of a U.N.-led consensus that such costs be internalized in economic decision making (Robert Goodland, senior World Bank environmental analyst, interview by Fox, Washington, DC, 3 June 1996).

29. Ernst B. Haas, *When Knowledge is Power* (Berkeley: University of California Press, 1990), p. 37. See also Peter M. Haas and Ernst B. Haas, "Learning to Learn: Improving International Governance," *Global Governance* 1 (1995). Haas's book surveys a wide range of international organizations and finds that the World Bank is one of the few that has shown a capacity for "true learning." This conclusion is based on the secondary literature on the World Bank's turn toward

"new-style" anti-poverty projects under MacNamara in the 1970s. However, it turns out that this literature does not provide empirical support for the view that the World Bank significantly changed the nature of its projects. Robert L. Ayres, for example, one of Haas' key sources, does not claim that the new-style projects had a significant impact (*Banking on the Poor* [Cambridge: MIT Press, 1983]). Rather than claiming that this shift involved a significant change of worldview, Ayres stresses the "tenacity of [the Bank's] ideology" and finds no "paradigmatic shift" (p. 90). In terms of the anti-poverty projects themselves, he refers to their "small, enclave nature" (p. 219), and his evidence does not disprove the critics who suggested that most of the new-style projects failed to have a significant anti-poverty impact and were often captured by local elites. See also the skeptical insider Art van de Laar's *The World Bank and the Poor* (Boston: Martinus Nijhoff, 1980), especially pp. 109–40. Perhaps most problematic for Haas's claim about the World Bank, which attributes anti-poverty shift in the 1970s to institutional learning, is Ayres's own underlying explanation. His extensive insider interviews led him to conclude that "the underlying political rationale for the Bank's poverty-oriented development projects seemed to be political stability through defensive modernization [aimed at] forestalling or preempting social and political pressures" (1983, p. 226). In other words, one could argue that the Bank's shift in the 1970s was driven more by adaptation (to the political threat of then-rising Third World nationalism) than institutional learning in Haas's sense. For then-contemporary radical critiques, see Payer, *The World Bank,* and Hayter and Watson, *Aid.*

30. Studies of change in large corporate systems have indicated that "frame-breaking" changes often required major turnovers in leadership, whereas the rarer "frame-bending" forms of learning required executives willing to undertake major change prior to organizational crises. See Nadler and Tushman, "Organizational Frame Bending." More generally, the patterns of change in organizations seem to follow a pattern of periods of relative stability punctuated by periods of great turbulence and rapid shifting of paradigms. See Connie Gersick, "Revolutionary Change Theories: A Multilevel Exploration of the Punctuated Equilibrium Paradigm," *Academy of Management Review* 16, no. 1 (1991).

31. Even the Bank's new president emphasized his frustration with the institution's resistance to change in an internal meeting with his entire senior management, where he said, "there is so much baggage. And yet there is a need, somehow, to break through this glass wall, this unseen glass wall, to get enthusiasm, change and commitment" (cited in Michael Holman, Patti Waldmeir, and Robert Chote, "World Bank Chief Accuses Staff of Resisting Reforms," *Financial Times,* 29 March 1996).

32. Samuel Paul examined some of the problems of public agency accountability in developing countries in "Accountability in Public Services: Exit, Voice and Control," *World Development* 20, no. 7 (1992). He draws on Hirschman's concepts of exit and voice to propose alternatives to hierarchical control for holding public agencies accountable. Most discussions of accountability, however, tend to be either very abstract and normative or extremely institution specific.

33. For an application of a related conceptual framework to the democratization of authoritarian regimes, see Jonathan Fox, "The Difficult Transition from Clientelism to Citizenship: Lessons from Mexico," *World Politics* 46, no. 2 (January 1994).

34. This approach is at the confluence of two distinct currents in the political science literature. As the discipline increasingly recognizes the importance of transnational political actors and processes and actors, the entrenched barriers between the subdisciplines of international relations and comparative politics are eroding. There is a growing consensus that the question is not whether to integrate the subdisciplines, but rather *how*. The most persuasive efforts so far are Keck and Sikkink, *Activists Beyond Borders;* and Peter Evans, Harold Jacobsen, and Robert Putnam, eds., *Double-Edged Diplomacy: International Bargaining and Domestic Politics* (Berkeley: University of California Press, 1994). Nevertheless, most international relations frameworks are still defined primarily in terms of state-to-state, state-international organization, or state-capital interaction, bringing in civil society actors on an ad hoc basis. Similarly, the conceptual frameworks of comparative politics that do deal systematically with state-society interaction bring in international actors on an ad hoc basis. Each subdiscipline still tends to treat the other's key actors as implicitly monolithic black boxes. To understand the inherently uneven process of transnational institutional change, however, the conceptual challenge is to unpack these black boxes and explain how internal conflicts within each one affect the prospects for change within and between the others.

35. As Peter Haas, Robert Keohane, and Marc Levy's comprehensive study of international environmental policy concludes, *"if there is one key variable accounting for policy change, it is the degree of domestic environmentalist pressure in major industrialized democracies, not the decision-making rules of the relevant international institutions"*. See their edited collection, *Institutions for the Earth* (Cambridge: MIT Press, 1994), p. 14, emphasis in original.

36. See the comprehensive comparative study by Paul Mosley, Jane Harrigan, and John Toye, *Aid and Power: The World Bank and Policy-Based Lending,* 2 vols. (London: Routledge, 1991). In principle, Bank capacity to oblige compliance should be highest in this area of macroeconomic policy, where performance is much simpler to measure than, for example, targeted poverty alleviation projects, whose effects are widely dispersed and difficult to monitor. On the limited effectiveness of "green" conditions on the use of international economic aid, see Robert Keohane, ed., *Institutions for Environmental Aid: Pitfalls and Promise* (Cambridge: MIT Press, 1996).

37. For an unusually detailed country case analysis of this process, see Robin Broad, *Unequal Alliance*, which documents "the World Bank's conscious strategizing to link forces with sympathetic Philippine transnationalists ... tilt[ing] the domestic power struggle in favor of transnationalist over nationalist factions" (p. 11). For a related approach that unpacks both capital and the state in terms of cross-cutting coalitions, see Sylvia Maxfield's *Governing Capital: International Finance and Mexican Politics* (Ithaca: Cornell University Press, 1990). See also

Jonathan Fox, "The World Bank and Mexico: Where Does Civil Society Fit In?" paper presented at the Universidad Nacional Autónoma de México, Facultad de Economía, Mexico, City, March, 1997.

38. For extensive documentation of the Bank's decision to proceed with the Amazon road in spite of detailed information about its social and environmental risks, see Price, *Before the Bulldozer*. On the internal Bank politics behind Narmada, see Wade, "Greening the Bank."

39. This process of taking advantage of the "political opportunity structure" in turn increases the influence of those NGOs that are the gatekeepers of access to political institutions that can be influenced. The U.S. lobbying context also encourages NGOs to focus more on issues where they have potential allies within the state, such as environmental assessment (in contrast to structural adjustment, for example). Under pressure from large environmental constituencies, U.S. congressional and executive leaders from both parties could push for World Bank environmental reforms without questioning the core elements of the dominant macroeconomic model. See Paul Nelson, "Transnational NGO Networks in Global Governance: Promoting 'Participation' at the World Bank," paper presented at the International Studies Association, San Diego, April 1996. On "political opportunity structures" more generally, see Sidney Tarrow, *Power in Movement* (New York: Cambridge University Press, 1994).

40. For analyses of World Bank support for authoritarian regimes, see, for example, the classic study by Walden Bello, David Kinley, and Elaine Elinson, *Development Debacle*; and more recently, Robert Browne and Caleb Rossiter, *Financing Military Rule: The Clinton Administration, The World Bank and Indonesia* (Washington, D.C.: Research Report by the Project on Demilitarization and Democracy, April 1994) and *In the Name of Development: Human Rights and the World Bank in Indonesia* (New York: Lawyers Committee for Human Rights, Institute for Policy Research and Advocacy (ELSAM), July 1995). For overviews, see the proceedings of the Conference on Human Rights, Public Finance and the Development Process, published in *The American University Journal of International Law and Policy* 8, no. 1 (fall, 1992). For case studies of the links between environmental movements and human rights, see Human Rights Watch, Natural Resources Defense Council, *Defending the Earth: Abuses of Human Rights and the Environment* (New York: HRW, NRDC, 1992). On the issue of linking official development aid to human rights issues, see Georg Sorenson, ed., *Political Conditionality* (London: Frank Cass, EADI, 1993).

41. The first major turning point toward increased direct voice for grassroots leaders was in 1987 when Brazilian rubber-tapper leader Chico Mendes lobbied the multilateral banks directly. See, among others, Margaret Keck, "Social Equity and Environmental Politics in Brazil: Lessons from the Rubber Tappers of Acre," *Comparative Politics* 27, 4 (July 1995); and Stephan Schwartzman, "Deforestation and Popular Resistance in Acre: From Local Movement to Global Network," paper presented at the American Anthropological Association, Washington, D.C., 1989.

42. Gopal Siwatoki, public presentation at the "Fifty Years Is Enough" conference, American University, Washington, D.C., 8 November 1995.

43. See "Arun III Campaign Update," unpublished report, 2 November 1995.

44. For examples of some of the World Bank's assumptions about advocacy groups, see World Bank External Affairs Department, *Setting the Record Straight: The World Bank's Response to Bruce Rich's* Mortgaging the Earth, unpublished ms., (Washington, D.C.: World Bank, March 1994), or the more extreme Brigid McMenamin, "Environmental Imperialism," *Forbes*, 20 May 1996. Several cases of international NGO involvement have provoked charges of "environmental imperialism" from grassroots groups and NGOs in developing countries, including the U.S.-NGO-threatened boycott of Scott Paper because of its activities in Indonesia and the conflict over the role of U.S. NGOs negotiating with oil companies over drilling in the Ecuadorian Amazon. On these two cases, see James Riker, "State-NGO Relations and the Politics of Sustainable Development in Indonesia: An Examination of Political Space," paper presented at the American Political Science Association Annual Meetings, Washington, D.C., September 1993, and Joe Kane, "With Spears from All Sides," *The New Yorker*, 27 September 1993. See also Cleary, "In Whose Interest?" For an example of an analysis that assumes rather than demonstrates NGO representation of grassroots interests, see Wapner, *Environmental Activism*. The problem of the lack of NGO accountability to grassroots groups is not limited to international NGOs, however. For examples of critical analyses of local NGO-grassroots relations, see Silvia Rivera Cusicanqui, "Liberal and *Ayllu* Democracy: The Case of Northern Potosí, Bolivia," in Jonathan Fox, ed., *The Challenge of Rural Democratisation: Perspectives from Latin America and the Philippines* (London: Frank Cass, 1990); Jonathan Fox and Luis Hernández, "Mexico's Difficult Democracy: Grassroots Movements, NGOs and Local Government," *Alternatives* 17, no. 2 (spring 1992); and Smitu Kothari, "Social Movements and the Redefinition of Democracy," in Philip Oldenberg, ed., *India Briefing* (Boulder, Colo.: Westview, 1993). For further analyses of NGO accountability issues, see Michael Edwards and David Hulme, eds., *Beyond the Magic Bullet: NGO Performance and Accountability in the Post-Cold War World* (West Hartford: Kumarian, 1996); David Hulme and Michael Edwards, eds., *NGOs, States and Donors: Too Close for Comfort?* (New York: St. Martin's, 1997); and John Farrington and Anthony Bebbington, *Reluctant Partners? Non-Governmental Organizations, the State and Sustainable Agricultural Development* (New York: Routledge, 1993).

45. Among recent studies of grassroots environmental movements, see Robert Bullard, ed., *Confronting Environmental Racism* (Boston: South End Press, 1993); Cleary, "In Whose Interest?"; Mark Dowie, *Losing Ground* (Cambridge: MIT Press, 1995); Al Gedicks, *The New Resource Wars: Native and Environmental Struggles against Multinational Corporations* (Boston: South End Press, 1993); Dharam Ghai, ed., *Development and Environment: Sustaining People and Nature* (Oxford: Blackwell, UNRISD, 1994); Dharam Ghai and Jessica Vivian, ed., *Grassroots Environmental Action* (London: Routledge, 1992); Richard

Hofrichter, ed., *Toxic Struggles: The Theory and Practice of Environmental Justice* (Philadelphia: New Society, 1993); Bron Raymond Taylor, ed., *Ecological Resistance Movements* (Albany: State University of New York, 1995); Jeremy Seabrook, *Victims of Development: Resistance and Alternatives* (London: Verso, 1993); and Andrew Szasz, *Ecopopulism: Toxic Waste and the Movement'for Environmental Justice* (Minneapolis: University of Minnesota Press, 1994).

46. It is worth emphasizing that there are often substantive political differences between coalition partners, not just differences of style. Coalition partners may perceive strategic choices very differently; some may see political trade-offs between means and ends where others do not. Most notably, some may be more willing to bargain for partial concessions than others. There may also be trade-offs between short-term, "elite" NGO lobbying opportunities to influence a particular project versus the more difficult long-term process of bolstering grassroots voice and representation within the transnational advocacy coalition. Recognizing these dilemmas is the first step to confronting them, but even then transnational alliance partners may well have very different ideas about how to balance them.

47. It would have been ideal to have included a chapter written from an insider, World Bank reformist perspective. Several candidates were invited but declined. This gap in the study is partially covered by Fox's chapter on internal Bank dynamics.

48. The one exception is Treakle's chapter on Ecuador's structural adjustment, where the World Bank and Inter-American Development Bank campaigns were closely linked. On the regional multilateral development banks, see Harris Mule and E. Philip English, *The African Development Bank* (Boulder, Colo.: Lynne Reinner, North-South Institute, 1996); Nihal Kappagoda, *The Asian Development Bank* (Boulder, Colo.: Lynne Reinner, North-South Institute, 1995); Chandra Hardy, *The Caribbean Development Bank* (Boulder, Colo.: Lynne Reinner, North-South Institute, 1995); Diana Tussie, *The Inter-American Development Bank* (Boulder, Colo.: Lynne Reinner, North-South Institute, 1995); Roy Culpeper, *Titans or Behemoths?* (Boulder, Colo.: Lynne Reinner, North-South Institute, 1996). On the Asian Development Bank, see also Antonio Quizon and Violeta Perez-Corral, *The NGO Campaign on the Asian Development Bank* (Manila: Asian NGO Coalition for Agrarian Reform and Rural Development, 1995).

49. The original set of campaign cases examined for this volume was larger, but the political sensitivity of the research questions led the leadership of at least two Southern anti-project campaigns to decide that they did not want to be studied by an international research network. We decided to respect their decisions, and those cases were not included in the field research.

50. In addition to sources already cited, see Silvina Arrossi et al., *Funding Community Initiatives* (London: Earthscan, 1994); John Clark, *Democratizing Development* (West Hartford, Conn.: Kumarian, 1991); Tom Carroll, *Intermediary NGOs: The Supporting Link in Grassroots Development* (West Hart-

ford, Conn.: Kumarian, 1992); R. Livernash, "The Growing Influence of NGOs in the Developing World," *Environment* 34, no. 5 (June 1992).

51. For one new analytical approach to the study of NGOs, see E. A. Brett, "Voluntary Agencies as Development Organizations: Theorizing the Problem of Efficiency and Accountability," *Development and Change* 24 (1993); Carrie Meyer, "Environmental NGOs in Ecuador: An Economic Analysis of Institutional Change," *Journal of Developing Areas* 27 (1993); and her "Opportunism and NGOs: Entrepreneurship and Green North-South Transfers," *World Development* 23, no. 8 (1995).

52. See Barbara Gray, *Collaborating: Finding Common Ground for Multiparty Problems* (San Francisco: Jossey-Bass, 1989), and David Brown, "Bridging Organizations and Sustainable Development," *Human Relations* 44, no. 8 (1991).

53. On this issue, see especially Edwards and Hulme, *Beyond the Magic Bullet.*

54. Some of the chapters included in this volume address issues of accountability and grassroots representation, but it will be up to other researchers to do the long-term, participant-observation research needed to provide convincing explanations of the determinants of degree of internal accountability. For a longitudinal study of the changing degrees of leadership accountability over time, see Jonathan Fox, "Democratic Rural Development: Leadership in Regional Peasant Organizations," *Development and Change* 23, no. 2 (April, 1992).

55. See Nelson, *The World Bank* and "Internationalising," and the official annual report, Operations Policy Group, World Bank, *Cooperation Between the World Bank and NGOs: FY 1995 Progress Report* (Washington: World Bank, March 1995).

56. *Indonesia: Kedung Ombo Multipurpose Dam and Irrigation Project (Ln. 2543-IND): Project Completion Report*, unpublished report (Washington, D.C.: World Bank, 1995). This study confirms that the "internationalization of the conflict" helped encourage the Bank to take "corrective action" about the resettlement and rehabilitation of the thousands of displaced families (p. 62). This internal report offers as a "lesson learned" that

NGOs contacted the Bank and the Government of Indonesia as early as 1987 with information that there were major problems with the resettlement operation, much of it subsequently confirmed, but the Bank did not respond to the NGOs or supervise the component for more than a year. This initial lack of interest in dialogue with NGOs contributed to the subsequent polarization of discussion. Had the Bank and project authorities participated in an early, informed discussion of Kedung Ombo with NGOs and others concerned groups, much of the embarrassment and cost suffered by the Bank and borrower subsequently could have been averted. (p. 63)

57. Comprehensive critiques are presented in the special issue of *The Ecologist*, "Banking on Disaster: Indonesia's Transmigration Programme," 16, nos. 2 and 3 (1986).

58. See the chapter by Gray, this volume, for further discussion of the transmigration project; also see Riker, "State-NGO Relations," for more on Indonesian NGO politics. See also Philip Eldridge, *Non-Government Organizations and Democratic Participation in Indonesia* (Kuala Lumpur: Oxford University Press, 1995).

59. Stephan Schwarzman, Environmental Defense Fund, interview by Fox, Washington, D.C., April 1996.

60. For a detailed explanation of the official policy, see Cindy Buhl, *A Citizen's Guide to the World Bank's Indigenous Peoples Policy* (Washington, D.C.: Bank Information Center, 1994).

61. Environment Department, *Resettlement and Development: The Bankwide Review of Projects Involving Involuntary Resettlement, 1986–1993*, Environment Department Papers, No. 32, Resettlement Series (Washington: World Bank, March, 1996 [originally published April 8, 1994]). P. 7

62. From Sidney Tarrow, "Fishnets, Internets and Catnets: Globalization and Transnational Collective Action," *Occasional Paper* (Madrid: Juan March Foundation, Center for Advanced Study in the Social Sciences, winter 1995).

I

Actors

2

Partnership Advocacy in World Bank Environmental Reform

David A. Wirth

The mistake the bank has paid the highest price for is not recognizing the importance of the environment.
—World Bank President Lewis T. Preston[1]

Uniti in vita e in morte
Entrambi troverà.
—Don Alvaro and Don Carlo in *La Forza del Destino*[2]

Notwithstanding the copious literature on the environmental impacts of the World Bank's[3] lending activities in borrowing countries,[4] little attempt has been made to analyze the methods by which nongovernmental organizations (NGOs) undertake to catalyze policy change at that institution. Since at least the early 1970s, public interest environmental groups in the United States have systematically and successfully promoted major policy changes on the domestic level on a wide-ranging environmental agenda, including restoration of air and water quality, exposure to environmental toxins, management of public lands, protection of endangered species, and regulation of pesticide use. But efforts to improve the World Bank's environmental performance, led by many of those same organizations, differ in significant respects from the model of domestic U.S. public policy advocacy on the environment. An examination of the mechanics of the campaign to encourage environmental reform of the World Bank is illustrative not only in revealing the dynamics underlying the international environmental NGO movement, but also in exposing the inner workings of the Bank as an institution. This chapter focuses on relationships between U.S. groups—particularly those with a presence in Washington, D.C., the headquarters of the World

Bank—and counterpart organizations in the Third World and Western Europe.

Partnership Advocacy and the Case Study Approach to Environmental Reform

Over time, the NGO-initiated World Bank environmental reform campaign has come to be characterized by an approach that might fairly be termed *partnership advocacy.* Cooperation among Washington-based advocacy NGOs and counterparts in other donor and recipient countries has become so deeply entrenched in NGOs' interventions with the Bank as to have become an essential component of the reform effort.

The *case study approach,* based on a critique of demonstrable problems with specific projects, has been the principal modus operandi of advocacy NGOs for influencing both specific projects and broader, generic policies. This approach has been perhaps most obvious in campaigns to influence particular Bank-financed infrastructure projects, such as large dams. In such cases, the necessity for a local perspective and accurate on-the-ground information is readily apparent. In such situations, the nexus between Bank-financed activities and the effects of those activities is often clear and immediate, the impacts are highly focused on identifiable individuals and communities, and the need for those voices to be heard is consequently often apparent. Case studies crystallize attention around a concrete, immediate problem, while raising broader issues of Bank policy and accountability.

Because of the importance of the case study approach to influencing the Bank, partnerships that focus on a particular Bank-financed project, as opposed to a generic Bank policy that affects all borrowers, can have an especially great impact both on projects and policies. Where campaigns start with efforts to influence broader policies, however, agreement on campaign goals often involves more players and usually much more diverse constituencies with potentially divergent agendas. Moreover, campaigns with clear time lines and goals (often imposed by project cycles) typically tend to be more coherent than more open-ended policy dialogues.

The case study method can nonetheless be expanded to address broader policies. Through a process of induction, analyses of specific project loans become particularized expressions of generic or systemic difficulties and can be used as a tool to leverage change on those more general policies. For example, in January 1989, the World Bank, under NGO scrutiny, indefinitely deferred consideration of a proposed $500 million electric sector loan to Brazil that would have supported that country's nuclear energy program and a power expansion agenda involving up to seventy-nine dams in the Amazon basin.[5] Here, a loan-specific campaign raised broader issues of policy. In the most notable precedent set by a project-focused campaign, an independent, ad hoc review of the Sardar Sarovar Dam projects in India "inspired" the creation of a new Inspection Panel with the power to report on the adequacy of Bank staff's compliance with the institution's own policies in the context of particular loans.[6] Before describing the various forms that partnership advocacy takes in practice, however, it is instructive to analyze the reasons why this strategy has evolved.

The Role of Partnership in Case Studies of Bank Lending

The most hotly contested debates surrounding World Bank lending have often been factual questions concerning anticipated, ongoing, or completed Bank-financed development projects, some of which are quite complex and all of which are physically located outside the United States. By contrast, many of the more contentious U.S. domestic environmental disputes concern not the integrity of empirical evidence or scientific data, but the policy implications of specific information more or less agreed upon by various stakeholders—such as the environmental community, industry representatives, and governmental agencies. Environmental advocacy organizations often have little in the way of financial or technical resources to undertake independent efforts to monitor air or water quality, for example. As a consequence, in domestic environmental controversies, factual and scientific information generated by governmental authorities tends to dominate the policy debate.

For members of the public in borrowing and donor countries alike, dealing with the World Bank requires a very different approach. Although the situation has changed gradually and incrementally for the better in recent years, the Bank is still notoriously parsimonious with hard data concerning its actual operations, especially those under consideration for future approval.[7] Indeed, many of the documents critical to NGO campaigns became available only through unauthorized disclosures or "leaks" from the Bank. Further, because the necessity for wholesale environmental overhaul has been less than thoroughly internalized by the Bank's management, there is a continuing necessity to establish the factual predicate for policy reform. The consequence is a situation in which detailed, fact-specific case studies of particular loans, most often supporting large-scale infrastructure projects, have been the engine that has driven the environmental reform campaign.

Accordingly, international NGOs working on issues in which the World Bank is a critical player have engaged in "basic research" in a way that is much less necessary in the U.S. domestic arena. The evidence, moreover, is virtually always to be found abroad; thus, partnership with local counterparts becomes both desirable and necessary merely to collect sufficient and minimally accurate information. As a result, the NGO reform effort more often than not has been characterized by disagreements with the Bank over such basic information as the details of project design, the performance of borrowing government agencies, and the severity of actual or anticipated harm to the environment, public health, and quality of life. Although their environmental effects are somewhat less directly attributable to a particular Bank action and consequently more difficult to document, structural adjustment and sector loans, which do not finance an individual, readily identifiable infrastructure project, have nonetheless been successfully analyzed in case studies.[8]

To be sure, NGOs have also made major challenges to broader Bank policies. One good example is the Tropical Forestry Action Plan (TFAP)—an $8 billion joint undertaking of the Food and Agriculture Organization of the United Nations (FAO), the United Nations Development Program (UNDP), and the Bank that was the subject of intense worldwide criticism.[9] Also, in 1989, the World Bank, after exposure in the press, abandoned a premature and ill-advised greenhouse gas policy that would have done nothing more than codify the status quo.[10] Even the Global Environment Facility (GEF), established to support environmentally beneficial activities, has come in for its share of criticism.[11] Environmental considerations in sector loans—which may support governmental activities in an entire sector, such as energy, and which do not necessarily finance specific infrastructure projects—and policy-based adjustment lending—which targets macroeconomic variables such as exchange rates, government deficits, and subsidies with the goal of fundamental economic reform—have also attracted attention from a policy perspective. International campaigns targeted at influencing generic policies, however, have rarely had the same intensity or impact as the controversies surrounding individual World Bank loans that support specific infrastructure projects, such as those discussed in the second section of this book.[12]

The Role of Partnership in Influencing Multilateral Institutions

The World Bank, simultaneously a multilateral organization and a bank, has an unusual governance structure that tends to insulate it from external NGO influence. The members and shareholders of the Bank are sovereign states, as represented by their governments. Each member state is represented by a governor, ordinarily that country's finance minister. The board of governors meets as a body only once a year and in practice gives only very general guidance to the Bank's professional staff. As of this writing, twenty-four executive directors, appointed or elected by member country governments, represent member nations in Washington on a day-to-day basis. They have offices physically located in the World Bank and make the final decisions on staff proposals for individual loans. As in the case of certain other donor states—Japan, Germany, the United Kingdom, and France—the U.S. executive director represents no other member states. Other executive directors represent groups of countries, some of them quite curious. For instance, one executive director represents the unlikely configuration of the Netherlands, Armenia, Bulgaria, Cyprus, Georgia, Israel, Moldova, Romania, and Ukraine. The individual generally identified as the "Canadian" executive director, a donor state, also represents most of the Caribbean countries, which are borrowers.

The board considers more than two hundred loan decisions each year, taking decisions by weighted majority voting. Votes are allocated in proportion to a member state's shares and its capital contribution to the institution. So, as of this writing, among the 177 current members of the International Bank for Reconstruction and Development (IBRD)—the Bank's "hard" loan window—the United States exercises somewhat more than 17 percent of the total voting power, nearly three times as much as the next largest shareholder, Japan. The situation in the International Development Association (IDA), the Bank's "soft" loan window, is similar.

The Bank's professional staff also has considerable autonomy in achieving a clearly identifiable, operational, on-the-ground mission: lending for economic development. The board typically operates by "consensus," implying unanimity. In the unusual situation in which a formal vote is taken, no single Bank member can defeat the approval of a loan proposal. Indeed, even the five largest donors to the Bank acting in concert could not, by themselves, assure that a particular loan proposal would be rejected. For the executive directors who represent groups of countries, particularly those configurations that include both donors and borrowers, the calculus before a negative vote is even more complex than for the large donors. Despite a handful of negative votes at the board level for environmental reasons,[13] effective and lasting change at the Bank requires structural and institutional changes among the Bank's management. Although a negative vote or an abstention by a member government can dramatically draw Bank management's attention to an issue such as the environment, that approach is rarely, if ever, successful in rejecting a staff proposal presented at the board level.[14] Instead, unilateral, informal representations from individual executive directors to Bank management at an early stage in the evolution of a controversy are often more effective in obtaining measurable progress than are formal board-level votes. Indeed, the timely threat of a "no" vote can be considerably more effective than a negative vote actually cast.

Because the Bank is formally accountable to its member governments, those policy interventions that come from a variety of countries are much more effective than those originating from a single Bank member. As an important corollary to this principle, demands from directly affected members of the public in borrowing countries carry a legitimacy that is difficult to deny. Communications from front-line, affected groups or individuals—who have a local perspective, often considerable firsthand factual information, and typically the most to lose—possess a particular credibility. When appeals from the public in recipient countries, in theory the intended beneficiaries of Bank lending, stand in opposition to the positions of their own governments, they create an uncomfortable kind of cognitive dissonance for the Bank's management by forcing a choice between what can appear to be the parochial interest of the borrowing country government, on the one hand, and the Bank's stated mission of sustainable development, on the other.

Partnership as a Legitimizing Factor

The Bank's management naturally views borrowing country governments as an important constituency. Representations strictly on behalf of private groups in the United States and other industrialized countries are vulnerable to accusations pointing to "environmental imperialism" on the part of the Bank and borrowing country governments.[15] Accordingly, Bank staff, implicitly or explicitly, may be unresponsive to interventions made purely by organizations in donor countries. Borrowing country governments, moreover, may dismiss environmental demands emanating from industrialized countries unless a direct connection with their own nationals is apparent. On the other hand, donor country advocacy NGOs have been quite successful in acting as intermediaries for directly affected local groups in borrowing countries, in part because of their greater leverage over their respective governments and therefore over the Bank's most powerful executive directors. Presumably for similar reasons, the Bank's resolution creating the new oversight mechanism, the Inspection Panel, specifies that only organizations in borrowing countries have "standing" to complain of the Bank's actions; in exceptional situations, Washington-based groups may nonetheless serve as local representatives on behalf of affected foreign groups.

Many of the substantive environmental problems confronted in the context of the Bank reform campaign are qualitatively different from those encountered in the United States and thus require skills and expertise that may be unfamiliar to NGOs based in industrialized countries. The World Bank loans of greatest concern can finance massive interventions in the natural environment, such as large dams and irrigation systems, of a scale that would be difficult to imagine domestically. In contrast to the situation in the United States, where issues of pollution are often paramount, questions of resource management are likely to be considerably more important in Third World countries. In the developing world, where many environmental hazards are exacerbated by the twin pressures of poverty and population that are rarely decisive in the United States, inappropriate choices with respect to resource management may have serious consequences for public health and people's ability to provide for themselves and their families. The environmental dilemmas raised by Bank lending can cut to the very heart of national development priorities in economies that are often heavily dependent on the resource base and that emphasize export-led growth and development strategies. More often than not, Third World organizations have more experience working in these contexts than do their Northern counterparts and consequently are more effective in identifying solutions to such local environmental dilemmas.

In addition to improving environmental quality, the World Bank reform campaign has the additional broad and deeply embedded aim of directly enhancing the democratic accountability to the public of a closed, technocratic institution. It is unthinkable that this goal could be achieved without the participation and support of those most directly affected by the Bank's activities.

The campaign to reform the World Bank's and regional banks' environmental practices has focused on both technical-substantive and process-accountability policy reforms. The former directly address substantive environmental issues such as pollution mitigation and forest conservation. The process agenda calls for greater public participation in the design and implementation of loans and improved access to information generated and obtained by the development banks. Realization of process-accountability goals implies a wholesale democratization of the development-lending process as undertaken by both the Bank and the recipient country government. Certain elements of the environmental reform platform incorporate both types of considerations. For example, public participation and transparent policy decision making are central components of the methodology of environmental assessment both in the United States and abroad, which in turn has influenced the Bank's environmental policy.[16]

NGOs can facilitate the democratic decision-making agenda among themselves by cooperating and sharing information with other groups. More specifically, by creating access and leverage for locally based borrowing country groups, donor country NGOs can successfully empower their counterparts to demand what they cannot obtain directly from their own governments or from the Bank. There is variation between those North-South partnerships that end up empowering Southern partners and those that do not or that do so to a lesser degree. Not all partnerships are equally empowering, nor are all equally accountable to the communities most affected by Bank projects or policies (see Brown and Fox, chapter 12, this volume). Nevertheless, the indirect route by which Washington-based organizations act as interlocutor with the U.S. government and the World Bank, although perhaps a second best alternative, is often a more effective mechanism for Third World activists to achieve access than is dialogue with their own governments.

Partnership Advocacy in Practice
The World Bank campaigns include politically diverse constituencies. Among advocacy groups, it is possible to identify two reasonably distinct subperspectives: (1) *reformists* who by and large accept the reality of the continued existence of the international financial institutions, at least for the medium term; and (2) *abolitionists* who tend to identify their goals as elimination of the multilaterals altogether.[17] This distinction may be more philosophical than practical, and in some cases the two approaches seem to coexist within the same organization. In fact, the abolitionists can create a radical pole that makes the reformists appear more moderate and "reasonable," thereby enhancing their effectiveness. (For distinctions among the more moderate NGOs and those participating in the official World Bank–NGO Committee, see Covey's chapter, this volume.) As the terrain of advocacy shifts beyond preventing or mitigating disasters to promoting the implementation of "reformed" social and environmental policies, it will be interesting to see if such mutually reinforcing dynamics are sustained.

Since the partnership advocacy approach emerged in the mid-1980s, Northern and Southern NGOs have gradually increased their levels of coordination and collaboration. Since 1986, NGOs from around the world have attended parallel convocations coinciding with the World Bank/International Monetary Fund (IMF) annual meetings.[18] In recent years, the NGO alternative annual meetings have expanded in participation and focus to include not just environmental organizations, but also church groups and development NGOs, and they increasingly have taken up broader policy issues, such as structural adjustment. These meetings have been a major opportunity to formulate joint agendas, often memorialized in formal statements. One outgrowth of these annual NGO assemblies was a broad advocacy campaign to mark the fiftieth anniversary of the founding of the World Bank and the IMF. Challenging the World Bank's institutional capacity to perform its stated mission, the "Fifty Years Is Enough" campaign highlighted the "strongly held belief by growing numbers of people around the globe that the type of development that the World Bank and the IMF have been promoting cannot be allowed to continue."[19] Structured as a coordinated cluster of national or regional campaigns, this global effort gained more than 140 organizational members in the United States and more than 170 foreign affiliates in about thirty countries. This endeavor triggered a direct response from the Bank, which issued its own vision statement.[20]

North-South NGO partnerships are not the only critical components of Bank reform campaigns. Coordination among advocacy groups in diverse donor countries is also essential for achieving policy change. As noted above, the Bank's governance structure requires coordinated advocacy efforts in North America, Europe, and Japan to pursue environmental change agendas to influence borrowing governments, which in turn are often skeptical of, if not resistant to, policy reform. Western European governments have only recently begun to become more responsive to environmental reform agendas driven by coordinated NGO efforts to target their finance and foreign aid officials as entry points into the World Bank's decision-making process. The Sierra Club developed a close partnership with Euronature, a German organization with offices in Bonn and Brussels that has been active on World Bank–related matters in eastern Europe. Similarly, the German groups Urgewald and World Economy, Ecology, and Development (WEED) have played influential roles in Germany, often coordinating World Bank campaign strategy with the Environmental Defense Fund in Washington. National NGO networks in Britain, France, Holland, Italy, Norway, and Sweden are also working to influence the policies of their national representatives at the World Bank.

A number of U.S. NGOs have created pools of money to be used for small grants to counterpart organizations in developing countries. Small amounts of money can go very far in the Third World, thereby substantially increasing the human resource base abroad. Starting in 1986, the Environmental Policy Institute (now Friends of the Earth–U.S.), the Environmental Defense Fund, the National Wildlife Federation, the Bank Information Center, and the Global Greengrants Fund each developed funds for awarding small grants, distributing at least several hundred thousand dollars to counterparts in developing countries for on-the-ground capacity building, travel, computers, operating expenses, and training.

Some NGOs by their very nature have a multinational character, which makes for natural partnership relationships. One of the most active in Bank reform efforts is the Friends of the Earth (FOE), currently a worldwide federation of fifty-two national affiliates (the Philippine affiliate is central to the discussion in Royo's chapter and the Italian branch is discussed in Keck's chapter, this volume). Approximately 40 percent of these sister groups are located in developing countries. The yearly assembly of FOE affiliates is an occasion for setting a joint agenda, building consensus, sharing information, updating the campaign, undertaking new initiatives, and adopting resolutions. The World Wide Fund for Nature (formerly World Wildlife Fund) is a family of twenty-eight national organizations (NOs) that share finances, information, and publications; it also coordinates multilateral development bank advocacy work among its affiliates as well as with other Northern advocacy NGOs. Greenpeace International, whose headquarters is located in Amsterdam, coordinates international, regional, and national campaign activities in forty-three offices in thirty-two developed and developing countries. The social change–oriented Oxford Committee for Famine Relief (Oxfam) international NGO network has also become very influential, especially on debt and structural adjustment issues.

Sardar Sarovar: The Paradigm Project Campaign

An excellent example of the case study approach to partnership advocacy is the campaign around the Sardar Sarovar Dam on India's Narmada

River, which focused on a 1985 World Bank loan of $450 million.[21] The Sardar Sarovar Dam, the centerpiece of a huge and complex multidam project, quickly became the subject of heated environmental and human rights controversy.

In many ways, the Narmada campaign became a paradigm of project-specific partnership advocacy. Indian nongovernmental organizations and citizens in the region banded together in an aggressive and vocal grassroots coalition known as Narmada Bachao Andolan (Save the Narmada Movement) and made repeated threats that they would perish voluntarily before accepting forced resettlement. The movement's chief spokesperson, Medha Patkar, has achieved the status of a global celebrity.[22] In December 1990 and January 1991, thousands of tribal peoples, oustees, and their supporters staged a "Long March" to the dam site that was interrupted by the police and military. The high level of organization at the local and national level, however, guaranteed a steady flow of both factual information and local perspectives from those affected by the project.

This solid, sustained, in-country base facilitated the efficacy of Northern partners on behalf of their Indian constituents. In most such campaigns, the first representations are typically made to the Bank's professional staff, which has the capability to affect appreciably the borrowing country's behavior.[23] On its own initiative, management may even discontinue disbursements for loans that have been approved by the board of executive directors or, in the case of proposed loans that have not yet been considered by that body, suspend or terminate negotiations with the borrowing country. Because the Bank's management has this significant degree of power, it is often desirable to communicate at the staff level in the beginning of a campaign to exhaust this remedy from a procedural point of view, even if the outcome is likely to be unsatisfactory. In the Sardar Sarovar case, sustained interventions with Bank staff and with upper-level management in 1988 led to a high-level mission to India.

A second strategy in the Sardar Sarovar campaign involved efforts to influence ministers, parliamentarians, and other representatives of national governments in key Bank member states, to whom the individual executive directors are ultimately accountable.[24] On October 24,

1989, U.S. Representative James H. Scheuer (D-N.Y.) chaired a special oversight hearing on the Sardar Sarovar Dam.[25] In April 1990, NGOs organized an International Narmada Symposium in Tokyo, an important venue because the Japanese government was cofinancing the project. More than twenty members of the Japanese Diet wrote to then–World Bank President Barber Conable demanding an end to Bank funding for the dam. Similar activities in Europe resulted in communications from members of the Swedish Parliament and the European Parliament.

A third strategy involved systematic attempts to influence the executive directors individually and, ultimately, to affect the decisions of the board of executive directors as a body. After direct communications from Indian nationals and Washington-based groups, the former and present (as of this writing) Dutch executive directors became personally interested in the potential adverse effects of the Sardar Sarovar loan. In the next stage, representations were made directly to the Bank's governors, individually or collectively. So, during the World Bank/IMF annual meeting in Washington, a full-page, open letter to World Bank President Lewis Preston, endorsed by 250 signatories from thirty-seven countries, appeared in the *Financial Times,* demanding that the World Bank "withdraw from Sardar Sarovar immediately" and challenging donor country contributions to the World Bank.[26]

A fourth strategy involved international media publicity and public education, both crucial elements in the NGO campaign on the Narmada projects. For instance, in 1992 the Environmental Defense Fund and the Bank Information Center created the Narmada International Human Rights Panel to monitor and publicize human rights abuses in the region. During the 1992 monsoon season, members visited the dam area and reported on human rights violations, including a fatal shooting of a tribal woman by local authorities. The Narmada campaign also worked with established human rights networks, such as the Lawyers Committee for Human Rights.

Key members of the Bank's board, faced with sharply conflicting "facts" on the project from Bank staff and NGOs, took the unprecedented step of encouraging Bank management to commission an independent expert review. Led by Bradford Morse, former administrator of UNDP, with Thomas Berger, a noted Canadian human rights lawyer, as deputy chair, the independent review confirmed much of the NGO critique and called for the Bank to "step back" from the project—a high point for the case study–based advocacy campaign approach.[27]

Bank management nonetheless decided to continue to fund the projects.[28] It subsequently decided to make continued disbursement of the 1985 loan conditional on improvements in resettlement before the end of March of 1993, a step that Morse opposed. The Bank's board voted on October 23, 1992, to continue funding the project despite the objections of executive directors representing 42 percent of the Bank's voting power—including directors from the United States, Germany, Japan, Australia, Canada, and the Nordic countries.[29] On March 30, 1993, the deadline for the implementation of the resettlement reforms, the Indian executive director requested the Bank to cancel the remaining disbursements of $170 million and announced that the government of India would finance the remainder of the project by itself. Apparently playing its part in a behind-the-scenes deal, the Bank reasserted its confidence that India had nonetheless satisfied the benchmark conditions precedent for the continued receipt of funds under the loan.[30]

The Narmada campaign gave the NGO campaign the leverage needed to encourage broader procedural reforms at the World Bank. The campaign revealed widespread flaws in the content or implementation of key Bank policies—such as those governing environmental assessment, forced resettlement, the treatment of indigenous peoples, and access to information. The Bank's abdication of responsibility with respect to human rights issues highlighted the multilateral institution's longstanding policy of asserting its legally constrained capacity to respond to human rights violations in its decision-making processes. The campaign also raised issues about the Bank's inconsistent treatment of violations of loan agreements.[31] Seen as but the most visible recent example of systemic deficiencies in project quality and implementation documented in the Bank's internal Wapenhans Report,[32] the Sardar Sarovar project created an opening for Washington-based NGOs to argue that the Bank's funding should be curtailed.[33] The history of the project also exposed the barren cynicism behind the Bank's argument of last resort, namely that the institution's continued presence can nevertheless leverage improvement in borrower performance. If, as in the Sardar Sarovar case, this

"We can do bad projects better than anyone else" philosophy is carried to its logical conclusion, there is no loan so flawed or fundamentally unsound that the Bank should refuse to fund it. As a result, the Narmada campaign generated the momentum needed to catalyze both the creation of the Inspection Panel and the adoption of a more open public information policy (see Udall chapter, this volume).

Policy Reform at the National Level in Donor States

Early on in the environmental reform campaign, the potentially pivotal role of the U.S. government, the World Bank's single largest shareholder, was well appreciated. It goes without saying that the U.S. government will likely be more responsive to input from U.S. citizens than from foreign nationals. Thanks largely to the work of U.S. public interest organizations and bipartisan support among lawmakers, the U.S. Congress has been quite active for the past decade in enacting legislation specifically targeted at encouraging environmental reform of the World Bank and other international financial institutions. The principal strategies employed in legislation to maximize U.S. influence in these multilateral institutions are (1) policy-based instructions to representatives of the United States, such as the U.S. executive director to the World Bank; (2) voting restrictions and mandates to those representatives; and (3) conditions on U.S. contributions to these multilateral institutions. Of the three, the "power of the purse" has generally been regarded as the most effective vehicle for transmitting directives and expectations in a form that will have the greatest impact on the Bank's management. In large measure, this perception accounts for the major role played by the Appropriations Committees in both houses of the Congress on the World Bank environmental reform issue.

A set of hortatory recommendations was the first conduit through which Congress articulated its expectations with respect to environmental reform at the World Bank and the regional development banks.[34] A series of binding appropriations measures have since been an annual feature of the appropriations cycle for the World Bank and other international financial institutions.[35] In 1987, permanent authorizing legislation was signed into law, directing U.S. representatives to the development banks to promote improved environmental performance by these institutions, including the adoption of environmental impact assessment (EIA) procedures. All these measures also require the involvement of the local public and health and environment ministers from borrowing countries in project design, improved access to information, and strict monitoring and enforcement.

A particularly noteworthy development on the legislative front was the enactment of the International Development and Finance Act of 1989, containing the so-called Pelosi Amendment, named after its principal congressional sponsor. This statutory authority requires the U.S. Department of the Treasury to promote the adoption of procedures that improve public access to EIAs at the World Bank, as well as at the regional development banks for Africa, Asia, eastern Europe, and Latin America. Effective December 19, 1991, two years after enactment, the Pelosi Amendment also prohibits the U.S. director of any multilateral development bank from voting for a proposed loan unless an EIA has been prepared 120 days before the vote and disseminated to the public. The purpose of this law is not primarily to trigger a negative vote by the United States or the rejection of a loan proposal, but to encourage the development banks to adopt improved policies. U.S. abstentions with respect to votes with MDBs because of the requirements of the Pelosi amendment have declined since the amendment's passage. In 1995, the United States did not abstain on Pelosi grounds from any projects, compared to its abstention from 17 percent of projects in 1992.

The U.S. executive branch, and particularly the Department of the Treasury—the executive department with principal responsibility for relations between the U.S. government and the MDBs—has been similarly supportive of the environmental reform agenda at the World Bank and the regional banks during the Reagan, Bush, and Clinton presidencies. For example, when James A. Baker III was Reagan's secretary of the treasury, he authorized intervention at the board level on a number of specific development bank loans, both those that had already been approved and those that were proposed for approval but had yet to be implemented. Baker also authorized the adoption of a series of specific generic standards for MDB projects that harm specific sensitive ecosystems—including tropical forests, tropical wetlands, and grasslands—

with the goal of leveraging further policy reform at these multilateral institutions through the exercise of U.S. voting power. Significantly, these voting instructions were drafted with a high degree of input from American NGOs.

The 1994 congressional elections brought the Republicans to power in Congress and reshaped the terrain for advocacy. The bipartisan basis of reform efforts in the Congress had relied, in part, on a coalition between environmentalists and Republican foreign aid critics, a coalition that became much more difficult for many environmental NGOs to sustain after 1994, particularly when the Republican majority started attacking domestic environmental programs. The dominant Republican attitude expressed itself primarily, if not exclusively, as an anti-aid agenda, rather than a strategy for using appropriations as a means of leveraging reform at the Bank. Without leverage in the Congress, partnership advocacy became more difficult in the United States.

Lessons Learned and Questions Unanswered

The partnership advocacy model has plainly been a successful strategy for NGOs to catalyze environmental reform at the World Bank. The teamwork of borrowing country groups and U.S. NGOs allows for a division of labor that synergistically magnifies the impact of each partner's contribution. In an ideal case, Southern groups bring legitimacy, local perspective, and on-the-ground information to the table, whereas their U.S. counterparts provide political access, leverage, media coverage, and an international policy perspective based on accumulated experience that is not confined to a particular country or project. In interactions with third parties—most notably the Bank's professional staff and borrowing country governments—the collaboration can take positions that are both representative of local interests and strategically and technically well informed. Within the international NGO movement, the partnership approach seeks to assure accountability to those with the most at stake and the true beneficiaries of the reform effort, namely the public in borrowing countries.

But precisely because of the accomplishments of the World Bank environmental reform campaign spearheaded by NGOs and the partnership advocacy model that has been a principal strategy for achieving those successes, a large number of second-order dilemmas have emerged. For whatever reason—because they have not arisen in a concrete setting, because they are not considered important, or because they have not been given much thought—these somewhat more sophisticated challenges have received little or no attention from the NGO community.

One of the more salient of these dilemmas concerns the authority of Northern groups to represent their Southern partners. The partnership advocacy model anticipates the formation among private actors of ad hoc, voluntary relationships. Although a standard based on accumulated practice and custom may emerge over time, there are now no explicit, objective ethical criteria applicable to these collaborations like those that govern the attorney-client relationship. For some groups, and certainly for U.S. groups, the role of representing the interests of others, as opposed to their own goals and those of their members, has been a somewhat uneasy one, although tensions may have abated through the years as groups have learned how to work together.

The Bank's board of executive directors has addressed this consideration directly through the mechanism of the World Bank's new Inspection Panel. The resolution creating the panel expressly provides that only "an affected party in the territory of the borrower" may initiate a request for inspection. The resolution anticipates that the complainant (i.e., the affected party) may have a local representative in Washington, or elsewhere if it contends that appropriate representation is not locally available and the executive directors agree at the time they consider the request for inspection.

Another significant concern is the appropriate role of Northern organizations in donor countries such as the United States, which do not borrow from the Bank but which have the most influence at the institution because they are the source of the Bank's capital. It is clear that most, if not all, environmental advocacy organizations in industrialized donor states have their own policy priorities and do not regard themselves as mere agents of their partners in World Bank borrowing countries. Donor country NGOs have their own worries about the propriety and efficacy of the use of their own tax monies to fund the World Bank's overseas operations. The priorities of potential Southern and Northern partners may overlap or coincide with those of counterpart groups in other nations, but ultimately each NGO decides its program for itself.

These considerations suggest that an NGO in a donor country should be free to refuse an overture from a potential collaborator in a borrowing country because of divergent perspectives. If differences arise after a more or less formal partnership relationship has been formed, however, there may be complications. In such a case, if the U.S.-based organization, acting both in a representative capacity and on its own behalf, were to withdraw or abandon the undertaking, the interests of the borrowing country organization may be severely compromised, especially when that organization cannot identify a new collaborator. This counsels both special attention to the identification of a common agenda at the partnership formation stage, as well as a shared understanding of how to sustain and, perhaps, eventually terminate the relationship.

Although this approach might seem suitable to a project-specific intervention, more serious conceptual difficulties arise in the formation of partnerships targeting broader Bank policies that affect the public in all borrowing countries. One difficulty concerns whether Washington-based groups should or must take into account divergent opinions among the NGO community worldwide, and if so, how. Indeed, there is some question whether a worldwide NGO community consensus can be identified, especially considering the diversity among potential public interest–oriented parters. But as NGO efforts become increasingly sophisticated and successful, as more is at stake in NGO dealings with the World Bank, and as the diversity of the NGO movement worldwide expands with an increasingly wide array of viewpoints on an ever-growing number of subjects, the issue of North-South NGO accountability on policy advocacy issues becomes more likely to arise.

Those U.S. groups that have contemplated the matter appear to have a "let a thousand flowers bloom" philosophy. In other words, each group, both in borrowing and donor countries, is free to choose its own collaborators. There is a tension between the obvious necessity for Northern activists to be anchored by alliances with their Southern analogues in borrowing countries and the practical impossibility of representing every NGO in the world. Moreover, some disagreements are bound to arise. In the environmental area alone, active organizations encompass both field-based groups with operational, on-the-ground missions and advocacy organizations whose goal is broad policy reform. Within the subset of environmental advocacy groups, some believe that reform of the World Bank is possible, whereas others express profound skepticism that the Bank and other international financial institutions will ever serve as affirmative vehicles for preserving and improving environmental quality.

This laissez-faire approach does have some drawbacks. In some situations, it may be difficult to identify a minimal level of consensus in the borrowing country—thereby inviting the question, Is a shared in-country perspective a desirable or necessary precondition to intervention by Northern groups? Identifying the "public interest" and a public-policy advocacy agenda is uniquely difficult under such circumstances. When there is no agreement among local groups, an ancillary issue is whether a Washington-based organization can or should avoid exacerbating factions or divisions among in-country NGOs. Yet another question that may arise in such turbulent circumstances is the extent to which representatives of any one group in fact have been deputized to speak for their own constituencies, let alone for any others. Widespread sharing of information with other groups—even those with divergent viewpoints—may at least improve the situation.

At the other end of the spectrum, the "thousand flowers" philosophy quite obviously invites the possibility of marginalizing dissident voices that cannot find counterparts in donor states.[36] Both dilemmas suggest that a necessary, although not always sufficient, condition for a U.S. group to become involved in an overseas controversy is a thorough knowledge of the country concerned. There are also indications from past partnership experiences that action does not always match rhetoric when highly technical matters are at stake, such as end-use energy efficiency in alternatives to Bank-financed power projects. Although partnerships have been highly successful on project campaigns, as in the Narmada case, when it comes to broader policy-level advocacy, NGO technical and information-gathering capacity is limited. These considerations counsel against the creation of "marriages of convenience" and instead suggest the advisability of long-term partnerships in which the collaborators learn to know each other well on a day-to-day basis before initiating potentially stressful confrontations with sophisticated institutions like the World Bank.

There is no longer any question but that North-South partnerships can and do work in practice, and the necessity for such collaborations as a matter of principle is now well accepted. But these lessons only invite yet another tier of trenchant second-order questions, themselves large in number and with, as yet, only incomplete answers. Although to a certain extent the nature of partnership relationships will vary depending on the context, several other unresolved generic concerns can also be identified:

• What attributes, if any, should the proposed partner organizations have before such a relationship is formed? For instance, how representative of local opinion should or must the borrowing country organization be?

• Is it relevant that either or both of the partners receive financial support from governments or have contractual relationships with the World Bank?

• How should a partnership relationship be established?

• What are appropriate terms for a partnership relationship?

• What, if any, objective indicia should there be to demonstrate that such a relationship has been established?

• What obligations do partner organizations have toward each other?

• Under what conditions should the partner be consulted before major decisions are made?

• Under what circumstances may the partners purport to represent each other?

• Is there a difference between active representation of a partner and sympathetic expressions of support for the partner's position?

• Should the partnership relationship imply a commitment of financial resources, and, if so, what kind of commitment?

• What sort of consultative process, if any, should donor country groups engage in before expressing their views on generic policies adopted by the Bank and other international financial institutions?

• What obligations, if any, do the partners have to third parties, such as in-country NGOs with divergent views, that may be interested in the partners' activities?

Based on previous experience, it is unlikely that the NGO community will address these issues in abstract terms. Rather, it will be fascinating to observe the partnership advocacy model as experience accumulates.

After a decade and a half of forming advocacy partnerships around World Bank issues, the NGO community has developed a significant capacity to identify destructive projects before they happen and sometimes even to stop them. The case study approach has succeeded in contributing to the change or creation of Bank policies in a range of areas. NGOs have developed a sophisticated network of trip wires to identify problem projects, a global communications network to link local grassroots activists and skilled lobbyists, and a well-developed capability to use the media.

The partnership advocacy model will need to adapt to a new set of "second-generation" challenges to NGO Bank campaigns. Having successfully pressured the Bank to adopt social and environmental reform policies, the main challenge now, in many cases, is to hold the Bank and borrower governments accountable for the full implementation of those commitments. This kind of advocacy requires not only informed and capable local communities throughout an area where a project is implemented, but also effective "real-time" monitoring capabilities, which in turn require greater organizational capacity at all levels, from the directly affected communities up through the provincial and national capitals to Washington, D.C. Making the transition from doing damage control to holding the World Bank accountable to its policy reform commitments poses a key challenge for future partnership advocacy.

Acknowledgments

This work was supported by grants from the Creswell Foundation and the Frances Lewis Law Center of Washington and Lee University. The author gratefully acknowledges the helpful comments on earlier drafts provided by Patricia Adams, Ian Bowles, Daniel Bradlow, Eric Christensen, Clif Curtis, Jon C. Cooper, Charles Di Leva, William M. Eichbaum, Edward G. Farnworth, Eugene Gibson, Andrew Gray, Michael Gregory, Leanne Grossman, Peter Haas, Faith Halter, Christopher Herman, Ellen S. Kern, Paul Kibel, Lee A. Kimball, Judith Kimerling, Jonathan Lash, Juliette Majot, Michael McCloskey, Alan Miller, Patti L. Petesch, Kal Raustiala, Walter Reid, Lutz Ribbe, Bruce Rich, Armin Rosencranz, Saleemul Huq, S. Jacob Scherr, Frances Seymour, A. Dan

Tarlock, Lori Udall, Stephen Viederman, Edgar Wayburn, Jacob Werksman, and Alexey Yablokov. Steven Patrick provided greatly appreciated research assistance. This chapter draws in part from the author's previously published writings and in part from a presentation by the author entitled "Representing Non-State Actors in International Environmental Controversies" at the "Lawyers' Ethics and International Human Rights Violations: Reconciling Professional Detachment and Moral Anguish" conference at the Fordham University School of Law, New York, New York, on October 20, 1993.

Notes

1. See Clay Chandler, "The World Bank Turns 50, and Promises Some Changes," *Washington Post*, 20 July 1994, p. F1 (quoting Preston); Thomas L. Friedman, "World Bank, At 50, Vows to Do Better," *New York Times*, 24 July 1994, p. 4.

2. Giuseppe Verdi, *La Forza del Destino,* act 3, scene 2 (1862 and 1869) (libretto by Francesco Maria Piave, with additions by Antonio Ghislanzoni).

3. The *World Bank,* as that term is used in this chapter, consists of two legally and financially distinct entities, each established by multilateral treaty: (1) the International Bank for Reconstruction and Development (IBRD), established in 1945, which lends to governments at approximately market rates of interest; and (2) the International Development Association (IDA), established in 1965, which lends at low interest rates to governments of the very poorest countries. The "World Bank group" also includes three other affiliates: the International Finance Corporation (IFC), which, in contrast to the IBRD and IDA, lends to the private sector; the Multilateral Investment Guarantee Agency (MIGA); and the International Center for the Settlement of Investment Disputes (ICSID)—none of which will be explicitly addressed in this chapter.

4. See, for example, Raymond F. Mikesell and Larry Williams, *International Banks and the Environment: From Growth to Sustainability. An Unfinished Agenda* (San Francisco: Sierra Club Books, 1992); Bruce Rich, *Mortgaging the Earth: The World Bank, Environmental Impoverishment, and the Crisis of Development* (Boston: Beacon Press, 1994); Patricia Adams, *Odious Debts: Loose Lending, Corruption, and the Third World's Environmental Legacy* (Toronto: Earthscan and Probe International, 1991); Robert E. Stein, *Banking on the Biosphere? Environmental Procedures and Practices of Nine Multilateral Development Agencies* (Lexington, Ky.: Lexington Books and International Institute for Environment and Development, 1979); Pat Aufderheide and Bruce Rich, "Environmental Reform and the Multilateral Banks," *World Policy Journal* 5 (1988); John Horberry, "The Accountability of Development Assistance Agencies: The Case of Environmental Policy," *Ecology Law Quarterly* 12 (1985); Zygmunt J.

Plater, "Multilateral Development Banks, Environmental Diseconomies, and International Reform Pressures on the Lending Process," *Denver Journal of International Law and Policy* 17 (1988), revised and reprinted in 9 B.C. *Third World Law Journal* 169 (1989); Bruce Rich, "The Multilateral Development Banks, Environmental Policy, and the United States," *Ecology Law Quarterly* 12 (1985), and "The Emperor's New Clothes: The World Bank and Environmental Reform," *World Policy Journal* 7, no. 2 (1990), pp. 305–29; David A. Wirth, "The World Bank and the Environment," *Environment* (December 1986), p. 33.

5. See "World Bank Shelves Vote On Brazilian Loan," *Financial Times*, 12 January 1989, p. 5. See also "Brazil's Future Tied to Bank Fight," *Insight*, 6 February 1989, p. 44; "Citing Environment, Wis. Senator Assails Loan Plan for Brazil," *Boston Globe*, 10 January 1989, p. 7.

6. David B. Hunter and Lori Udall, "The World Bank's New Inspection Panel," *Environment* 36, no. 9 (1994), p. 2. See IBRD res. no. 93-10; IDA res. no. 93-6 (22 September 1993) and Udall chapter, this volume.

7. The World Bank's information policy in principle is guided by a presumption in favor of disclosure. See Udall chapter, this volume, and World Bank, "Disclosure of Operational Information," Bank procedure 17.50 (September 1993). The Bank's most commonly cited reason for refusing to release documentation—protection of confidential relationships with borrowing country governments—is quite plainly not the only operative principle because it has refused to release country-specific documents even with the approval of the borrower. See David A. Wirth, "Legitimacy, Accountability, and Partnership: A Model for Advocacy on Third World Environmental Issues," *Yale Law Journal* 100 (1991), pp. 2645, 2653 n. 31. Although the Bank's current information policy is considerably more forthcoming than before, detailed documentation is still available only for approved loans and not for those under development or "in the pipeline."

8. See, for example, Wilfrido Cruz and Robert Repetto, *The Environmental Effects of Stabilization and Structural Adjustment Programs: The Philippines Case* (Washington, D.C.: World Resources Institute, 1992); David Reed, ed., *Structural Adjustment and the Environment* (Washington, D.C.: Wordwide Fund for Nature, 1992) with case studies of Côte d'Ivoire, Mexico, and Thailand.

9. With the guidance of the World Resources Institute (WRI), a Washington-based policy research organization, the basic outlines of TFAP were set out in 1985. See their *Tropical Forests: A Call for Action* (Washington, D.C.: World Resources Institute, 1985). In 1987, relying on the TFAP infrastructure, then-president of the World Bank Barber Conable promised to double Bank investments for tropical forestry projects within the TFAP infrastructure (Barber B. Conable, president, the World Bank and International Finance Corporation, address to the World Resources Institute, Washington, D.C., 5 May 1987). Worldwide criticism of TFAP alleged that the plan was failing to respond to the deforestation crisis and in some cases exacerbating it by emphasizing export markets and commercial logging; by opening forests to colonization, agriculture, and ranching; by not involving local communities and NGOs in planning and

implementation; and by neglecting the significance of and effects on forest-dwelling peoples. A high-level review team in June 1990 recommended considerable changes, including greater attention to policy reform and to local capacity building. Several unofficial critical reports made additional recommendations: making the TFAP process more open and accountable to the public, developing a new management structure as an alternative to that now housed in FAO's Forestry Department, and improving quality control. By 1990, WRI had concluded that *"TFAP as currently implemented is not achieving many of the plan's original objectives*. Moreover, it seems unlikely that the present TFAP planning process will ever be able to achieve some of them" (Robert Winterbottom, *Taking Stock: The Tropical Forestry Action Plan After Five Years* [Washington, D.C.: World Resources Institute, 1990], emphasis in original). In 1991, TFAP was characterized as "all but disavowed by some of its original sponsors" and then-president of WRI James Gustave Speth described the plan as the "biggest disappointment of my six years at W.R.I." (Eugene Linden, "Good Intentions, Woeful Results: How an Ambitious Environmental Program Ended Up Damaging the Tropical Rain Forests," *Time,* 1 April 1991, p. 48). According to a leading observer of the Bank, "by 1991, the TFAP was for all practical purposes dead" (Bruce Rich, *Mortgaging the Earth,* p. 164 and generally pp. 160–66). See also Mikesell and Williams, *International Banks and the Environments,* pp. 131–34.

10. See Larry B. Stammer, "World Bank May Ignore Climate Fears: Predicted Warming Doesn't Justify Loan Limits," *Los Angeles Times,* 31 August 1989, p. 1.

11. Although this chapter does not address the GEF directly, one could argue that it was created as a multilateral response to growing nongovernmental concern about the environment. After its establishment in 1990, the GEF began a three-year pilot phase with approximately $1.3 billion in funds for distribution on a grant or highly concessional basis for situations in which the recipient country bears the cost of environmental protection, but the benefits accrue to the global community. The GEF addresses four global environmental issues: (1) stratospheric ozone depletion; (2) the "greenhouse" effect; (3) the degradation of international water resources; and (4) loss of biodiversity. It is a joint undertaking of the Bank, the United Nations Environment Program (UNEP), and United Nations Development Program (UNDP), but the Bank administers the funds and manages project implementation. See Andrew Jordan, "Paying the Incremental Costs of Global Environmental Protection: The Evolving Role of GEF," *Environment* (July/August, 1994), p. 12; World Bank, "Documents Concerning the Establishment the Global Environment Facility," 30 *I.L.M.* 1735 (1991); United Nations Development Program, United Nations Environment Program, and World Bank, *Global Environment Facility: Independent Evaluation of the Pilot Phase* (Washington, D.C.: UNDP, UNEP, World Bank, 1994). NGO criticisms have focused on the exclusion of nongovernmental organizations as observers at GEF participants' meetings, lack of access to GEF documentation, and failures in consultation with local communities. See Ian A. Bowles and Glenn Prickett, *Reframing the Green Window: An Analysis of the GEF Pilot Phase Approach to Biodiversity and Global Warming and Recommendations for the Operational*

Phase (Washington, D.C.: Conservation International and Natural Resources Defense Council, 1994); David Reed, ed., *The Global Environment Facility: Sharing Responsibility for the Biosphere* (Washington, D.C.: World Wide Fund for Nature International, 1993); David Reed, "The Global Environment Facility and Non-Governmental Organizations," *American University Journal of International Law and Policy* 9 (1993); and David A. Wirth, *Environment and the International Financial Institutions: The Next Steps* (Washington, D.C.: Sierra Club Center for Environmental Innovation, 1992).

12. See David A. Wirth, *Environmental Reform of the Multilateral Development Banks* (Flint, Mich.: C. S. Mott Foundation, 1992) (quoting World Bank staffer as observing that "[e]veryone in Washington has a recommendation about how we can do better. These case studies are unique in that they concretely demonstrate the need for change" [p. 11]).

13. See, for example, Hobart Rowen, "World Bank, U.S. Dispute Brazil Loans, Agency Approves Funding Despite Officials' Objections," *Washington Post*, 20 June 1986, p. 11 (the first U.S. negative vote for environmental reasons in the World Bank); "U.S., Citing Ecology, Avoids Botswana Loan Vote," *New York Times*, 19 August 1987, p. 2. More common is a scenario like the one surrounding a $20 million forestry loan to Sri Lanka considered by the World Bank in June 1989. Based on the representations of a Lankan group channeled through a U.S. NGO to the U.S. government and then to the Bank, the government of Sri Lanka announced a moratorium on logging operations on forty-five thousand acres of natural forests until ecological surveys were completed. At the World Bank board meeting, the U.S. executive director supported the proposed loan, subject to the full implementation of the Sri Lanka government's promises. See Larry Stammer, "Environmental String Tied to Sri Lanka Loan," *Los Angeles Times*, 17 June 1989, p. 1.

14. See U.S. General Accounting Office, *Multilateral Development: Status of World Bank Reforms*, no. 19 (Washington, D.C.: GPO, 1994). (The GAO noted that from October 1990 through December 1993 the United States voted "no" on twenty-two loans and abstained on ninety-eight more, of which ten negative votes and seventy-five abstentions "stemmed primarily from legislative requirements on the observance of human rights and the performance of environmental impact assessments," but that all 120 loans were nonetheless approved by the Bank's board.)

15. See, for example, "World Bank: Greener Faces for Its Greenbacks," *Economist*, 2 September 1989, p. 41.

16. However, a recent Bank report recognizes that this is one of the weakest parts of the EIA provisions in practice. World Bank, *Mainstreaming the Environment* (Washington, D.C.: World Bank, 1995).

17. See Wirth, *Environmental Reform* and *Environment and the International Institutions*.

18. See Iain Guest, "The World Bank Is Going for the Green," *Washington Post*, 13 October 1991, p. 3; Philip Shabecoff, "World Bank Stressing Environmental

Issues," *New York Times*, 24 September 1990, p. 13; Art Pine, "World Bank, Greens Ready for Faceoff," *Los Angeles Times*, 17 September 1989, p. 1, col. 5; Philip Shabecoff, "Global Bankers and Ecology: Amazon Rain Forest Tells the Story," *New York Times*, 29 September 1987, p. 8.

19. For information on the World Bank/IMF "50 Years Is Enough" U.S. campaign platform (1994), see wb50years//igc.apc.org. For more on the campaign, see regular coverage in *Bankcheck Quarterly*; and Kevin Danaher, ed., *Fifty Years Is Enough: The Case against the World Bank and the International Monetary Fund* (Boston: South End Press, 1994). The quote comes from Clay Chandler, "The Growing Urge to Break the Bank: A Chorus of Critics Say the World Bank Has Hurt the Poor and the Environment," *Washington Post*, 19 June 1994, p. 1.

20. See World Bank Group, *Learning from the Past, Embracing the Future*, 1994 p. 7 for the statement of World Bank President Lewis T. Preston that "[i]nstitutional change is already under way. The themes of human resource development, environmental sustainability, and private-sector development—which are essential to reduce poverty—are being given higher priority." Similarly, the Bank took the unusual step of issuing a formal response to Bruce Rich's *Mortgaging the Earth*. See Andrew Steer, "Errors and Misrepresentations: The World Bank Files a Rebuttal," *Earth Times*, 14 May 1994.

21. This analysis draws largely from Lori Udall, "The International Narmada Campaign: A Case Study of Sustained Advocacy," in William Fisher, ed., *Towards Sustainable Development?* (Armonk, N.Y.: M. E. Sharpe., 1995). See also her chapter in this volume.

22. A cursory search disclosed mention of Medha Patkar's name in the English-language Western press no fewer than 130 times since 1987. See Patrick McCully, "Why I Will Drown: Medha Patkar Talks to Patrick McCully," *Guardian* (London), 16 April 1993, p. 18. Representations from internationally renowned individuals may be especially important in countries that are less tolerant of the formation of associations such as NGOs to represent local interests. For example, Dai Qing, a celebrated journalist and noted advocate of democratic political reform, has become a principal spokesperson representing those opposed to the massive proposed Three Gorges Dam on the Yangtze River in China. See Dai Qing, ed., *Yangtze! Yangtze!: Debate Over the Three Gorges Project*, trans. Patricia Adams and John Thibodeau, (Ann Arbor, Mich.: Association for Asian Studies, 1994). Compare "Saving China from Itself," *Boston Globe*, 1 June 1992, p. 10, an editorial opposing potential construction of Three Gorges Dam, which notes that "[t]here are no opportunities for a Medha Patkar in China, but the World Bank and other outside lenders should treat the Three Gorges project as if there were a Chinese Medha Patkar." In 1992 Professor Wangari Maathai, an internationally recognized Kenyan environmental activist, was charged as a criminal for her supposedly political activities (See letter from Denis D. Afande, ambassador of the Republic of Kenya to the United States, to Michael L. Fischer, executive director, Sierra Club, 22 May 1992, Sierra Club files). All three women

have received the prestigious Goldman Environmental Prize, sometimes called the "Nobel Prize for the environment."

23. Although presented sequentially for the sake of narrative clarity, the strategies described in this analysis can quite obviously be employed simultaneously, as they were in the case of the Sardar Sarovar Dam.

24. There is an ongoing legal controversy concerning the legality of direct instruction of executive directors by the governments of Bank member states. An opinion of the Bank's general counsel asserts that member governments "are under an obligation not to influence the Bank's President and staff in the discharge of their duties, and Executive Directors are under the duty not to act as the instrumentality of members to exert such prohibited influence." See Ibrahim F. I. Shihata, *The World Bank in a Changing World: Selected Essays* (Boston: M. Nijhoff, 1991), p. 107 (quotation and analysis by World Bank general counsel of his own internal memorandum). According to this view, the Bank's executive directors are "officers ... of the Bank" who "owe their duty entirely to the Bank and to no other authority." "Articles of Agreement of the International Bank for Reconstruction and Development," 27 Dec. 1945, art. 5, sec. 5, para. c, 60 stat. 1440, T.I.A.S. no. 1502, 2 U.N.T.S. 134. Consequently, the executive directors may not "interfer[e] ,in the political affairs of any member [or be] influenced ... by the political character of the member or members concerned" in the exercise of their voting rights because "[o]nly economic considerations shall be relevant to [the] decisions" of the Bank and its officers (Article IV, sec. 10). The memorandum consequently concludes that "[t]he Chairman of the Board [of Executive Directors] is entitled to rule out of order a political debate or statement which does not have a clear relevance to the economic considerations related to the subject matter under discussion." (Shihata, *The World Bank,* p. 107). Nonetheless, the U.S. Department of the Treasury, citing the negotiating history and text of the Articles of Agreement, regularly instructs its executive director to the Bank. See Bartram S. Brown, *The United States and the Politicization of the World Bank: Issues of International Law and Policy* (London and New York: K. Paul International, 1992), p. 236.

25. House Committee on Science, Space, and Technology, *Sardar Sarovar Dam Project: Hearing before the Subcommittee on Natural Resources, Agricultural Research and Environment,* 101st Cong., 1st sess., 1989.

26. *Financial Times,* 21 September 1992, p. 6.

27. Bradford Morse and Thomas Berger, *Sardar Sarovar: The Report of the Independent Review* (Ottawa: Resource Futures International, 1992).

28. See *India: The Sardar Sarovar (Narmada) Projects: Management Response,* IBRD. doc. sec. M92-849 (Washington, D.C.: IBRD, 23 June 1992).

29. See "Executive Directors Approve Funding to Complete Controversial Dam in India," *International Environment Reporter Daily,* 27 October 1992.

30. See Steven Holmes, "India Cancels Dam Loan from World Bank," *New York Times,* 31 March 1993, p. 5; Stefan Wagstyl, "India to Drop World Bank

Dam Loans: Government Refuses to Meet Stiff Conditions on $3 Billion Project," *Financial Times*, 30 March 1993, p. 6; "India Asks World Bank to Cancel Loan for Controversial Dam, Canal Project," *International Environment Daily Reporter* (BNA), 1 April 1993.

31. On human rights, see Shihata, *The World Bank in a Changing World*, pp. 97–134, and "The World Bank and Human Rights," *Proceedings of the American Society of International Law* 82 (1988), pp. 60–2. Loan agreements between the Bank and borrowing countries have a status in public international law similar to that of treaties and are enforceable under international law.

32. *Report of the Portfolio Management Task Force: Effective Implementation: Key to Development Impact* (Washington, D.C.: World Bank, 1992).

33. See, for example, House Committee on Banking, Finance, and Urban Affairs, *Reform Measures Adopted by the World Bank: Hearing before the Subcommittee on International Development, Finance, Trade and Monetary Policy*, 103d Cong., 2d Sess. 1994 (for the statement of Lori Udall, International Rivers Network, using Sardar Sarovar as example in urging withholding authorization of the U.S. contribution of $1.25 billion to the third year of the tenth replenishment of IDA).

34. House Committee on Banking, Finance, and Urban Affairs, Subcommittee on International Development Institutions and Finance, *Multilateral Development Bank Activity and the Environment*, 98th Cong., 2d Sess., Committee Print, 1984 (for the recommendations to the executive branch and MDBs concerning improved environmental performance, including consultation with affected public in project preparation).

35. See Ian A. Bowles and Cyril F. Kormos, "Environmental Reform at the Multilateral Development Banks: The Role of the U.S. Congress," *Virginia Journal of International Law* 35, no. 4 (summer 1995), pp. 777–839.

36. See, for example, Wirth, *Environment and the International Institutions*, p. 22, quoting World Bank staffers describing "environmental imperialism" on the part of private groups in the United States and other industrialized countries, characterized as "commissars" putting pressure on Third World organizations "for politically correct purposes."

3

Is Critical Cooperation Possible? Influencing the World Bank through Operational Collaboration and Policy Dialogue

Jane G. Covey

Most nongovernmental organizations (NGOs) around the world rarely interact with the World Bank even though they may recognize important consequences that it has on their nation's development and their work at the grassroots. Depending on an NGO's own mission and development theory, the political environment, and government attitudes, NGOs choose to be more or less active in trying to influence policy and practices of state and international agencies. Many interested in changing development systems prefer to achieve influence through means other than advocacy and direct lobbying.

This chapter looks at two modes of interaction between the World Bank and NGOs to provide a broader view of the range of these modes of engagement. Its focus on project-level collaboration and policy dialogue complements the previous chapter's analysis of the dynamics of an environmental advocacy reform campaign by tracing two other patterns of relationship between NGOs and the World Bank. The chapter looks at trends in World Bank–NGO project cooperation commonly referred to in the World Bank as *operational collaboration,* and reviews the evolution of the NGO–World Bank Committee, one of several channels for policy dialogue with NGOs.

By examining the nature of these alternative influence strategies, the chapter seeks to contribute insights about the range of potentially effective means NGOs can employ for institutional reform. As the World Bank, largely due to outside pressure, adopts policies and practices that create space for greater engagement with civil society, NGOs need a repertoire of influence capacities effective in a wider variety of situations. Specifically, the chapter asks two questions: (1) To what extent are NGO policy dialogue or operational collaboration characterized by a critical perspective? and (2) Can policy dialogue or operational collaboration lead to significant change in the World Bank?

Operational collaboration is defined as direct involvement of NGOs in any phase of Bank-funded projects. Project-level engagement is rapidly increasing and offers interesting opportunities for NGOs to promote on-the-ground implementation of many broad policy reforms that advocacy groups have struggled to achieve. The chapter reviews the history of World Bank–NGO operational collaboration, along with examples of recent experience.

Policy dialogue is defined here as an attempt to influence policy through information exchange and debate that explores both differences and areas of agreement between NGOs and the Bank. It may occur formally, through membership in advisory bodies, for example, or through informal interactions. Policy dialogue is often viewed as "soft," or "insider" advocacy. Its goals are often directed toward incremental reform rather than the fundamental restructuring that often characterizes more confrontational advocacy. Policy dialogue seldom includes high-pressure tactics such as public protest.

Policy dialogue will be examined through the experience of the NGO–World Bank Committee, a nearly fifteen-year-old formal relationship between NGOs from five regions and Bank management. The committee is acknowledged to have made important contributions to the evolution of the Bank's policy on popular participation, but at the same time it has struggled to define a viable policy role.

Operational collaboration and policy dialogue are distinct approaches, although some organizations engage in both. The design and implementation of Bank-funded projects are the locus of operational collaboration; policy dialogue may include NGOs that do not collaborate in projects and may involve policies that govern these operations, whether related to broad development approaches (structural adjustment programs [SAPs]), sectoral concerns (women in development), or operational issues (the project cycle).

Finally, the chapter explores *critical cooperation*, an influence strategy that embodies significant levels of both cooperation and conflict. It proposes that there may be significant opportunities to use a high-cooperation/high-conflict approach to institutional reform, particularly in cases where there are shared interests between civil society and the World Bank.[1]

Operational Collaboration: NGOs in the Project Cycle

The Bank involves NGOs[2] in *operational collaboration*—cooperation in the identification, design, implementation, and evaluation of Bank-funded projects. NGO collaboration is sought when Bank staff believe NGOs can make important contributions to achieving project goals, usually as defined by the World Bank and the borrowing government. NGO involvement spans a wide variety of project types. Bank figures for fiscal year 1995 report that NGOs are associated with more than 70 percent of projects in health, agriculture, and social sectors; 60 percent in industry; 50 percent of all projects in education; and 40 percent in environment, mining, and urban and water projects.[3] In this work, NGOs may play very minor roles (e.g., meet once with Bank staff to give their opinion on some aspect of a project), or they may be significantly involved in defining project goals and methods as well as implementing project activities.

Through this collaboration, NGOs bring operational skill and expertise in grassroots development work. They also potentially channel the views of the grassroots and offer perspectives and analyses that can improve project decisions and operating practices. NGOs have the potential to bring badly needed perspectives into the collaboration, and along with Bank staff and government agencies they can influence project design, implementation, or evaluation. However, to the extent that NGOs merely endorse decisions or implement preestablished designs, they risk being "coopted"—that is, cooperating without influence.

Project design and evaluation by their nature offer more opportunity for influence on projects than implementation. Historically, however, the Bank has rarely involved NGOs "upstream" in the project cycle, where they might influence project definition or design. More commonly, it has invited NGOs to participate in project implementation.[4] Bank-funded projects typically use NGOs to deliver services or involve beneficiaries in project activities. Recently, however, under pressure from advocacy

NGOs, the Bank has changed operational policies to create greater opportunity for more substantial NGO collaboration and greater community participation in shaping Bank policy and projects. For example, the information disclosure policy, participation action plan, and environmental impact assessment (EIA) requirements all demand early project involvement of NGOs and grassroots organizations. Experiments with NGO involvement in World Bank country poverty assessments, for example, create the possibility for civil society influence on the design of far-reaching country assistance strategies (CASs). However, a World Bank Operations Evaluation Department (OED) study shows that participation in poverty assessments and other "building blocks" of a CAS does not actually influence the final Country Assistance Strategy document.[5] The Bank is poised to make some significant changes in its practices, but progress is uneven.

This review of trends in operational collaboration and the nature of NGO involvement in the project cycle is mindful that the Bank is in a state of transition as Bank staff change their practices or resist new policies. Innovations are taking place, but the long-standing culture in the Bank remains a strong force in favor of the status quo.[6] Documented NGO–World Bank operational collaboration experiences reviewed here come primarily from projects designed prior to these new policies. The issues these experiences raise exemplify the changes needed to implement new policies effectively, but they may not fully reflect innovations in current practice.

Trends in Operational Collaboration

Annual Bank reports trace trends in operational collaboration from 1973. They show that NGO involvement was minimal until 1989. It increased dramatically after the Bank issued an operational directive that instructed staff to involve NGOs in projects.[7] Between 1973 and 1988, NGOs were involved in only 6 percent of Bank projects (see figure 3.1). In 1989, this number rose to 20 percent and then averaged about 30 percent up until 1993. In 1994, fully one-half of all projects approved by the executive directors are reported to have some type of NGO involvement. The Bank staff who authored this 1994 report voice some skepticism about the meaning of that figure, however, noting that "NGO

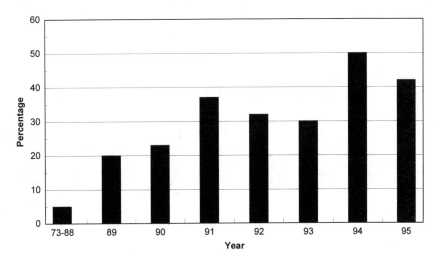

Figure 3.1
Projects with NGO involvement as a percentage of total projects

involvement may be modest and ... Staff Appraisal Reports (SARs) do not always contain detailed or extensive descriptions of such collaboration."[8] The *Fiscal Year 1995 Progress Report* on Bank-NGO cooperation documents NGO participation in 41 percent of all Bank-funded projects that year. This decline from the 1994 percentage is not explained. It could be that NGOs are less involved, or it may be an artifact of more accurate reporting.

Analysis of NGO involvement in the design, implementation, and evaluation stages of the project cycle show a discernable increase in all three areas in 1989 and then again in 1991.[9] Figure 3.2 illustrates the percentage of projects that involve NGOs in project design, implementation, and evaluation as reported by the Bank in 1995.

In the period since 1988, a year when there was minimal involvement of NGOs in implementation and virtually none in design or evaluation, there has been a substantial increase in all three areas according to Bank figures.[10] Participation in implementation increased to a high in 1994 of 43 percent of all projects approved that year. Levels of NGO involvement in project design and evaluation, however, are much less spectacular. They remained well below 20 percent until 1995, when they are reported at 50+ percent and 20 percent respectively. These numbers

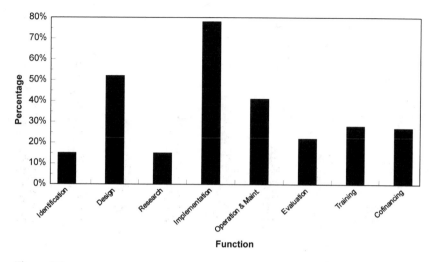

Figure 3.2
Function of NGOs in Bank-financed projects, as a percentage of total NGO involved projects FY 1995

suggest that the Bank is beginning to encourage more NGO involvement in project design and/or that NGOs are successfully demanding greater inclusion.

Other observers and NGOs themselves offer further analysis. Using Bank data in his study of NGO-Bank cooperation for the 1973–1990 and 1991–1992 periods Paul Nelson categorizes NGOs as having "major roles" in one quarter of the projects in which they were involved. "Major roles" include NGO participation in project design, receipt of direct funding, or conflict with the Bank about the impact of a project.[11]

Nelson observes that the dramatic increase of NGO involvement in project design in 1991–1992 reflects an increase in NGO participation in two specific project types. The first type is subprojects, which are NGO activities "funded by a project-financed authority." He found no data in Bank project reports to suggest that NGO design of subprojects influenced the design of the larger project. The second type of NGO involvement in design in 1991 and 1992 was in projects such as Social Investment Funds created to compensate for the impacts of structural adjustment. Nelson concludes that participation in adjustment-related project design may have softened the impact of structural adjustment, but he does not address adjustment policy itself—a major area of concern for NGOs. When Nelson factored out these "marginal" forms of cooperation, he found that NGO involvement in project design remained constant in the 1973–1990 and 1991–1992 periods.[12] Because 1989 begins a period of major growth in NGO participation in design, according to Bank figures (see fig. 3.1), Nelson's analysis raises serious questions about the quality of the World Bank indicators of NGO involvement.

Nelson also looks at the scope of NGO involvement in project design as measured in project dollars. He notes that NGO participation in design of the Ghana transportation and Philippine education subprojects, for example, involved less than 1 percent of these project's funds.[13] If NGO participation touches such a small component of the project portfolio, what kind of meaningful influence do they actually have on development program design?

Although reported NGO involvement in Bank-funded projects is rising quantitatively, there is little, if any, evidence that their involvement is making a substantial difference. Overall, NGOs continue to become involved most often during implementation, when there is little chance to shape the nature of the project. When they are involved in design, they typically participate in "compensatory" projects or marginal components of large projects.[14]

Participation of Local NGOs

Involving affected people in the processes of their own development is a hallmark of current development approaches. This is highlighted in the terms of reference given to a World Bank Task Group on Social Development, which "stated that the building of 'social infrastructure' had 'increasingly eclipsed the building of physical infrastructure as the central challenge of development.'"[15] NGO participation is valued because NGOs are often more effective than government agencies in eliciting community participation and building social infrastructure. "Indigenous" rather than "foreign" (i.e., international) NGOs are also valued because they are part and parcel of the civil society of the nation. As these NGOs grow in number and strength, they build the institutional fabric of this sector and create a force for democratization and improved governance.

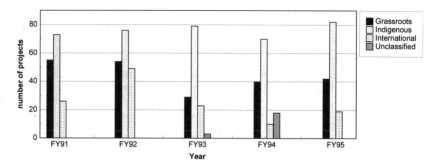

Figure 3.3
Type of NGOs involved in Bank-financed projects FY 1991–1995[21]

Many officials in the Bank value NGOs (especially local NGOs) as lower-cost service providers; one Bank official observed that the best type of NGOs to work with are local groups, such as church NGOs that do not need overhead.[16] Local NGOs are also valued for their local knowledge and contacts helpful in both project design and implementation. Increasingly, the World Bank has also begun to value local NGOs for their capacity to create pressure for institutional change.

As stated in official reports, the Bank "has placed emphasis on engaging local NGOs in the operations it finances."[17] Its definition of "local" NGOs includes a range of groups—from NGOs with national scope to small grassroots groups (often called community-based organizations). The Bank reports concerted efforts to identify local NGOs with whom it can work, in part by recently assigning NGO liaison staff to resident missions. Over time, the involvement of local NGOs in Bank-funded projects has increased and the presence of international NGOs has decreased.[18]

Bank analysis represented in figure 3.3 shows that international NGO operational collaboration has decreased from 34 percent of total projects during the 1973–90 period to 12 percent in 1995.[22] In the light of current development thinking, this is a positive trend. One Bank staff member observes, however, that international NGO involvement is rising from year to year in absolute number of projects even as it declines as a percentage of total NGO involvement. By virtue of their size and institutional capacity, international NGOs can manage larger projects and therefore a large proportion of total project dollars. There are trade-offs for the World Bank in choosing to work with international NGOs, however. Although many of these NGOs are large agencies with the capacity to mount national-level projects and help finance project operations, they often carry higher overhead costs than local groups, including support of "home offices" outside the borrowing country.

Further analysis is needed to disaggregate NGO involvement. How large, in dollars or numbers of people benefited, are the projects involving local NGOs compared to those involving international NGOs? Where international NGOs cooperate in projects, to what extent do they work with local NGOs and act as intermediaries to strengthen local civil society involvement?

The Quality of Operational Collaboration

World Bank statistics on operational collaboration show positive trends in the overall quantity of NGO activity in Bank-funded projects and in both "upstream" and, to some extent, "downstream" NGO involvement in the project cycle. However, Bank officials responsible for the annual progress reports on cooperation express caution in assessing this reported increase in operational collaboration: "Bank-NGO operational collaboration is increasing and becoming a feature of Bank-financed operation. However, "head counts" cannot fully or adequately represent the depth and quality of the NGO involvement"[19]

The following section discusses issues in the quality of operational cooperation from both Bank and NGO perspectives. It is based on a review of selected World Bank documents, NGO project evaluations, and interviews with Bank and NGO staff.

Project Implementation   Much of the documented NGO experience in Bank-funded project implementation predates the recent policy changes and so reflects the "old rules of engagement."

NGOs have quite specific concerns about operational collaboration under these old rules, as do some Bank officials. NGOs commonly perceive that the Bank treats NGO and community involvement as a tool to facilitate project completion, preferably at a low cost, rather than as a way to improve project effectiveness. In the Karnataka Rural Water Supply and Environmental Sanitation Project (KIRWSES), for example, "community participation was invited by the government of Karnataka after the project was approved by IDA [the International Development Association]. [Decision-making] processes were not transparent." As a consequence, "selection of villages was done on grounds other than genuine needs. Some villages did not need the project, certainly not in the form of the package presented to them."[20] Stakeholder participation in decision making during implementation is also problematic from the NGO perspective. In the KIRWSES project, primary stakeholders chose from predetermined options, which led to "squandering of physical installations, inadequate utilization, and lack of local ownership and involvement."[21]

NGOs also commonly perceive that logistical and financial constraints thwart effective operational collaboration and sometimes create hardships for their organizations. In an example from Africa, a large international NGO helped the government negotiate with the Bank for IDA funds for a natural resource project. The international NGO's role in the project was to provide technical assistance to the government and local implementers. Once the loan was negotiated, both Bank and government administrative practices presented problems. According to a staff member of an international conservation NGO, start-up was "very slow due to chaos and delays by the Bank and the government." The Bank task manager was replaced five times during the course of the project start-up, and the project began one year behind schedule. Ultimately, it was difficult for the NGO to get reimbursed for work done, although funds were available in the responsible government ministry. The NGO implemented the project with partial reimbursement and incurred debt to cover its costs. In phase two of this project, however, the NGO negotiated a separate account to alleviate this problem.[22]

Project Design and Evaluation   A 1991 Bank review of thirty-nine projects in Asia discusses upstream collaboration.[23] The author of the review observes that the Bank tends not to involve NGOs in project identification and design in most regions, but that NGOs participated in those activities in more than 50 percent of the Asia projects he reviewed. In several of the examples that illustrate this point, NGOs provided "expert" knowledge of environmental issues or insights about local service demands. In one, an NGO surveyed affected populations to assess needs and prepare a resettlement plan. NGOs were also invited to help the Bank learn from NGO projects that "stimulated demands and enabled the poor to articulate these demands."[24] A number of well-established, internationally known local NGOs worked on Bank projects in Asia, but the report does not comment on whether these NGO recommendations conflicted with Bank opinion, nor whether NGO advice affected design decisions.

The author notes, however, that the record in Asia for NGO involvement in monitoring and evaluation "has been insignificant." He implies that in the small number of instances in which NGOs have been involved in monitoring and evaluation, they have done so unofficially. He comments that NGOs

have contributed to Bank-supported projects by involving and self-advertising themselves rather than being asked up front by the Bank to monitor and evaluate projects. For example, several NGOs involved themselves in the Sardar Sarovar Project in India by critically evaluating the Gujarat Resettlement Scheme.[25]

He recommends that the Bank take a proactive posture toward involving NGOs in monitoring and evaluation, and states that "the proverbial first step in the right direction was taken when two NGO-Bank Committee (India) members were recently contracted to evaluate the Population VII project."

Undoubtedly, the intention of the author was to encourage greater involvement of NGOs in design and evaluation of Bank-funded projects. However, juxtaposing an independent NGO critique of a highly contested project with the suggestion that the Bank "contract" NGOs to undertake project evaluations can reinforce NGO perceptions that the Bank wishes to "use" or coopt NGOs through project collaboration.

Tripartite Relations.   Finally, implementation of Bank-financed projects involves NGOs in a tripartite relationship with government bureaucracies and the Bank. World Vision's experience as an implementing agent for a component of the Ugandan structural adjustment compensatory program exemplifies the dilemmas and challenges in such cooperation.[26] The World Vision evaluation reports:

Experience in this program after 18 months suggests that working in a government partnership with funding from the World Bank is problematic, with unique challenges for achieving effective community development.

There are differences between World Vision's philosophy and approach to ministry [i.e., development] and the World Bank's procedures for supporting development through government agencies.

The report discusses challenges of many kinds: an overambitious project design, bureaucratic inflexibility in the Ugandan government and World Bank, government interdepartmental conflicts and poor capacity, and lack of political will to implement the program. It portrays an NGO project in which work was hamstrung by poor planning, bureaucratic inefficiency and ineptitude, rigid contractual requirements, and inadequate funding flows. It describes classic frustrations of NGOs trying to work with government agencies.[27]

In countries where the state grants limited space for civil society to operate (e.g., Vietnam, Indonesia, many African states), it also often perceives NGOs as a political threat and engages with them ambivalently, if at all. Few opportunities for NGO involvement in Bank-financed projects exist under these circumstances, especially when governments create agencies (government NGOs or GONGOs) to undertake Bank-funded projects. World Bank resident missions can play a constructive role in assisting governments to learn about the benefits of civil society engagement in development processes and to accept NGO roles in projects. They can also enable NGO participation by affirmatively following Bank policy on information disclosure despite government discomfort. The proactive stance of the Vietnam resident mission exemplifies the positive role the Bank can play.

NGOs are giving greater attention to the challenges in government-NGO-Bank cooperation, especially where uneasy relationships exist between the state and civil society. They call on the Bank as a powerful development actor to use its influence to enable productive government-NGO interaction and greater opportunity for the growth of an independent NGO sector. Staff within the Bank who agree with this analysis are helping it to be responsive to these issues to some extent. Its *Participation Sourcebook* offers practical guidance on ways Bank staff can approach constraints in the legal framework governing participation. It addresses the right to information, the right to organize, and the impact of financial and other regulations on NGOs.[28] Experiments with participatory CAS formation bring NGOs and government into dialogue. The Bank has also commissioned a handbook on the legal framework for NGOs[29] in order to assist Bank staff in promoting an "enabling environment." This handbook surveys "best practices" in NGO law and identifies principles for "model" NGO legislation. A Bank consultant, an expert in nonprofit law, assisted Palestine in creating its NGO laws. However, Bank staff are also reluctant to become involved when they feel their efforts to promote NGO-state cooperation might elicit counterproductive state action.

Promising Innovations

Some Bank officials acknowledge that it often does take an instrumental approach to cooperation. However, the Bank's Participation Action Plan[30] sets forth a different standard for interaction with NGOs. It defines participation as "a process through which stakeholders *influence and share control over* development initiatives, decisions and resources which affect them."[31] The spirit of this definition is exemplified by one Bank official's statement: "Through participation we lost 'control' of the project and in doing so gained ownership and sustainability, precious things in our business."[32] These positive examples of NGO-government-Bank cooperation are more common as some Bank staff actively embrace civil society participation as a legitimate and necessary component of successful development. For example, a review of a new ecodevelopment project in India notes that "more than 40 conservation and rural development NGOs participated in the NGO and public consultations."[33] and describes further NGO and community involvement that will occur through "participatory micro-planning" at the village level during implementation.[34] At a recent regional meeting of the NGO–World Bank Committee in Manila the task manager responsible for this project plan attested to the important role NGOs had in project design.[35]

With funding from the U.S. Agency for International Development (USAID), one NGO working in a newly independent state piloted a micro-projects component of a Social Investment Fund. The Bank joined this project in its early phases with the intention of funding later phases by contracting local women's organizations to undertake a component of the social needs assessment. According to the NGO, it works "hand-in-hand" with the government and the World Bank on this project. It credits the Bank task manager with "making the difference" and characterizes her behavior as "Excellent, she shed light on Bank processes and had expertise, while we had cultural knowledge. She was very consistent and dedicated; attended cultural events; minimized the use of outside consultants; and coordinated financing, personnel, and negotiations." The NGO also credited the success of this project to its prior working relationship with the government. With this relationship in place, the NGO felt it could take critical stands when needed because it had earlier established trust with government officials.[36]

NGO Influence through Operational Collaboration

Since 1989, the Bank has been increasing the involvement of NGOs in all aspects of the project cycle. Official Bank figures attribute greater progress and importance to this collaboration than do many NGOs and Bank critics, however. Many observers inside and outside the Bank are skeptical that greater contractual involvement by NGOs in projects will produce greater influence in decision making. The emphasis on NGO involvement in the implementation phase, the Bank's proclivity toward "instrumental" relations, and NGO self-imposed restrictions on engaging in critical dialogue can all limit NGO influence.

However, recent reforms create an opportunity for greater influence through project cooperation. Anecdotal evidence suggests that NGOs can bring critical perspectives into operational collaboration. When they have no opportunity to influence project design, they chafe at the implementation projects that are poorly structured and administered. Close involvement in Bank-funded projects may in fact teach NGOs to be more critical. Many NGOs worry about the gap between Bank rhetoric and practice regarding its ability to reduce poverty and prevent environmental degradation, its will to implement operational policy reforms (e.g., social impact assessment, information, and resettlement policies), and its ability to tolerate loss of operational control even if that loss produces better performance. They also worry that the Bank does little to create conditions for effective government-NGO collaboration at both political and bureaucratic levels.

Through operational collaboration, NGOs have the responsibility to press the Bank to "walk its talk," though not all NGO partners live up to that responsibility. Unfortunately, some NGOs do not have a well-developed capacity for involving primary stakeholders themselves and therefore are not in a position to cooperate critically with the World Bank on these questions. Others avoid conflict with their funding source in order to preserve the relationship and maximize future project opportunities. Where their experience is at odds with project design or implementation, NGOs may need to take forceful initiatives to advocate for project changes.

Policy Dialogue: The NGO–World Bank Committee and Establishing Cooperation

Founded in 1981 to facilitate dialogue and collaboration, the NGO–World Bank Committee has evolved in response to changing development concerns, interactions between the Bank and NGOs inside and outside this forum, and its own dynamics as a global NGO body.

Originally, the committee's purposes were to improve information exchange and to develop and encourage new approaches for World Bank–NGO cooperation. The committee initially served as a mechanism that helped position the Bank and NGOs favorably with one another. It offered the Bank an opportunity to educate NGOs on its approach to economic and sectoral issues, for example, and NGO members were eager to identify ways to expand operational collaboration with the Bank. Fourteen NGO members from Europe, North America, and Japan, and fifteen Bank sectoral and area managers met together twice a year, alternating between Washington, D.C., and a Third World location. Over time, committee NGO membership expanded to include five organizations each from Africa, Asia, and Europe; four from Latin American and the Caribbean; four from the North America/Pacific region; and two international NGOs. Currently, the committee is exploring the inclusion of members from eastern Europe and Middle East regions as well.

Membership has evolved from operationally oriented NGOs to a group predominantly composed of development policy–oriented NGOs and NGO federations or networks. Individual members have ranged from those who are apolitical in their relations to the Bank (especially in earlier years) to a few who are active in the "50 Years Is Enough" campaign (which seeks radical restructuring of the Bretton Woods institutions). One NGO representative active from the mid-1980s to 1993 commented that during his tenure there was a swift change from the right to the left along political lines and that by 1993 the group was relatively balanced. (See the appendix for a list of NGO members in 1986 and 1996.)

The committee elects its members from a roster of self-nominated NGOs. Theoretically, any NGO can seek membership, but knowledge of the committee and its functions has generally not been widespread. Over time, the recruitment and election process has become more open and democratic through procedural changes and the committee's wider visibility to development NGOs worldwide. No formal mechanisms ensure the level of member "representativeness" or the committee's accountability to NGO communities. Nevertheless, the committee expects members to be a conduit for channeling information to and from their respective regions.

Initially, the committee was fully funded by the Bank. NGO members make in-kind contributions (staff, travel, research) to support their own participation, and they underwrite some of the work of the committee. As of 1993–1994, members began to share meeting costs as well. In 1994 they also agreed to pay nominal annual dues and in 1995 set the goal of increasing non-Bank funding to 50 percent of the total budget— in part an effort to establish a more balanced relationship between the two parties and in part to support a growing program because Bank contributions remain constant.

Observers inside and outside the Bank viewed the commitee in its early years simply as a public relations effort. Some NGOs, including committee members, viewed it primarily as a means for expanding project collaboration and identifying funding opportunities. However, as early as 1982, NGO members proposed in-depth discussions of policy issues. By 1984 NGO members had established the NGO Working Group on the World Bank (NGO Working Group) as an autonomous parallel body that included only NGOs. Since then, the group has had a dual identity. The NGO Working Group establishes its own program priorities, agenda for meetings with the Bank, and its operating policies and procedures. The joint NGO–World Bank Committee is the forum for dialogue between Bank staff and the NGO Working Group.

At the time of the 1987 Bank reorganization, responsibility for the committee was moved from External Relations to the Strategic Planning and Review Department. This change appeared to signal greater interest by the Bank in working seriously with NGOs.

In its 1987 spring meeting (in Santo Domingo), the committee issued a *Consensus Document* that recognized the importance of including Southern knowledge and experience in program and policy formulation, and that identified policy dialogue as an explicit purpose of the committee. This document also refined the committee's operating mechanisms in order to expand the role of Southern members. For example, it proposed regular member-organized consultations at national and regional levels to enhance direct Southern consultations with the Bank. An Indian NGO member, AVARD (Association of Voluntary Agencies for Rural Development), took this charge seriously and formed a consultation committee in India that held regular meetings with Bank representatives. Other members, however, were not as successful in setting up local consultation processes.

NGOs Define a Policy Advocacy Agenda
Several related themes have characterized the NGO–World Bank Committee dialogue throughout much of its life: including civil society (especially Southern)voice in Bank policies and projects; ameliorating the negative development and social impacts of structural adjustment; and monitoring the nature of NGO-Bank relationships themselves.

Beginning with discussions on operational collaboration, the committee explored how NGO involvement in projects could improve their effectiveness and influence policy formation. By 1987 the NGO members of the committee introduced structural adjustment into the dialogue, arguing that the costs of adjustment fall inequitably on poorer sectors of society due to program design flaws. This debate drew particularly on the work of Northern NGO members who had in-house research and advocacy capacity. Finally, NGO members of the committee raised the question of popular participation and information disclosure as preconditions for meaningful civil society participation. In spite of its many challenges, the NGO Working Group made contributions in part by simply putting an issue on the agenda of the formal Bank-NGO forum and in part by discussing issues in the spirit of cooperation.

At the Bangkok meeting in 1989, the NGO Working Group wrote a position paper[37] that articulated a broad-based critique of the Bank's development approach. It charged that there was little change in the Bank's approach despite high-level statements about concern for the environment and poverty, called for basic changes in structural adjustment policies, and demanded that local populations have better access to information and more voice in decisions affecting them. It warned that too much Bank emphasis on project cooperation could foster "bogus NGOs" or coopt others. This analysis was accompanied by a set of recommendations for poverty reduction based on an alternative development paradigm that challenged the Bank's neoclassical economic policies.

The Bank's written response disputed specific points in the *Position Paper*, especially those related to structural adjustment, and it criticized NGOs for proposing an "idealistic alternative approach" without defining it. The response also presented a record of Bank responsiveness to NGO concerns, challenged NGOs to acknowledge their responsibilities for effective development, and suggested that the *Position Paper* reiterated "old debates and stereotypes."[38]

If the 1987 Santo Domingo meeting's *Consensus Document* represents a high point of cooperation from the Bank's perspective, the 1989–90 exchange of position papers reflects a split between Bank and NGO Working Group views about the nature of an appropriate relationship. NGO members of the committee believed that their *Position Paper* was important in order to establish Working Group independence and to focus dialogue on critical flaws in Bank policy, thus bringing the committee more into the mainstream of policy dialogue. The paper critically assessed the Bank's policies and performance regarding poverty, gender, and the environment, as well as its record on grassroots participation. Finally, it recommended fundamental changes in the Bank's development model and specific program approaches. As indicated by its formal response, the Bank believed the Working Group had overstepped the boundaries of its competence and of appropriate dialogue.

In this same period, the Bank was preparing its operational directive for relations with NGOs. The NGO Working Group was invited to provide input and did so, but later felt betrayed when the directive was submitted to the board for approval without incorporating the group's suggestions. Bank members' consistently poor meeting attendance and disregard of NGO input signaled the low degree to which the Bank valued and respected the role of the NGO Working Group.

The NGO Working Group continued to formulate an autonomous advocacy agenda that was articulated in the "Saly Declaration" of 1991. It made structural adjustment and popular participation the focal points of the agenda and developed a work plan for the next several years.[39]

About this time, NGO and Bank staff members of the committee agreed that more formal case studies of structural adjustment might facilitate dialogue with the Bank. The Working Group raised funds to undertake independent studies in Sri Lanka, Senegal, and Mexico. The Bank staff supported this project by encouraging government cooperation. Presented at the 1993 Washington meeting, these studies provided substantive points of departure for policy dialogue among NGOs, World Bank officials, and (in the Sri Lanka case) government representatives. In the Mexico case, Bank staff challenged the methodology and accuracy of the case analysis and agreed to supply data to the case writers for further work. Overall, this meeting opened channels for dialogue. Although the quality of the cases varied, the Working Group demonstrated capacity to develop and debate analytic (as well as moral) positions. Agreements to follow up these Washington discussions with meetings in Sri Lanka and Senegal were made, but the meetings never took place. Explanations for this lack of follow-up vary. Some Bank staff believed the quality of the cases did not warrant continued dialogue; NGOs in at least one of the countries believed that resident Bank mission staff were adverse to any engagement with NGOs. In the Mexico case, however, the Working Group paper led to further dialogue and refinement of the analysis, which was subsequently published.[40]

At the global level, the dialogue on structural adjustment was difficult to sustain, and one year later this topic was not on the meeting agenda of the NGO–World Bank Committee, perhaps in part because the terms of some of NGO members most active on this issue expired. The Working

Group's lack of success in substantially influencing structural adjustment policy may stem not only from the complexity of the issues and the centrality of structural adjustment policy to World Bank strategy, but also from the distant relationship between the Working Group and "line management" decision makers. The Working Group was able to produce empirically based analyses of structural adjustment, but could not mount a sustained effort to debate their findings with Bank staff.

During this period the Working Group did contribute to the development of the Bank's policy on popular participation. The Operations Policy Department (OPR), which was responsible for developing participation policy, also oversaw Bank-NGO relations, so the Working Group's contributions were facilitated by this relationship. Bank members of the committee, for example, found the 1989 Bangkok dialogue on participation quite fruitful. In particular, they appreciated NGO frankness and good advice based on experience at the grass roots. Participation was an issue on which Bank staff believed NGOs had substantive knowledge to share. The NGO Working Group continued to pursue the "popular participation" theme through a subgroup that followed up on efforts to alter Bank policies on participatory development and now focuses on implementation of the 1994 board-approved Participation Action Plan.[41]

By 1992–1993, the NGO Working Group had established itself as an actor in policy dialogue on two issues: popular participation and structural adjustment. Through formal channels it brought the perspectives and concerns of development NGOs to the NGO–World Bank Committee and to informal individual contacts between NGO Working Group members, Bank officials, and national government representatives. However, the committee lacked clout, had difficulty sustaining the dialogue, and had severely limited resources with which to improve its substantive expertise on structural adjustment. The Bank representatives on the committee attended meetings once or twice a year, helped NGO members gain access to various Bank officials, and solicited and used NGO advice from time to time. But efforts to engage the Bank in joint analysis on the two NGO policy priorities—structural adjustment and participation—failed. As one NGO member said, "they [Bank staff] were not willing to risk their political capital in this way."[42]

A Split with the MDB Reform Campaign

Toward the end of 1992, NGO advocacy groups outside the NGO Working Group were pressuring the United States to significantly cut funding of the tenth IDA replenishment unless the Bank became more accountable, especially in the Sardar Sarovar project in India (see Wirth and Udall, this volume). The Bank cochair of the Committee called this campaign to the attention of NGO committee members during a Steering Committee meeting. Subsequently, a group of six Southern NGOs from Africa, Asia, and Latin America, including four NGO Working Group members,[43] met with Bank officials in Washington. At least one African representative also met with a U.S. legislator who was a central figure in the debate.

These Southern NGO representatives took a position favorable to IDA replenishment, but expressed concern about how IDA funds were being used. A letter to Bank President Lewis Preston—and signed by the NGO cochair of the NGO–World Bank Committee—summarizes this position.[44] It calls for full IDA 10 replenishment and future funding at increased levels; asserts that IDA funds should not be used for structural adjustment and that popular participation should become a Bank priority; and supports the broad NGO campaign for greater accountability and participation in the use of IDA funds, but cautions that protection of the global environment should not be accomplished by diminishing efforts to reduce poverty.

The Working Group made no specific demands for action on the part of the Bank. Led by Southern NGO perspectives, it supported IDA 10 replenishment without explicitly requiring a quid pro quo. Subsequently, the group discussed ways to monitor progress on IDA reforms, but took no concerted action to implement a systematic monitoring process. This lack of follow-up exemplifies the Working Group's limited capacity to monitor and hold the Bank accountable to a reform agenda. One Southern NGO representative active in the debate acknowledges that NGOs did not have solid proposals for how the Bank could reverse the objectionable features of IDA lending.

The Working Group's expression of support helped the Bank respond to the U.S. Congress's threat to cut off IDA funds and was clearly important to the Bank. President Preston thanked those NGOs personally at the 1993 fall committee meeting, and expressed the hope that it would stay involved in dialogue on IDA 11. Members of the Working Group continued to lobby for IDA with member governments and to discuss IDA funding and conditionalities in regional meetings. As Bank donor (especially U.S.) commitment to IDA eroded following the 1994 U.S. congressional elections, the fight to maintain IDA funding continued to be a priority of the Working Group, but discussion of IDA reform has also continued. The alliance between the Working Group and the Bank on IDA replenishment has been strengthened by increasing threats to IDA from donor countries.

Until the Working Group made the intervention in 1992, activists in the multilateral development bank campaign (the "MDB campaign") for Bank reform had largely dismissed the NGO Working Group as an ineffectual "nonplayer" because as a body it had not participated actively in the reform campaign. In retrospect, some Washington-based activists acknowledge that they had not sufficiently communicated with the Working Group or tried to gain their support for the campaign. Many feel that the IDA 10 episode exacerbated international and intercoalitional NGO tensions,[45] and some NGO advocates also interpret it as a "divide-and-conquer" tactic by the Bank in which the NGO Working Group position was held out as "more legitimate" because it was rooted in Southern NGO views. Generally, Southern NGOs with a poverty focus feel that the episode simply revealed the diversity of NGO interests and concerns. Certainly, it highlights the fact that NGOs working on World Bank reform have different priorities and different strategies that are sometimes in conflict (at least in the short term) with one another.

In many ways the IDA 10 campaign was a watershed event in relations among transnational NGO alliances. The incident demonstrated that the NGO Working Group (employing the relatively weak intervention of policy dialogue) could use its access to Bank officials to influence decision makers on contested issues. Some advocacy activists paid more attention to the NGO Working Group as a result, occasionally inviting it to support their positions, such as the demand for common disclosure and information policies in other World Bank units like the International Finance Corporation (IFC). Some however, expressed concern that NGO Working Group participation in various NGO-Bank fora might actually dilute their own efforts.

Tensions between NGO Working Group members and campaign activists continue to surface from time to time. For example, when the Working Group perceived that Southern voices were excluded from important advocacy NGO-Bank debates, or when the Working Group did not carry out agreed-upon advocacy actions that were the priorities of more critical groups, the two camps were likely to challenge each other.

Expanding Access, Relevance, and Effectiveness

In the early 1980s, when the NGO–World Bank Committee was formed, the Bank had few channels for policy dialogue with NGOs. The committee became a forum where Bank officials could engage with NGOs on topics of mutual interest. However, by the early 1990s the Bank had many more opportunities for such interaction and had gained considerable sophistication in engaging with NGOs. Now that Bank staff are talking with NGOs in a broad variety of fora in Washington and the regions, the purpose and unique contribution of the NGO Working Group are being questioned. No one is fully satisfied with the committee's past role or performance. Bank staff who value NGO participation in Bank reform are sometimes disappointed that the committee doesn't fulfill its potential. Others wonder what the appropriate role for a "generalist NGO body" is at a time when the Bank has begun to establish advisory bodies with specialist NGOs, such as the Consultative Group to Assist the Poorest (CGAP) to advise on microenterprise lending. Some Bank staff wonder about the relative return on investment in underwriting the generalist Working Group rather than other fora,[46] and some NGO members are concerned that the body lacks collective capacity to play a significant role in policy dialogue.

NGOs and Bank staff agree that there are limitations to the NGO Working Group's effectiveness. The group is essentially a voluntary body that relies on member resources to support its activities, meeting attendance, and basic secretariat functions. Few members have dedicated internal resources to the committee's work, and few have the ability to sustain ongoing in-depth dialogue with the Bank (especially in Washington). These resource limitations constrain the committee's ability to build competence and implement strategies for effectively linking microperspectives to macroissues.

A second limitation to the NGO Working Group's effectiveness is its lack of connection to the broad base of national NGO communities. The Working Group as a whole is not formally representative of any constituency. Historically, few members have demonstrated the capacity or the commitment to ground their participation in broad-based dialogue among their colleagues in a national/regional context. At best, members speak on behalf of themselves and their immediate constituencies, not a larger NGO community. Since the spring of 1995 that constraint has been partially ameliorated by regional meetings to which larger numbers of NGOs have been invited. Participants and observers agree that the fall 1995 annual meeting was substantially strengthened by the regional dialogues. NGO positions were better grounded and analyzed, and they conveyed the collective wisdom of a greater number of NGOs worldwide.[47] In their second year these regional fora further expanded in size. They now seem to be a useful way for both NGOs and the Bank to increase the inclusiveness of the dialogue.

The NGO Working Group membership is geographically and institutionally diverse, which can be both a strength and a limitation. The Working Group includes up to twenty-six country perspectives and organizations with operational, policy advocacy, networking, and grant-making capacities. Some have program units working on key policy issues such as multilateral debt, Social Investment Funds, and participatory evaluation. Some advise government and donors, whereas others work directly at the grass roots. One Bank staff member observes, "[The committee] is an effective way to get messages out, and an effective way to bring together, consolidate and prioritize NGO observations."[48] This diversity can result in thoughtful, nuanced analyses that bring important insights into the debate, or it can lead to a least-common-denominator formulation of views. Where the NGO Working Group has sustained substantive dialogue (e.g., its work on popular participation), its diversity seems to be an asset.

Assessing the Policy Impact of the NGO Working Group

Among Bank staff and NGOs familiar with the Working Group, there is substantial agreement on its contributions. Its role in promoting the Bank's development of a participation policy is consistently named as its most valuable contribution. The committee did not put the issue on the Bank agenda singlehandedly, but it was a key catalyst by raising the issue, by recommending a learning approach to policy change, and by consistently challenging the Bank's attitude toward participation in its discussions with the Bank. It also has been an ally to internal Bank reformers as the policy was developed. The committee's decision to make this issue one of its priorities, its competence in the area, and its official standing with the Bank all helped get and keep participation on the Bank reform agenda. In this instance, a critically cooperative dialogue contributed to Bank policy reform. NGOs and the Bank had a mutual agenda, and NGOs brought recognized, substantive expertise to the table.

There is less consensus on the Working Group's impact on the structural adjustment debate. Members who were involved in the group believe that its work in the 1980s helped increase the Bank's recognition of the issues surrounding structural adjustment and shift the focus of policy debate from a parallel strategy for mitigating social impacts of adjustment to incorporating these concerns into structural adjustment policy itself. Although the Working Group joined the debate early and through several members invested considerable resources and time, it lacked the combination of expertise, access, clout, and persistence to sustain a concerted long-term advocacy effort in this important policy area. Although not strongly influential itself, the Working Group's role in the structural adjustment debate may illustrate the view that "the Working Group has been strategic in terms of reinforcing (if not setting) policy issue areas."[49]

The Role of the NGO Working Group

The Working Group's story is one of attempting to establish an independent agenda, searching for common ground, and seeking to raise critical policy questions. As its history shows, a loosely knit, globally spread group of NGOs can encounter difficulty in establishing and maintaining effective policy dialogue with the Bank. The committee has sometimes met the Bank on its own analytic terms in debate, but it has seldom employed dramatic power-balancing tactics to gain influence. In its fifteen years of existence, the Working Group has evolved in a number of important ways. Most significant perhaps is the shift of power within the Working Group itself. Originally dominated by Northern NGOs, its leadership now is squarely in the hands of its Southern members. Operational NGOs that were interested in securing project funding dominated membership in the early years; in contrast, today's membership is predominantly intermediary NGOS and NGO networks concerned about policy reform. Members tend to agree on the priority issues, yet they are diverse in their analysis, their specific goals for Bank reform, and their influence style. The Working Group has also evolved from a strictly self-reproducing "club" to an emerging catalyst for broad-based Bank-NGO discussions by means of its annual regional meetings and a growing sensitivity to its responsibility to champion inclusiveness.

However, the role of the Working Group is not without controversy. NGOs critical of its structure and its role in World Bank reform cite its lack of financial independence from the Bank, its limited technical competence, and its lack of accountability to NGO constituencies as barriers to its effectiveness and legitimacy. Concerned, in part, that the Working Group is a tool with which the Bank can blunt strong advocacy, these critics particularly worry that the NGO Working Group is too responsive to Bank requests for assistance (e.g., IDA 10) and that it does not sufficiently support NGO interests as expressed through advocacy groups and campaigns.

The Working Group has been controversial among Bank officials on several dimensions. They too criticize its lack of substantive rigor and its inability to deliver analyses that meet Bank standards, a criticism leveled especially when the NGO Working Group oversteps Bank views of the proper boundaries of the policy dialogue in either substance or style as exemplified in the polemics of the 1989 *Position Paper*. For some Bank staff, the question is whether the Working Group delivers products worth its costs, particularly when they deem its meetings unsatisfactory and see other more specialized NGO advisory bodies as a productive alternative. Bank officials working for a more people-centered World Bank perhaps express the greatest frustration. Needing strong external allies, they lament the Working Group's unrealized potential.

Critical Cooperation in World Bank Reform

The struggles for accountability described and analyzed in this book attest to the need for grassroots groups and NGOs to challenge World

Bank policies and practices that cause rather than ameliorate poverty and environmental degradation. Clearly, advocacy and even civil disobedience have been useful in influencing institutions like the World Bank. This chapter asks if the strategies of operational collaboration and policy dialogue can be carried out with a critical perspective and if they can promote significant changes.

If one looks at World Bank–NGO relations prior to 1981, critical cooperation as a strategy for change was not a practical option at either policy or project levels. The Bank "used" NGO policy dialogue for public relations purposes and contracted with NGOs to deliver services. In this era, Bank decisions were neither transparent nor negotiable: the Bank paid for cooperation; it largely ignored criticism.

Similarly, NGOs interested in accessing World Bank projects and resources were for the most part not interested in or skilled at influencing World Bank policies and programs. They frequently eschewed criticism at either micro- or macrolevels because they agreed wholeheartedly with Bank practices or because economic and institutional considerations pressed for smooth relations.

The decade-long NGO struggle to make the World Bank more accountable for the impacts of its policies and practices has opened the Bank to more public scrutiny and has substantially changed the situation in which civil society organizations and the Bank interact. By 1995 the World Bank had adopted a series of policy reforms that enable or even require more stakeholder participation and Bank accountability. It has also raised standards for its own performance by creating mechanisms to monitor and evaluate consequences (e.g., environmental and social impact assessments). These policies, if fully implemented, have the potential to transform many Bank practices, increase its responsiveness to grassroots voices, and result in more sustainable development programs.

Protest and advocacy continue to be essential reform strategies, particularly where the Bank remains closed to outside influence.[50] But as the Bank begins to implement policies with which NGOs essentially agree, there is greater space for cooperation that can reinforce positive change, monitor progress, and maintain outside pressure to help bring about fuller implementation of desired Bank policies.

The history of NGO-Bank relations suggests to many that conflict (advocacy) or subordinate cooperation (cooptation) are the only

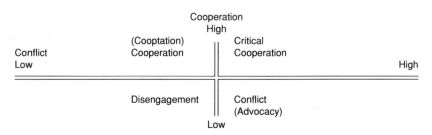

Figure 3.4
One-dimensional Bank-NGO relations

Figure 3.5
Two-dimensional Bank-NGO Relations

viable options for NGO engagement with the Bank. Informed by conventional wisdom, this perspective arrays conflict and cooperation along a single continuum (see fig. 3.4). One is either highly critical or highly cooperative—at either end of the continuum—or a little of both (wishywashy) and in the middle. To be both highly cooperative and highly critical is not conceivable within this unidimensional framework.

But cooperation and conflict are not necessarily poles of the same dimension.[51] If they are treated as two separate dimensions, as in figure 3.5, then it becomes possible to have interaction marked by both conflict and cooperation—the *critical cooperation* of the upper-right quadrant of figure 3.5. *Disengagement* is the fourth strategy for NGO-Bank relations in this model, involving low conflict and low cooperation.

Critical cooperation is civil society–Bank engagement that includes elements of both conflict and cooperation and so reflects a mix of differing and converging interests. It also involves a relationship component, expressed by one NGO as "a willingness to work with another institution while maintaining independence and critical distance."[52] In the Tondo Foreshore Development Project in the Philippines, for example, a federation of grassroots organizations challenged plans for urban renewal by the Bank and the Philippine government. For several years this engagement included both fierce conflict as well as cooperation and re-sulted in the redefinition of the initial project from redevelopment (slum clearance) to rehabilitation (improvement of existing residences), with much of the improvement to be carried out by local residents.[53] The evolution of the role of the NGO Working Group is another example of critical cooperation. Its successful dialogue on popular participation was carried out after it had (1) committed itself to principles of cooperation (the *Consensus Document* of 1987), and (2) asserted its independence (the *Position Paper* of 1989). Its emerging dialogue with the Bank on the design of Social Investment Funds offers fertile ground for critical coop-eration as well. Although NGOs and the Bank agree on the broad out-lines of these programs, they nevertheless still need to discuss many areas of fundamental differences.

The research literature on negotiation and conflict management sug-gests that critical cooperation is an appropriate strategy when parties desire both substantive outcomes (e.g., improving flow of benefits to women or improving portfolio performance) and a constructive rela-tionship with each other (e.g., continued civil society participation in project design). In contrast, when the substantive outcome (e.g., pre-venting forced resettlement) is much more important than preserving the relationship (e.g., NGOs do not see any long-term need for good relations with the Bank), then more adversarial strategies that achieve substantive gains without worrying about costs to the relationship are appropriate.[55]

Challenges of Critical Cooperation

Establishing and maintaining both high cooperation and high conflict is difficult. Some activists who have engaged in negotiation with the World Bank believe that prevailing conditions constrain or even preclude critical cooperation. They contend that the forces of a powerful institution such as the World Bank compel unconditional cooperation on one hand or advocacy and protest on the other. They also often believe that by ac-cepting a role in Bank-funded projects, NGOs are necessarily constrained by professional, institutional, and economic interests to forego their in-dependence and therefore their capacity for conflict. According to this view, collaborating NGOs that access Bank funding may think they are making a difference, but in reality self-interest clouds their ability to assess their situation.

Conditions where critical cooperation is a potentially viable alternative may, however, be found in a variety of situations: when NGOs and Bank staff share mutual interests, such as engaging local communities in the design of Social Investment Fund projects; when NGOs endorse institutional reforms, such as revamping the project cycle; or when NGO and Bank goals overlap, such as implementing debt relief for highly indebted countries.

Neither NGOs nor Bank staff have much experience in using critical cooperation, though. Where opportunities for this type of influence in fact exist, NGOs may miss them because they are accustomed to using purer forms of either cooperation or conflict, or they have been disengaged from World Bank affairs altogether. Bank staff, with many years of experience in making decisions without much mutual influence, may be inept at cooperation and unwilling to accept a new paradigm.

The choice to use critical cooperation requires good analysis of the situation in order to understand one's own and the other parties' interests, assess constraints and opportunities, and realistically appraise the larger forces at work in complex development policy and project contexts.

In addition, critical cooperation demands a learning process in which both parties (World Bank and NGOs) accept the notion of mutual influence and evolve the capacities necessary to achieve this goal—such as tolerance, the ability to confront differences, and the willingness to listen to one another. Critical cooperation demands a sophisticated blend of criticism and confrontation with understanding and problem solving. Finding and managing a good balance of these different elements in a dynamic situation at either project or policy levels are quite demanding tasks.

Additionally, the terms of engagement for critical cooperation need to be negotiated. NGOs who seek influence through critical cooperation must be prepared to challenge the Bank to ensure that they are jointly setting the agenda and defining the terms of operational collaboration or dialogue. For example, extended negotiations were involved in setting up the sites, the structures, and the methods for a joint World Bank–civil society study of structural adjustment in about eight countries. Throughout the process of initiating and designing the study, *both* NGOs and Bank staff developed the conditions and the capacity for critical cooperation.[56]

At the project level, however, there is reason to be concerned that the momentum of Bank-NGO relations favors Bank-dominated terms of engagement and limits space for critical cooperation. As the Bank puts vast resources into NGO participation and as it seeks closer ties with NGOs through direct funding mechanisms and contracts, cooptation is inevitable unless NGOs consistently push the boundaries of cooperation by being "advocates" of those affected most by Bank projects *as well as* "partners" with the Bank.

Efficacy of Critical Cooperation

As the cases in this volume illustrate, much NGO experience with the World Bank raises questions about the efficacy of cooperative influence strategies in dealing with an agency that historically has been well insulated against public scrutiny and public accountability. Gains in opening the Bank to public influence have been achieved, but many NGOs that have struggled with the Bank over project and policy reform fear that the Bank's growing enthusiasm for contracting and consulting with NGOs will at best result in pseudo-influence and at worst be counterproductive to the gains made through more traditional advocacy approaches.

Critical cooperation is a strategy suited to incremental reform rather than big wins or losses. Negotiation, problem solving, and mutual influence are its hallmarks, so measuring its effectiveness in terms of outcomes is always difficult. In cases where NGOs are trying to move the Bank to adopt positive processes (e.g., citizen participation in national policy dialogue), concrete results will inevitably be only short term and partial.

A Plurality of Influence Strategies

New opportunities are emerging for NGOs (especially in the South) to participate in both operational collaboration and policy dialogue. There is reason to believe that critical cooperation is an appropriate strategy for promoting policy refinements and policy implementation when Bank and NGO goals converge. In the case of the Participation Action Plan, for example, NGOs must vigorously hold the Bank to its commitments, push for full implementation of participation guidelines (e.g., influence not just consultation), responsibly participate in the project cycle, and continuously press for accountability through independent monitoring and other means. Elements of cooperation, critical cooperation, pressure, and pure advocacy will undoubtedly be required over time.

In many respects, the ball is now in the NGO court: they must recognize key issues on which their activity is particularly relevant, and they must build and use the skills of critical cooperation where they believe this strategy of "insider" influence is appropriate.

The growing number of instances in which the Bank and NGOs have agreed to undertake joint activities offer critical cooperation "learning laboratories." In the spring of 1996, for example, President Wolfensohn and Bank staff agreed to an NGO-initiated joint study of impacts of structural adjustment in ten countries. In a letter to NGO organizers dated April 9, 1996, Mr. Wolfensohn agreed that structural adjustment has been problematic in some respects. In the spirit of learning how to improve, he wrote

what I am looking for—and inviting your help in—is a different way of doing business in the future. My objective is to ensure that economic reform programs make maximum contribution to poverty reduction, that we fully appreciate the impact of reform on disparate population groups, that we promote measures which narrow income differentials, and that we encourage governments to consult and debate with civil society on policy reforms.[57]

Mr. Wolfensohn's expressed desire to work with NGOs to achieve shared goals is certainly welcome, but not all NGOs will want to accept such opportunities for critical cooperation. Nor should they, for there is continual need for "outsider" advocacy and protest, especially in those many situations where parts of the Bank do not yet see the value of serious dialogue, where it is slow to implement its reform agenda, and where its policies and operations are destructive to human well-being and environmental integrity.

The pluralism of the NGO sector itself is both a strength and a weakness. Dialogue with the World Bank becomes richer and potentially more reflective of legitimate interests of primary stakeholders (the persons directly affected by World Bank programs) as diverse groups bring their constituents' views to the table. Emerging debates among NGOs over the legitimacy and efficacy of different policy positions and influence approaches are but one manifestation of greater inclusion and the dynamism that accompanies it.

NGOs perhaps need to give greater attention to building new alliances among themselves that take full advantage of new opportunities within the Bank and with some borrowing country governments. A spirit of critical cooperation among NGOs could go a long way toward ensuring that future Bank reform campaigns are effective and accountable.

Appendix

Composition of the NGO Working Group on the World Bank 1986 and 1996

1986 Membership[a]	1996 Membership
Asia	
International Institute for Rural Reconstruction (Philippines)	Asian NGO Coalition for Agrarian Reform and Rural Development (Philippines)
Association of Voluntary Agencies for Rural Development (India)	Freedom from Debt Coalition (Philippines)
	Society for Participatory Research in Asia (India)
	Proshika (Bangladesh)
	Lanka Jathika Sarvodaya Shramadana Sangamaya (Sri Lanka)
Africa	
CONGAD (Senegal)	Inter-Africa Group (Ethiopia)
	Kagiso Trust (South Africa)
	Organization of Rural Association for Progress (Zimbabwe)
	Integrated Social Development Centre (Ghana)
	Institut Africain pour le Développement Economique et Social Centre (Cote D'Ivoire)

Composition of the NGO Working Group on the World Bank 1986 and 1996

1986 Membership[a]	1996 Membership
Latin America and the Caribbean	
Grupo de Tecnología Apropiada (Panama)	Centro de Estudios y Promoción del Desarrollo (Peru)
SOLIDARIOS (Dominican Republic)	Latin American Association of Development Organizations (Ecuador)
	Equipo Pueblo (Mexico)
	Caribbean Policy Development Centre (Barbados)
	Brazil Network on Multilateral Financial Institutions (Brazil)
Europe	
NOVIB (Netherlands)	APRODEV (represented by Christian Aid, U.K.)
Oxfam (U.K.)	
Union Nationale des Maisons Familiales Rurales d'Éducation et d'Orientation (France)	Eurostep (represented by Swiss Coalition of Development Organizations, Switzerland)
Deutsche Welthungerhilfe (Germany)	Centre de Recherche et d'Information pour le Développement (France)
World Ort Union (U.K.)	Swedish Mission Council (Sweden)
	Society for Improvement of Quality of Life (Croatia)
North America and the Pacific	
Technoserve (U.S.)	Inter-Church Coalition on Africa (Canada)
InterAction (U.S.)	
Canadian Council for International Cooperation (Canada)	Community Aid Abroad (Australia)
Aga Khan Foundation (U.S.)	Institute for Development Research (U.S.)
OISCA International (Japan)	Pacific Asia Resource Centre (Japan)

Composition of the NGO Working Group on the World Bank 1986 and 1996

1986 Membership[a]	1996 Membership
International NGOs	
CIDSE (Germany)	CIDSE (Belgium)
World Council of Churches (Switzerland)	International Save the Children Alliance (U.K.)
Lutheran World Federation (Switzerland)	

[a] Some geographic categorization for this year may be incorrect due to ambiguity in source data.

Note: Acronyms and abbreviations that appear in table, in alphabetical order
APRODEV—Association of Protestant Development Organizations in Europe
CIDSE—Cooperation Internationale pour le Développement at la Solidarité
CONGAD—Conseil des Organisations Non-Gouvernementables de Appui au-Développement
NOVIB—Netherlands Organization for International Development Cooperation
Proshika—Proshika Manobik Unnayan Kendra
SOLIDARIOS—Consejo de Fundaciones Americanas de Desarrollo

Notes

1. I have used several methods for collecting information: (a) interviews with Bank staff and NGO members who have been active in operational collaboration and policy dialogue—to understand their perspectives, analysis, and evaluation of NGO-Bank cooperation; (b) World Bank– and NGO-authored documents that chronicle operational collaboration and the NGO-Bank Committee; (c) studies and analyses of various examples of NGO-Bank interactions; and (d) various World Bank policy documents relevant to the chapter's focus. Finally because the Institute for Development Research has been a member of the NGO-Bank Committee since 1993, I have had firsthand participant observation experience of NGO-World Bank relations.

2. For simplicity, the World Bank definition of NGOs is used. The definition includes both nongovernmental organizations (NGOs) that assist community groups and community-based organizations (CBOs) made up of community members.

3. Poverty and Social Policy Department, World Bank, *NGOs and the Bank: Incorporating FY95 Progress Report on Cooperation Between The World Bank and NGOs* (Washington, D.C.: World Bank, 1996). p. 15.

4. Lawrence F. Salmen and A. Paige Eaves, "Interactions between Nongovernmental Organizations, Governments, and the World Bank: Evidence from

Bank Projects," in S. Paul and A. Israel, eds., *Nongovernmental Organizations and the World Bank: Cooperation for Development* (Washington, D.C.: World Bank, 1991), pp. 94–133.

5. Operations Evaluation Department, World Bank, *Poverty Assessment: A Progress Review*, report no. 15881 (Washington, D.C.: World Bank, August 1996).

6. Many authors have written of the nature and impact of World Bank culture. Among them are Susan George and Fabrizio Sabelli, *Faith and Credit* (Boulder, CO: Westview Press, 1994); Paul Nelson, *The World Bank and Non-Governmental Organizations: The Limits of Apolitical Development* (New York: St. Martin's, 1995); and Willi Wapenhans, *Report of the Portfolio Management Task Force* (Washington, D.C.: World Bank, 1992).

7. World Bank, *Operational Directive 14.70: Involving Nongovernmental Organizations in Bank-Supported Activities* (Washington, D.C.: World Bank, 1989).

8. Operations Policy Group, World Bank, *Cooperation between the World Bank and NGOs: FY 1994 Progress Report* (Washington, D.C.: World Bank, 1995), p. 2.

9. Poverty and Social Policy Department, *NGOs and the Bank* (1996), p. 16.

10. Figures are reported only through fiscal year 1994. The *FY 1995 Annual Progress Report* published by the Operations Policy Group of the World Bank does not include this analysis.

11. Nelson, *The World Bank*, p. 72.

12. *Ibid.*, pp. 78–9.

13. *Ibid.*, p. 79. Design participation by NGOs affected .13 percent of $230 million Ghanian and .07 percent of the $410 million Philippine projects.

14. In 1996 the Operations Evaluation Department of the World Bank undertook a study of the impact of NGO involvement in World Bank projects to determine the extent to which NGO involvement increases the quality of project outcomes.

15. Task Group on Social Development, World Bank, *Draft, Final Report* (Washington, D.C.: World Bank, 1996), p. 1.

16. Bank official, interview by author, April 14, 1995, Washington, D.C.

17. Poverty and Social Policy Department, *NGOs and the Bank* (1996), p. 16.

18. *Ibid.*, p. 6

19. Operations Policy Group, *Cooperation* (1995), p. 6.

20. Society for Participatory Research in Asia, *Issues Concerning Participation in World Bank Projects* (New Delhi, India: Society for Participatory Research in Asia, 1995). p. 3.

21. *Ibid.*, p. 10.

22. Staff member of an international conservation NGO, interview by author, May 9, 1995. Washington, D.C.

23. Bhuvan Bhatnagar, *Non-Governmental Organizations and World Bank–Supported Projects in Asia: Lessons Learned* (Washington, D.C.: World Bank, 1991).

24. *Ibid.*, p. 5.

25. *Ibid.*, p. 6.

28. Samuel J. Voorhies, *Working with Government Using World Bank Funds* (Monrovia, CA: World Vision International, 1993), p. 1.

27. See L. David Brown and Darcy Ashman, "Participation, Social Capital, and Intersectoral Problem-Solving: African and Asian Cases," *World Development* 24:9 (1996), pp. 1467–1479.

28. See World Bank, *The World Bank Participation Sourcebook.* (Washington, D.C.: World Bank, 1996), pp. 174–80.

29. The International Center for Not-for-Profit Law, *Global Standards and Best Practices for Laws Governing Non-Governmental Organizations* (forthcoming).

30. The Participation Action Plan approved by the World Bank board of directors in 1994 mandates mechanisms for engaging stakeholders in World Bank operations. However, it does not mandate stakeholder participation per se. NGOs continue to advocate for this standard for all projects and development policy processes such as the country assistance strategy. See World Bank, *The World Bank and Participation* (Washington, D.C.: World Bank, 1994), p. i.

31. *Ibid.*, emphasis added.

32. *Ibid.*, annex VI, p. 6.

33. World Bank, S. Asia Country Department *Draft Staff Appraisal Report: Ecodevelopment Project* (Washington, D.C.: March 1996), World Bank, p. 4.

34. *Ibid.*, p. 7.

35. Presentation by Ms. Jessica Mott, Asia Regional Meeting, NGO–World Bank Committee, Manila, Philippines, April 1996.

36. NGO staff member, interview by author, April 7, 1995.

37. NGO Working Group on the World Bank, *Position Paper of the NGO Working Group on the World Bank* (Washington, D.C.: NGO Working Group on the World Bank, 1989).

38. NGO working groups on the World Bank and the World Bank, "The World Bank and Development: An NGO Critique and a World Bank Response," *Trocaire Development Review* (1990), p. 21.

39. To implement the work plan, subcommittees were established and various members agreed to take the lead on different tasks. A full-time coordinator was hired to assist the committee in coordinating communications and with the committee's analytic work. He was based in the ICVA (International Council of Volumary Agencies) Secretariat in Geneva.

40. See Carlos A. Heredia and Mary E. Purcell *The Polarization of Mexican Society: A Grassroots View of World Bank Economic Adjustment Policies* (Washington, D.C.: The Development Group for Alternative Policies, 1994).

41. The World Bank, *The World Bank and Participation.*

42. Northern NGO member of the Working Group, interview by the author, February 28, 1996, Washington, D.C.

43. From the NGO Working Group, this group included; Mazid N'Diaye (FAVDO); Jalal Abdul Latif (Inter-Africa Group); Iqbal Asaria (Third World Network); and Allan Criton (Caribbean Council of Churches). Others were: Hugo Fernandez (Centro de Investigación e Promoción del Campesino [CITCA], Bolivia) and Shripad Dharmakidary (Narmada Bachao Andolan, India).

44. Letter to President Preston from the NGO Working Group, 30 October 1992, NGO Working Group Secretariat Files.

45. Interviews and discussion with several Washington-based NGO staff, February-March, 1995, Washington, D.C.

46. Interviews with several World Bank staff knowledgeable about the NGO–World Bank Committee, February 27, 28, 1995, Washington, D.C.

47. Several important changes in committee operations may help expand the dialogue. In 1994 the NGO Working Group decided to move its secretariat from Geneva to be housed with the Steering Committee chairperson to improve operations and reduce cost. The 1995 spring meeting was decentralized to the Africa, Asia, Latin America, and Caribbean regions to address regional concerns and involve a broader group of NGOs in the meetings. Somewhere between two and three hundred NGOs participated in these dialogues, expanding NGO participation in the committee's deliberations ten-fold. Some regional plans call for additional subregional meetings to prepare for or follow up to these spring regional meetings, which are now expected to be a permanent feature of the committee's operations. In the 1996 spring meetings NGO participation further expanded in numbers and range of countries represented.

48. World Bank staff member, interview by the author.

49. World Bank staff member, interview by the author, February 27, 1995. Washington, D.C.

50. Challenge (i.e., conflict) remains the strategy of choice in instances where Bank policies and practices are neither on the reform agenda nor aligned with NGO values and goals, as in the information disclosure policy of the International Finance Corporation (IFC). IFC is exempt from IBRD disclosure standards on the grounds that it must protect the private sector competitive playing field.

51. The possibility of simultaneously high levels of conflict and cooperation has been articulated in a number of contexts. W. F. Whyte, for example, found that rural villages in Peru could be sorted into the four quadrants of figure 3.5 and that the sorting indicated quite different patterns of development within the villages. See "Conflict and Cooperation in Andean Communities," *American Ethnologist* 2 (1995), pp. 373–92. The research literatures on conflict management and negotiations have also focused extensively on a similar topology, which suggests that differing interests may be handled by competition (conflict), accommodation (cooperation), withdrawal (disengagement), or problem solving (critical

cooperation). When the parties place a high value on both the substantive questions at stake and the quality of their continuing relationship, critical cooperation is the most effective strategy. See K. Thomas, "Conflict and Conflict Management," in M. D. Dunnette, ed., *Handbook of Industrial and Organizational Psychology* (Chicago: Rand McNally, 1976); L. D. Brown, *Managing Conflict at Organizational Interfaces* (Reading, Mass.: Addison-Wesley, 1983); and G.T. Savage et al., "Consider Both Relationships and Substance When Negotiating Strategically," *Academy of Management Executive* (1989), pp. 37–47.

52. See S. Commins, "NGOs and the World Bank: Critical Engagement," *World Vision Discussion Papers* 3 (1996), p. 4.

53. Global Research Consortium, *Multiparty Cooperation for Development in Asia* (New York, N.Y.: The United Nations Development Programme, 1992).

54. See, for example, Proceedings of the Spring Meeting of the LAC NGO Working Group on the World Bank, Santafe de Bogotá, 24–25 April 1995, Files of NGO Working Group Secretariat and T. Abbott and J. Covey "Social Investment Funds (SIFs): An Expanded Critique," *IDR Reports* (Boston: IDR, 12:4, 1996).

55. See Thomas, "Conflict," and Savage, "Consider Both."

56. This initiative, commonly known as the Structural Adjustment Participatory Review Initiative (SAPRI), was first proposed by NGOs to the World Bank president, Mr. Wolfensohn, in early 1996. Joint planning began in late 1996, and the study launch took place in the summer of 1997.

57. Letter from James Wolfensohn, President World Bank, to Doug Hellinger, Development Group for Alternative Policies, April 9, 1996, regarding a proposal to undertake a joint World Bank and Civil Society study of the social impact of Bank-financed Adjustment Programs.

II

Bank Projects

4

Indonesia: The Struggle of the People of Kedung Ombo

Augustinus Rumansara

In January 1989 the gates of the Kedung Ombo Dam closed, flooding the fields and villages of some five thousand families in the Serang River Valley in Java, Indonesia. The World Bank had financed $156 million of the cost of building Kedung Ombo to provide irrigation and control floods for about 125,000 families in the valley, improve municipal water supplies, and generate up to 22.5 megawatts of power.[1] However, even some of those for whom these facilities were built resisted Kedung Ombo.[2] Farming small plots, they complained that the compensation paid for land taken for irrigation canals was too low and that local officials cheated them out of money they should have received. In the rainy seasons they experienced flooding due to problems with the irrigation canals and dikes, and, of course, they had to pay for the irrigation water that they got. But whatever these farmers' concerns, those who were living in the roughly 45 square kilometers submerged by the reservoir were not even intended to enjoy benefits from the dam.

Instead, the government's plan was that they should move to South Sumatra, thousands of kilometers from their villages, as part of Indonesia's transmigration program. This controversial effort, also financed by the World Bank, sought to shift people from densely populated central islands to outer islands, where there is more land, but where they would be far from home, have a hard time building a new life, and often displace native peoples.

In the case of Kedung Ombo, the government's transmigration plan was based on a gross misconception. Conducted in 1984 by a state university in Bandung, the environmental impact assessment (EIA) for Kedung Ombo concluded that 75 percent of the people in the proposed reservoir area were willing to transmigrate. The accuracy of the figure was clearly dubious, however, because of the misleading nature of the survey question.[3] The local people might have wished to have a better living but not to leave their land without any guarantee that conditions would improve. They did not willingly transmigrate. When the dam was nearly complete in 1988, Indonesian officials informed the World Bank that some 3,500 families had transmigrated,[4] but reports from Satya Wacana Christian University stated that only 376 families, or 8 percent, had transmigrated. The latter figure was supported by the Kedung Ombo project leader from the Directorate General for Water Resource Development (DGWRD) of the Indonesian government, who pointed out that a large number of people were still living in the dam area.[5]

Despite these findings and based on the EIA, those who lived on land that was to be flooded were pressured to transmigrate. Threats and intimidation were applied, particularly to former members of the forbidden Communist Party.[6] For example, in January 1986 the head of Kemusu subdistrict forced ex-tapols (ETs), former political prisoners who had belonged to the Communist Party, to attend a meeting of the District Assembly Council, where they were warned not to refuse to transmigrate or they could be charged under the criminal code. In August, the same subdistrict head told the people of Kemusu that if they did not give up their land for the project, they would be regarded as enemies of the state and anti-development.[7] In September, however, seven hundred families who said that they had been intimidated went to the legal aid foundation (LBH) in the nearby city of Jogyakarta.[8] The Bank advised the Indonesian government to withhold information about possible livelihoods in the area and said instead to emphasize that those who stayed would likely suffer income losses.[9]

In the end, only a little more than one thousand households transmigrated; somewhat more than three thousand moved by themselves to other cities or villages on Java; and practically one thousand families[10] retreated only when the waters began to rise and settled by the dam.[11] Close to seven hundred of this last group[12] now live in villages that the government of Indonesia was ultimately pressured to establish in nearby forest lands, and another three hundred live in the greenbelt, a strip about 200 to 500 meters wide around the reservoir that, in order to prevent erosion, was not supposed to be inhabited. In fact, it is not possible to be too precise about where people displaced by the dam now live. For example, officials of the village of Mlangi, part of which is in the greenbelt area, say that a great number of young people go to cities to find jobs, mainly as seasonal workers.[13] This change requires them to change their residence identity card, so they are deleted from the list of inhabitants of their old villages, even though they always come back. The officials estimate that such migrants constitute approximately 60 percent of the population of the village. Only old people stay behind permanently.

As compared with other efforts against dams in Indonesia, the people of Kedung Ombo mounted an unusually forceful and consistent protest, in spite of pressure and intimidation from the military and government. The attitude behind such a protest was at least partly shaped by their beliefs that they are descendants of a mythological hero, Nyi Ageng, who fought for truth and against injustice.

The protest began late—about two years before the dam was flooded—and did not contest whether the dam should be built. The main concern the people of Kedung Ombo raised was that they receive fair compensation for their property. Recently, some nongovernmental organizations (NGOs) have said that asking for fair compensation was an indirect way to stop the dam, but these NGOs may be asserting that because their attitudes are now relatively anti-dam. It does not appear, however, that the local people actually asked that the dam project be stopped.

At the time the land in the area was expropriated in the late 1980s, it was selling for about 3,000 rupiah (Rp.) (about $1.50) per square meter. The minister of Home Affairs of the government of Indonesia in Jakarta said in 1987 that owners would receive Rp. 3,000 per square meter, but the governor of the province of Central Java said they would get only Rp. 700 per square meter. In 1990, after the fields and villages had been flooded, fifty-four villagers brought their case to the Provincial Court of Semarang, the capital of Central Java. They lost this case, but thirty-four of them appealed to the Supreme Court, which in 1994 awarded them the extraordinary sum of Rp. 50,000 per square meter in compensation for land and nonmaterial losses. In the end the government managed to get the award overturned. The people of Kedung Ombo never received more than Rp. 250 to 300 (less than fifteen U.S. cents) per square meter—a tenth of the market price and a hundredth of what the Supreme Court ultimately said they deserved. Total compensation for property fell about $90 million (U.S.) short of the market value of the land—a cost that the displaced people themselves bore on top of the social and cultural costs of dislocation.

The people of Kedung Ombo were first told that they would receive this small amount per square meter when they were called into their village heads' homes in 1987. Some accepted the compensation, inadequate though it was, and left. Some flatly refused it. The largest group took the money but continued to insist that it was inadequate and to struggle for better compensation.

In their struggle for fair compensation, beginning a couple of years before the dam was flooded, the people of Kedung Ombo reached out at first to local NGOs, primarily legal aid foundations, and later to national NGOs based in the capital, Jakarta. The Jakarta-based NGOs in turn reached out to partners in Europe, the United States, Canada, Japan, and Australia, forming an international alliance that brought the concerns of the people of Kedung Ombo to the media in numerous countries, the offices of the World Bank, the U.S. Congress and Treasury Department, and many other governments—not least, the government of Indonesia in Jakarta. The struggle spread from the issue of compensation to questions about where the people of Kedung Ombo could resettle and how they could regain livelihoods, to concerns about their human rights as the villages where they were still living were flooded, and to disputes about NGOs' rights of association. Although the Kedung Ombo campaign never made progress on the issue of compensation, it did on other matters.

The campaign found some sympathy within the government of Indonesia, notably from the Supreme Court justice who issued the favorable ruling, as well as from the minister of Home Affairs, who established a productive working relationship with many NGOs. The World Bank continued to maintain that compensation was the responsibility of the government of Indonesia, but did raise the issue directly with President Suharto. It also improved its procedures for project planning, opening dialogue with NGOs despite opposition from the Indonesian government. Moreover, the Kedung Ombo campaign may have contributed to new Bank policy guidelines on resettlement, disclosure of information, and independent review panels of the World Bank, as well as expanding the space for NGO participation in domestic policy making, at least regarding dam projects.[14]

In its project completion report on Kedung Ombo, the World Bank acknowledged a number of failures. It did not respond to "major problems with the resettlement operation" until too long after NGOs raised them. Had the Bank and local authorities "participated in an early, informed discussion of Kedung Ombo with NGOs and other concerned groups, much of the embarrassment and cost suffered by the Bank and [the government of Indonesia] subsequently could have been averted." Moreover, the Bank report noted that more open planning could have alleviated the "fear and uncertainty" of the people of Kedung Ombo, prevented the "steep income drop" for some, and helped others find "new, local opportunities" for work.[15]

Not least of the accomplishments of the Kedung Ombo campaign is the fact that a substantial minority of the displaced villagers were able to remain near their old homes. In 1988, as the protest over Kedung Ombo spread to international groups and the World Bank put pressure on the Indonesian government, the government gave up on the idea that virtually everyone was going to transmigrate. It began building three nearby villages on state forest land that it officially designated for resettling people from Kedung Ombo who still did not want to transmigrate. In addition, as mentioned, several hundred families are living near the dam on sites that are technically illegal but implicitly recognized by the government. Approximately fifty of these families, from the flooded village of Kedung Pring, squat in a settlement that they continue to call by the same name, though the land is poor and they are not well off. Others moved into the greenbelt, with its fertile agricultural soil, and are actually doing quite well—in many cases now cultivating more land than they had before. The Indonesian government appears to prefer letting these families live in technically illegal sites rather than risk additional domestic and international protest by trying to move them.

The following account describes how the Kedung Ombo campaign grew and reasons for both its successes and its failures. The account will consider causes for the divisions that arose among some groups involved in the campaign, but also for the essential success of the international alliance in representing the actual interests of the people of Kedung Ombo, as they stated them. Of course, the alliance was often unable to actually secure what the local people wanted. This account also considers the sources of the alliance's strengths or weaknesses in achieving its goals.

The Legalistic Strategy

A key weakness of the Kedung Ombo protest—and a reason why it did not achieve more—was that it got started very late. To begin with, it was thirteen years after planning began on the dam before the people of Kedung Ombo even got an inkling as to what was going on. Netherlands Development Consultants (Nedeco), an engineering company from the Netherlands, and Snowy Mountains Engineering Company from Australia, carried out a feasibility study for the dam for the Indonesian Department of Public Works in 1969. However, news of Kedung Ombo did not come out until 1982, and only then in a roundabout way. The governments of three districts in Central Java—Boyolali, Grobogan, and Sragen—instructed village heads in areas that would be flooded to register ownership of the land—that is, to ascertain precise information as to which villagers had exactly what rights to what parcels. The village heads invited villagers to their homes to do this. They did not mention the dam: all the people knew was that suddenly somebody wanted to find out who owned the land. They suspected something was going on and started asking questions. The village heads never were able to register the land, but the news came out that the government wanted to build a dam.

In 1982 construction of roads and other infrastructure began, and actual dam construction commenced in mid-1985, but still there was relatively little communication among the villagers, who lived in three different districts, and they had no clear idea about the pace of the construction or the specific plans for the dam. Though people were called to village heads' homes and pressured to sign papers agreeing to accept compensation for their land, this was done village by village, so the process did not link the people together and may actually have been a clever way to avoid stirring up a mass opposition. From 1986 on, the press in

Semarang and in Jakarta gave cautious reports about the compensation issue, but one had to read between the lines to try to figure out what was really going on. Indonesian newspapers require licenses to operate, and the government may use its ability to revoke a license in order to pressure the editor in chief to fire journalists who report too much on sensitive issues, which results in continual self-censorship.[16]

For five years after they first began to learn about the dam in 1982, the only strategy the people pursued was legalistic. In response to requests from disparate groups of villagers, various legal aid foundations sought through the courts to ensure that the government of Indonesia would abide by its own agrarian laws, which required that the government discuss and agree on compensation with those whose land was expropriated—something it had not done and indeed never did.

Through 1987 no attempt was made to form pressure groups within the country to back up the people of Kedung Ombo. Even though they protested strongly, the Kedung Ombo people did not choose a pressure-group strategy for a long time for several reasons. They were subject to intimidation, so surely protest by pressure groups would have been more dangerous, particularly for the people at the local level. National NGOs are often formed by dissident members of the elite—they may have educational, social, or family ties with government officials or the military—so they face somewhat less danger than local people, such as the people of Kedung Ombo. Also, obstacles to communication often existed between Kedung Ombo and national groups: the area is relatively isolated from the media, and the people mostly speak Javanese, not Bahasa Indonesia, the language of groups in Jakarta. Moreover, the people of Kedung Ombo did not know that the World Bank was financing the dam, only that it was being built by the government of Indonesia.

A further reason for following the legal strategy was simply that one key individual disseminated information about the possibility of using this particular strategy. An activist at an NGO in the nearby city of Solo had learned about the dam and from 1982 to 1985 sought to inform villagers of their legal rights—a process called *legal awareness training*. The activist expected that the villagers would thus be able to defend themselves and secure adequate compensation. They would discuss compensation with the government, as required by law, to agree on a fair price. If project managers did not follow the laws, the people would resort to the courts. In 1990, in interviews with thirty-seven people, 46 percent stated that they had decided to follow this legal strategy because of the workshops conducted by the NGO activist, and another 27 percent stated that their leaders had advised them to take this approach. All the leaders who were mentioned stated that they had been advised by the NGO activist.[17]

The people in Kedung Ombo were not at all organized when they began the legal strategy. Each group, representing perhaps only one village, went to a preferred legal aid foundation that someone in the group knew. If people in a group doubted that one foundation or NGO was serious about their problem, they went to others. For example, some of those who had gone through the legal awareness training formed the Committee of Eight to help villagers in Boyolali District secure their legal rights. (Sometimes one individual left the committee, or another joined, so at various times it was called the Committee of Seven or the Committee of Nine.) Different committee members put villagers in touch with different legal aid foundations (LBH) in nearby cities. Initially, in 1986 many from Boyolali went to the legal aid foundation in the nearby city of Jogyakarta (LBH Jogya), where the activist from Solo had links. However, in February 1987 the same group went to Semarang to meet with members of Provincial House of Representatives, and in June 1987, while the case was still being handled by LBH Jogya, the same group went to Jakarta to seek help from the Indonesian Institute for Legal Aid (YLBHI) there. This irritated LBH Jogya, which withdrew from defending the group. For their part, the people accused LBH Jogya of being too close to the district head of Boyolali because instead of fighting for better compensation, it tried to persuade them to accept what the government had offered. Some in the group accused LBH Jogya of failing to defend them when they were accused of being communist.[18] Others also went to legal aid foundations in Solo, Salatiga, and Semarang as well.

In spite of the Indonesian context, where the judiciary is weak, the legal strategy could have been more effective if it had been better organized and particularly if it had been combined with broader campaigns by human rights or student groups. However, by creating the expectation that the people would win in the court, the legal strategy *per se* led to a wait-and-see attitude, thus weakening other efforts.

The International Campaign

By 1987 it was evident that the legal strategy was not making progress, and with the dam only a little less than a year from completion, some groups decided to launch a broader protest at national and international levels. In September 1987 the Indonesian Institute for Legal Aid (YLBHI)—the Jakarta NGO that, as mentioned, some people of Kedung Ombo had contacted—sent an account of the events surrounding the dam to the World Bank. *Inside Indonesia,* an Australian magazine, ran a detailed English-language report on Kedung Ombo in December 1987, which alerted international NGOs such as Tapol in Britain and the Canadian-based group Probe International; from then on, *Inside Indonesia* and *Tapol Bulletin* provided a stream of news about Kedung Ombo in English.[19]

International NGO alliances began to organize around Kedung Ombo. The main international alliance that focused on Kedung Ombo, the International NGO Group on Indonesia (INGI), was established in 1985 and was founded in response to the Inter-Governmental Group on Indonesia (IGGI), a consortium including the World Bank, the Asian Development Bank, and fourteen nations—the United States, Canada, Japan, Australia, and many in Europe—that provided billions of dollars a year in aid and loans to Indonesia. In 1985 YLBHI and the Netherlands Organization for International Development Cooperation (NOVIB), along with other Indonesian and foreign NGOs, organized INGI as a consortium of NGOs to raise issues of human rights and sustainable development pertinent to Indonesian projects financed by the members of IGGI. Indonesian members of INGI encompassed a wide range of NGOs, regionally from North Sumatra to Irian Jaya and sectorally from charity and religion-based NGOs to political advocacy and human rights groups. They included the thirteen so-called "big" NGOs, such as YLBHI and the Indonesian Environment Forum (WALHI).[20]

After having notified the World Bank in 1987, YLBHI raised the Kedung Ombo issue at INGI's March 1988 conference in the Netherlands.

INGI in turn described the problems of Kedung Ombo to the Dutch minister for development, who brought them up at the June 1988 IGGI meeting in the Hague, thus informing the World Bank and major donor nations to Indonesia of those problems. The Indonesian Documentation and Information Center (INDOC) in the Netherlands also raised the issue at the June 1988 IGGI meeting. For the next few years INGI, along with other Indonesian and international NGOs, continued to press the claims of the people of Kedung Ombo for better treatment.

Indonesian NGOs chose to seek support from the international community because they understood that the World Bank had guidelines with regard to compensation and involuntary resettlement on projects it finances. They also knew that Indonesia was dependent on foreign funding for development projects, so international pressure, especially by shareholder nations of the World Bank, could have an effect in the country and on World Bank development policies. The political atmosphere in Indonesia was closed, so Indonesian NGOs "borrowed" the more democratic atmosphere in donor countries where NGOs have broader political space to pressure governments and the World Bank.[21]

The international campaign proved significant. If there had been no international publications and pressure, efforts at the national level would have died out. As compared with the local and national efforts, those at the international level were more coordinated. The target of the advocacy was clearly defined—namely, the World Bank and governments of IGGI member countries, particularly the United States, the biggest shareholder in the World Bank. Information flow was also relatively more organized than at the local and national levels.

INGI's main strategy for gaining leverage was to point out the disparity between Bank policies regarding what is supposed to happen on projects it finances and what actually happened in Indonesia. On paper, World Bank policies are generally consistent with concerns for promotion of broad public participation and sustainable development.[22] As an international alliance, INGI had access to information both about World Bank policies on resettlement and related matters and about what was going on at Kedung Ombo. INGI sources on Indonesia were NGOs in the country, particularly its own members, which even World Bank missions, like the one featured in a report by Dennis Purcell on April 14,

1989, acknowledged as the best sources of information as to what was really happening.[23] Thus, INGI could demonstrate when actual results contradicted Bank policies.

Following the 1988 IGGI meeting, various groups continued to raise the Kedung Ombo issue with the World Bank. The Environmental Defense Fund (EDF) in Washington sent a letter to the Bank in October, and Tapol in the U.K. sent a letter to the Bank in December concerning two thousand families in Kedung Ombo who had not received fair land compensation. In addition, INGI asked members to raise the Kedung Ombo issue with their respective governments, which in turn were asked to reform Bank policies or to ensure that existing policies were implemented. Each participant in INGI was also asked to discuss Kedung Ombo at international meetings related to the environment, development banks, human rights, or democratization. Thus, Lori Udall from EDF and George Aditjondro, one of the founders of INGI, raised the issue in a seminar of the International Rivers Network in San Francisco in June 1988, and Agus Purnomo from WALHI raised it in a seminar on the World Bank in Berlin in September. In May 1988 the Bank sent David Butcher, a consultant, to study issues surrounding the dam.

The Indonesian government thus faced international pressure as well as the refusal of many villagers to transmigrate. Wanting to clear the Kedung Ombo area for flooding in early 1989, it gave up trying to pressure nearly all the families in the area to transmigrate, and in August 1988 the Directorate General for Water Resource Development (DGWRD) developed a plan to build three new villages on state-owned forest land near the reservoir where 850 families could live. The first village was Kayen, planned to accommodate four hundred families, each of which was offered 1,000 square meters of land and was promised the title. In April 1989 a group of thirty-five families moved there, but as late as February 1990 only forty-seven families lived in this village.[24] To attract more people from Kedung Ombo, the government offered 2,000 square meters per family, and in May 1989 it provided two other locations, Kedungmulyo and Kedungrejo. Approximately seven hundred families now live in the three villages.

The government's change of attitude caused a split among the people of Kedung Ombo. Some, especially from Kedung Pring, remained suspicious of the sudden offer and continued to press their case in court. They suspected the Group of Eight who represented them of having been bought off by the government because they had asked for fair compensation but were ultimately offered resettlement land still owned by the state forest company. This split in some measure weakened the movement.

Human Rights and Mass Protest

In January 1989, when the dam gates were closed even while people in Kedung Ombo were still living in their homes, the questions of compensation and observance of Indonesian law shifted to concerns about the human rights of those whose homes were being flooded and about broader issues of democratization. Students from cities not far from the dam site—Salatiga, Jogyakarta, and Surabaya—formed the Solidarity Group for the Victims of Kedung Ombo (KSKPKO), and in February 1989 they held demonstrations in Semarang, Kedung Ombo, and Jakarta. They also sent a letter signed by 961 people to the chair of the National Assembly, Kharis Suhud, requesting that land prices be raised, corrupt government officials who had pocketed compensation money be punished, filling of the dam be halted until the land compensation issue was settled, and an independent body be established to supervise construction.

Of particular importance in the campaign was Father Mangoen Wijaya, a Catholic priest. In February he called for solidarity with the people in Kedung Ombo and sought especially to help children from the flooded villages who were physically and mentally hurt by the situation, having lost their homes, schools, and playgrounds. The governor of Central Java called Father Mangoen's effort an attempt to support broad political resistance and forbade it, in spite of the fact that Moslem leaders, academics, and even some high-ranking military officials supported it. On March 11, Father Mangoen wrote an open letter to the minister of Home Affairs in Jakarta, who actually supported him. This letter received a lot of attention and sympathy for the people of Kedung Ombo from the public and from inside the bureaucracy and the military, as well as from abroad.

International NGO involvement in the campaign increased in response to these events. In addition to those already mentioned, foreign NGOs involved in the campaign included Tapol (U.K.), Initiative for Human Rights in Asia (IMBAS, Germany), the Dutch Indonesian Committee, Japan Tropical Forest Action Group (JATAN), International Working Group of Indigenous Affairs (IWGIA), Friends of the Earth–Japan, Reseau Solidaire (France), APPEN (Malaysia), Third World Network (Malaysia), ILSA (Colombia), Canada Asia Working Group, Trocaire (Ireland), and South East Asia Information Center (Germany). There had been no direct relationship among these NGOs before the campaign, except through international meetings. Involvement in the Kedung Ombo campaign arose because of common concerns, such as violation of human rights and environmental degradation caused by projects financed by development banks.

In April 1989 there occurred what came to be known as the Brussels incident. When homes of people in Kedung Ombo were being flooded by the dam, the INGI conference that month (actually held in a city near Brussels) focused on the human rights of those people and concluded with a decision to write a letter to Barber Conable, president of the World Bank, pointing out that what was happening violated the Bank's own resettlement policy just published the previous year.[25] Various NGOs also followed up on the INGI letter to the Bank. For example, in the fall of that year the Environmental Defense Fund in Washington arranged for Peter Van Tuijl, secretary of INGI, to testify before the U.S. Congress about Kedung Ombo. With NGO staff from the United States and Indonesia, Van Tuijl also presented the case to the U.S. Treasury. It was hoped that these efforts would influence the Bank, mainly because the United States is its biggest shareholder.

In its letter to the Bank, sent April 26, 1989, INGI described the problems that had occurred in Kedung Ombo. In violation of Bank policies, affected communities had been excluded from meaningful participation in decision making about resettlement and compensation: the Bank had not consulted them directly or through community leaders. The Bank did not conduct regular supervision of resettlement operations, and there was no access to information about the project. Because the Kedung Ombo families had "totally lost trust in local government officials," INGI requested that the Bank, with full involvement of NGO representatives of the community, should (1) seek to promote fair compensation for all families displaced and locate resettlement land equivalent to what they had lost; (2) ensure the restoration of income through agriculture or fish farming for the nine hundred or so families still in Kedung Ombo; (3) independently monitor progress on these matters in Kedung Ombo and other dam projects in Java; (4) discuss Kedung Ombo with the government; (5) before financing any other Indonesian projects involving purchasing land and resettling people, request a full review of policies on these matters; and (6) develop a regional environmental plan for Kedung Ombo.[26]

On March 13, 1989, before receiving that INGI letter, the Bank had sent a mission, consisting of Dennis Purcell and Sahulata, Bank officials, and William Partridge, a consultant, to assess the situation in Kedung Ombo. Their report of April 14 was sent to Thomas E. Davies, the Bank's chief agricultural staff resident in Indonesia, as well as to the Indonesian minister of Home Affairs, minister for National Development Planning, minister of Public Works, and secretary of state. In an annex of the report on lessons from Kedung Ombo, Purcell states,

[T]he involuntary resettlement component of a development project must be given as much importance as the directly productive project components.... Valid representatives of the populations should be involved developing or modifying the resettlement plan. Bank missions should openly encourage this.... All the anticipated costs of resettlement should be included as a project cost, and, wherever possible, loans provided to any components which are likely to suffer from inadequate funding from local resources.... Bank missions should supervise disbursement of annual budgets for compensation and resettlement.

Further, Bank supervisors must "work with a permanent, responsible agency or group ... capable of responding to changing circumstances in the complicated task of involuntary resettlement.... NGOs can have a special role in monitoring implementation of resettlement programs." And, though compensation levels are always a contentious issue, "if there is reason to question the adequacy of [legal] process [for setting them], this should be addressed as a general policy issue on a country-wide basis before proceeding with the project."[27]

Officially, the World Bank kept strictly to the loan agreement, which gave responsibility for land compensation—the chief issue that the people of Kedung Ombo had raised—to the Indonesian government.[28] Even after INGI wrote a letter on November 20, 1989, to Bank President Barber Conable raising the issue of compensation, the December 15 Bank response only alluded to it in terms of resettlement. However, the Bank did ask the Indonesian government if INGI's charges were true and requested it to deal with the compensation problem. Indeed, the Bank raised the matter with President Suharto himself, which resulted in Suharto becoming angry and holding a cabinet meeting. After the meeting, the chief of the armed forces (later vice president) made a statement in an interview with a Jakarta newspaper to the effect that Indonesian NGOs should not hang dirty linen outside the country, and the state minister for social and political affairs warned NGOs and students not to act for foreign interests.[29] Further, President Suharto instructed Retired General Rudini, the minister of Home Affairs, to handle NGOs so such issues would not continue to be aired outside Indonesia.

Rudini was one of most progressive ministers in the government during this time, though he did not have a lot of power. He told NGOs not to go to the World Bank but said that Indonesia and the Bank were partners in development and, if they had concerns, to discuss them with him. A dialogue allowing NGOs to talk to him not only about Kedung Ombo but about other issues as well continued until another Home Affairs minister was appointed in 1992. Moreover, Rudini actually had a lot of sympathy with Father Mangoen's solidarity action to help the children in Kedung Ombo.

Relations within the Campaign

Relations among local, national, and international levels grew complicated and in some ways conflictive. As mentioned, the people of Kedung Ombo—in part based on myths about the hero Nyi Ageng, who had fought for truth and against injustice—held beliefs about human rights and environmentalism that met in a fertile way with those of the national and international groups. However, because of their lack of knowledge of Bahasa Indonesia and their vulnerability to intimidation, grassroots groups played a very limited direct role in the legal strategy and the effort to lobby officials in the bureaucracy, parliament, and the military;

moreover, they were not involved at all in the international campaign. Instead, they were represented by staff from local and national NGOs, for example by YLBHI at INGI conferences and by WALHI at other international NGO events. A staff member of an NGO based in Solo joined the INGI lobby team in Washington in 1989 to meet with World Bank and U.S. government officials.

The legal and social position of NGOs in Indonesia affected relations within the Kedung Ombo campaign. The official Indonesian term for NGO is *Lembaga Swadaya Masyarakat* (LSM), or community self-reliance organization. The English term *nongovernmental organization* was translated into Indonesian as *Organisasi Non-Pemerintah* (OR-NOP), which the government sees as having a noncooperative or anti-government connotation, so it prefers the term LSM. This linguistic fine point has broader ramifications: an NGO can be established by the government as a GONGO or government-organized NGO. There is today a wide range of NGOs, organized by citizens, the government, or even political parties.

The government tends to separate NGOs according to whether they are *advocacy* or *development* groups. Development NGOs are considered to provide services or technical support, such as clean water supplies or training, without getting involved in broader political matters. Legal aid or environmental groups that talk about people's rights and challenge development policies are considered advocacy NGOs. In reality, the distinction between advocacy and development is not always sharp. Genuine NGOs engage in both. Most adopted a development strategy in the 1970s, but following student demonstrations against government corruption in the mid-1970s, outspoken university students were dismissed, and many joined NGOs. When some of these activists became NGO leaders in the 1980s, they adopted advocacy as an important part of their work.

The key sanction against NGOs that are deemed advocacy groups is to take them off the official list of NGOs. To be on this list, NGOs register at the offices of the Department of Home Affairs. If an NGO is not on this official list, it is prevented from getting a recommendation to receive foreign funding. Because Indonesian NGOs are not allowed to raise funds from the public, if an NGO is denied official status, it effectively cannot secure money to operate and to get permits to organize meetings with other NGOs or seminars.

The government attempted to keep internationally connected NGOs from politicizing Kedung Ombo by cutting off or threatening to cut off their ability to get funds, but there was no way to stop the students by cutting off funds because they raised their own funds from members. (Membership fees are allowed by the government if they serve only as a temporary measure of fund raising. If an NGO has many members with a permanent membership fee, it is categorized as a mass political organization and strictly controlled by the Mass Organization Act.) Instead, the government moved in the military to stop the students after they started their campaign at the dam site in February 1989,[30] an intervention that succeeded to a certain extent in slowing down or even suppressing the movement.

Differences among NGOs and between them and grassroots groups became a problem in part through lack of planning or any clear division of labor, especially at local and national levels. INGI had its approach, the students had theirs, Father Mangoen had his, the legal foundations had theirs. In general, NGOs had low credibility with people at the grassroots, as evident in the people's suspicion toward some NGOs and their efforts to seek support from one NGO after another and even from several at the same time. Some NGOs had the impression that each NGO was merely seeking to "raise its own flag," as when one legal aid foundation felt angry because people from Kedung Ombo went to other NGOs to seek support.

There was also suspicion between local NGOs and the broader national and international coalition. Student and legal groups in Central Java saw Kedung Ombo as their issue, now being taken over by INGI, which had access to the World Bank, Washington, D.C., and the Hague. Local NGOs did not understand that INGI took up Kedung Ombo based on proposals from Indonesian members. They also suspected that INGI was getting funding for Kedung Ombo and accused it of "selling the people of Kedung Ombo to travel abroad" when Indonesian members of the group did international lobbying. In fact, this was not the case: a check of INGI records showed that there was no special budget for

Kedung Ombo, and the group received no special funding for it. Nevertheless, this suspicion continued for a number of years.

A further division occurred when the government got angry at INGI for making Kedung Ombo an international issue, and local NGOs accused INGI of getting them into trouble. The national-level Jakarta NGOs could go to the relatively progressive minister of home affairs and, as mentioned, often had social or educational ties with the elite, but NGOs in remote areas had no such place to go. Local groups told the press that INGI, not they, were responsible for "airing dirty linen abroad," so it appeared that INGI was losing its base inside the country. As a result, group leaders who had signed INGI documents became more vulnerable.

These problems arose even though the international coalition did reflect stated grassroots interests. When INGI wrote the World Bank, it was in fact substantively focusing on the issues that local people had raised: compensation money, resettlement, and adequate consultation with the government. Moreover, these issues were brought to INGI by Indonesian groups, and the people who went to the World Bank and the U.S. Congress to testify were from NGOs working with local communities. However, it is not clear that the students did represent the local people. A feeling of solidarity was the driving force behind their movement, and in their demonstrations they placed increasing emphasis on broad issues of democracy and the constitution—not just on the interests of people in Kedung Ombo.

The national and international campaign may not, in the end, address the objectives and hopes of the people of Kedung Ombo who still expect compensation. National and international NGOS may not necessarily "represent" the interests of the people affected by the dam, whose main concern was and is compensation rather than the broader national and global issues.

To understand the campaign process more completely, we need to differentiate between actors and issues at local, national, and international levels. At the local level, the main issue was fair compensation for land taken by the government for the dam. People from the affected communities approached NGOs specifically to enlist their assistance in filing a court case to obtain just compensation.

At the provincial and national levels, however, students in solidarity with the affected communities and other NGOs broadened the issues beyond compensation to include demands for clean government, transparency in development policy decision making, human rights, and democratization. Land compensation was used by NGOs and student groups as an entry point for these broader issues.

At the international level, Kedung Ombo was packaged to fit within the context of a global environmental NGO campaign theme: an "anti–big dam campaign," encompassing issues of resettlement policy, indigenous peoples' rights, access to information, and destruction of the biological and cultural environment.

It is not clear that the shift from the issue of "compensation" to "democratization" and "no more big dams" directly benefited the people in Kedung Ombo, nor is it clear to what extent NGOs "represented" the people of Kedung Ombo in the campaign. When the international campaign was over, the people of Kedung Ombo returned to their original legal strategy, working through the legal aid foundation to try to obtain fair compensation for their land. The issues of compensation *are,* however, linked to the broader issues of transparency and democratization of development policy decision making. NGO advocacy has a legitimate role in promoting these issues, even if NGOs do not directly represent the affected communities in doing so. The international campaign actually linked and framed issues in terms of the goals and objectives of actors facing different targets at different levels. The lines of accountability within the international campaign are complex and are not easily drawn from international NGO to local NGO to affected community.

Lessons of Kedung Ombo

In December 1993 the World Bank officially declared the Kedung Ombo project closed: the funds were disbursed, the construction had been finished, and the project was thus considered complete. The problems of resettlement and compensation had not been resolved, but the Kedung Ombo campaign could be seen as effective in important respects, especially as the broad protest began only a little more than a year before the dam gates closed.

Indonesian NGOs are far more aware now of the potential problems involved in dams and are able to get information about large projects from EIAs. (The Indonesian government may say that EIAs belong to it and are not to be released, but NGOs can in practice get them from the World Bank or other funding agencies.) The question now is how NGOs actually make use of EIAs: Do they understand the technical language used in the EIAs? Can they check whether the claims in the EIA are correct?

As a result of both the broad protest and the legal strategy, Indonesian government officials may now be more careful in dealing with this type of project. Many Indonesians interpret the Supreme Court ruling for the thirty-four families in Kedung Pring, even though ultimately overturned, as the judge's indication that the people would have received very substantial compensation if there had been a fair trial. The staff of LBH Semarang, which represented the Kedung Pring families, say that both the governor and the army commander of Central Java warned the consultant for social impact assessment to be careful in dealing with land compensation.[31] When INGI wrote the Department of Energy about consultation with people living in the area of the Kuto Panjang Dam being built in North Sumatra, the director general responded directly.

As mentioned, the government has provided official resettlement villages near the dam and is not displacing those who live in technically illegal areas. However, the people in these villages and areas will not get irrigation water because their land (including the greenbelt) is higher than the water level of the dam. The villagers were told that an aquaculture project would allow them to earn income, but it does not look as if this will happen because they are supposed to be working for a private firm that was not operating as of May 1994. The villagers say it would be much more sensible if they themselves could decide what they wanted to do, rather than having projects contracted to another party.

The World Bank officially continues to maintain that, although its own guidelines require adequate compensation and resettlement, these matters must be handled by the borrowing government. Informally, however, Bank officials have hinted that if NGOs raised the issue of improving the livelihoods of people around the Kedung Ombo Dam rather than compensation per se, the Bank might be able to do something. Moreover, the

Bank says it will be wary about financing construction of more big dams, particularly in Indonesia. The Kedung Ombo campaign surely contributed to this decision, although, of course, protests about other big dam developments were currently being waged around the world, such as Narmada in India and Itaparica in Brazil. Thus, as a result of this overall decision, the World Bank did not finance the Kuto Panjang Dam in North Sumatra. Although it is not clear whether it asked the Japanese Overseas Economic Cooperation Fund (OECF) to finance the dam, OECF executives did state that in financing it they would avoid "another Kedung Ombo"[32]; they would minimize environmental damage, disclose EIAs, provide proper compensation, and respect cultural and historical sites. The World Bank itself is involved in a small dam in South Sumatra, where, according to Bank reports, local communities were consulted with the help of NGOs. The results of the consultations are not yet clear because the dam project is still in the planning phase.

Several international factors created leverage for the Kedung Ombo campaign. Global awareness of the importance of human rights, the environment, and participation in development made a difference. The international alliance made some progress in enforcing Bank policies that address these matters—although perhaps more for future projects than for Kedung Ombo—by demonstrating policy violations to the Bank and officials of governments that are shareholders in the Bank. In the case of Indonesia, falling oil prices in the early 1980s may have helped the campaign by making the economy more dependent on international loans, which put the government in a weak position vis-à-vis development banks and international pressure from donor countries.

Local resistance and information about it is crucial to an international strategy. The location of Kedung Ombo Dam itself helped: in Central Java, about one hour by air from Jakarta, and hence accessible to journalists, academics, international NGOs, and World Bank staff, who could observe and disseminate information.

Several factors seemed to influence the level of resistance—or at least information about it—raised in affected areas. Though parts of three districts in Central Java were flooded by the dam, and though there may have been resistance elsewhere, virtually all the reported resistance came from Boyolali District. Boyolali is, indeed, more accessible from urban centers. Also, because, the district leadership is more open, NGOs and the press are more active, so more information was available in Boyolali about what was going on. The university in Solo is actually closer to Sragen District, but students have to get a permit to do field work there, so they tend to work in Boyolali instead, thus making more information about this district available. Finally, the NGO of the particular activist who initiated the legal campaign had other projects in Boyolali.

NGOs and grassroots groups face general difficulties in seeking to influence multilateral development banks (MDBs), mainly because they do not have direct access to these banks. Agreements about project loans are between the MDBs and borrowing governments, so the banks are more receptive to what the governments have to say. In the case of Kedung Ombo, the World Bank always relied on official reports from the Indonesian government and even had to consult with the government about what it found in its own missions to Kedung Ombo. Because the Bank always argued that matters of compensation and resettlement were the responsibility of the government—even though its own missions to Kedung Ombo, such as Dennis Purcell's, showed that there were problems[33]—in the end, the people had to accept what the provincial government offered.

Some of the particular weaknesses of the Kedung Ombo campaign are as follows. Especially if a future NGO effort sought to change or stop a dam, it would have to avoid these problems:

• The campaign started at a very late stage, in part because the government of Indonesia did not let the public know about plans for the dam from 1969 through 1982. In spite of new policies of the World Bank and Asian Development Bank, such governmental silence will likely continue in the future if there is no pressure from NGOs and the community on a project site.

• The campaign never asked for the project to be stopped. The people asked only for fair compensation for their land and other assets. This approach was furthered by efforts of legal aid foundations to follow a legalistic path and ask for compensation.

• The legal-awareness training begun in 1982 and legal actions begun in 1986 were not very effective. However, the strategy was not without merit. At least it showed that there were sympathetic people in the government, such as the Supreme Court justice. In general, the effectiveness of a legal strategy would depend on the type of regime.

• In the initial phase, no attempt was made either in or out of Indonesia to form pressure groups to support the people of Kedung Ombo, and thus to supplement the legal strategy. The lesson that members of the international alliance could draw is that they started their campaign too late—just when the dam was almost finished. Although INGI began its campaign earlier than other advocacy networks, for a long time it too missed what was going on at Kedung Ombo.

• The international alliance got started late in part because the Indonesian government did not inform the people about the dam in a timely way, but even when information started coming out, dissemination was very slow. The legal aid foundations, working through official channels, did not seek to disseminate information or to form pressure groups. Also, until the end of 1987, most of the available information was in Bahasa Indonesia, and thus the issue was restricted to the national level. Only after information became available in late 1987 through INGI, *Inside Indonesia*, and other English sources did Kedung Ombo become an international issue, prompting international NGOs to pose questions about it to the World Bank.

• The Indonesian government badly needed this project to maintain the momentum of its economic growth, so stopping the dam project would have been very unlikely. In addition to the problems posed by government resistance, lack of planning and organization in the campaign was a problem. The student demonstrations seemed mainly to be part of the pro-democracy movement against the ruling regime, a politicization of the Kedung Ombo project that aggravated the government's tendency to suppress the movement. People in Kedung Ombo said they did not want to go that far and that it was better to take the legal approach. The international coalition to some extent did generally represent the actual goals of the people of Kedung Ombo, and it did have positive effects. It too, however, tended to politicize issues, even if not deliberately. For example, when the government felt that international efforts were designed to offend it rather than deal with development problems, it retaliated by repressing Indonesian NGOs.

These parallel effects of an international campaign would seem to be inevitable: on the one hand, it can raise the interests of grassroots groups with major institutions such as the World Bank and donor governments, but on the other hand, the resulting international pressure can prompt the government to repress national and local groups. However, if international coalitions can come up with reliable information about a project from the field as well as from donor governments and multilateral development banks, they can at least mitigate the repression of national NGOs and grassroots groups.

One necessity for strengthening a campaign, which might also mitigate repression of national and local groups, is that actors at each level should be involved in planning: grassroots groups with local NGOs, local NGOs with national NGO networks, and national NGO networks with international NGO networks. This planning should include the kind of strategies to be taken and a detailed task description of who is doing what. Of great importance are informational links among these levels. In the case of Kedung Ombo, some planning took place at the international level, but lack of planning at local and national levels prior to taking action reduced NGOs' credibility with the grassroots, as demonstrated by people's efforts to seek support everywhere and to feel suspicious of some NGOs. Suspicion is particularly a danger in the Indonesian context if a campaign is taken up by an NGO or network as a project for which it receives money, which suggests that the procedure of giving NGOs funding for particular advocacy projects makes them more vulnerable. To avoid suspicion, there must be continuous, direct communication, in addition to planning, between NGO activists and people at the grass roots, even after NGOs are no longer active at the site of the troublesome project. A visit to the site of Kedung Ombo in June 1994, revealed, however, that the local people had raised the complaint that only one or two NGOs had tried to maintain such a continuous contact.

Notes

1. According to World Bank's staff appraisal report, there were 570,600 people or 125,000 households in the project area, whereas according to a study sponsored by INGI there were 5,268 families, or approximately 26,000 people, who had to be removed to fill the reservoir. See World Bank, *Staff Appraisal Report: Indonesia—Kedung Ombo Multipurpose Dam and Irrigation Project,* report no. 5346a-IND (Washington, D.C.: World Bank, April 1985).

2. See, for example, George J. Aditjondro, "Covering Kedung Ombo: How the Media Determine What We Should Not See in This Part of the World," mimeo, Department of Government, Cornell University, Ithaca, New York, 1992.

3. The source is a World Bank evaluation report, stating that the formulation of the interview question was not accurate. The question was whether the people want a better life. If transmigration can be considered one way to a better life,

then the question is an indirect one. The study was conducted by a state university as part of an EIA of the project.

4. See Daniel R. Gibson, "The Politics of Involuntary Resettlement: World Bank-Supported Projects in Asia," Ph.D. thesis, Duke University, Durham, North Carolina, 1993, pp. 247–48.

5. See David Butcher, *A Review of the Land Acquisition and Resettlement under Four World Bank Financed Projects in Indonesia* (Washington, D.C.: World Bank, June 1988).

6. See Stanley, *Seputar Kedung Ombo* [*Around Kedung Ombo*] (Jakarta: Lambaga Studi dan Advokasi Masyarakat (ELSAM), 1990/1994?), p. 90.

7. *Sinar Harapan* (Jakarta newspaper), 30 August 1986.

8. See INDOC (Leiden, Netherlands: Indonesian Documentation and Information Centre, 1988).

9. See "The Politics of Involuntary Settlement," Gibson, pp. 249–50.

10. See Stanley, *Seputar Kedung Ombo*, p. 265, which mentions 529 families, but the World Bank figure is more recent.

11. See World Bank, *Indonesia: Kedung Ombo Multipurpose Dam and Irrigation Project (Ln. 2543-IND): Project Completion Report* (Washington, D.C.: World Bank, April 1995).

12. *Ibid.*

13. Mlangi officials, interview by the author and the staff of Yayasan Geni (an NGO in Salatiga established by students from the Christian University of Satya Wacana who are concerned with human rights, the environment, and democracy), April 1994, Mlangi, Indonesia.

14. In March 1996, the National Planning Body (BAPPENAS) invited leaders of thirteen NGOs for advice on how BAPPENAS could ask for financing for large dams in Indonesia. Abdul Hakim Garuda Nusantara, personal communication to author, March 1996.

15. See World Bank, *Indonesia: Kedung Ombo*, pp. 63–4.

16. See Aditjondro, "Covering Kedung Ombo."

17. These interviews were conducted by Yayasan Geni Salatiga in 1990 in the region.

18. Stanley, *Seputar Kedung Ombo*, p. 124.

19. John Paterson, "The Price of Dam Development in Central Java," *Inside Indonesia* (December 1987). YLBHI, "Pengantar Siaran Pers Yayasan LBH Indonesia tentang Kasus Kedung OMBO [Introduction to Press Release on Kedung Ombo]," Jakarta, 1991. For more on the Kedung Ombo campaign, see George J. Aditjondro, "Large Dam Victims and Their Defenders: The Emergence of Anti–Large Dam Movement in Indonesia," paper presented at the Asian Studies Association of Australia Biennial Conference at Murdoch University in Perth, July 1994; G. N. A. Hakim, "Membangun Waduk dan Masa Depen Sebuah Komunitas: Studi Kasus Kedung Ombo, Diskusi: Dimensi Sosial dan Kemanusiaan

dalam Pelaksanaan Pembangunan," YLBHI, Jakarta, 22 March 1989; George J. Aditjondro, "Mengapa Kampanye Kedung Ombo Berhasil [Why the Kedung Ombo Campaign Succeeded]," mimeo, Salatiga, Indonesia, 1990; George J. Aditjondro, "Setelah Ramai-ramai Kampanye Kedung Ombo [After the Kedung Ombo Campaign]," mimeo, Salatiga, Indonesia, 1990; Lawyers Committee for Human Rights and Institute for Policy Research and Advocacy (ESLAM), *In the Name of Development: Human Rights and the World Bank in Indonesia* (New York: Lawyers Committee for Human Rights, July 1995); Y. A. Toorop, "Kedung Ombo door de Wereld Bank Genomen" [Kedung Ombo Taken by The World Bank], M.A. thesis, University of Amsterdam, Amsterdam, 1990; Y. P. Widyatmadja, "Peaceful Atmosphere, Not Aid that Kedung Ombo Needs," unpublished notes of dialogue with John Clark from the World Bank, Solo, Indonesia, 1993; J. B. Mangoen Wijaya, "Permasalahan dan Saran-saran Tentang Resettlement Penduduk Kedung Ombo [Problems and Recommendations for Resettlement of Kedung Ombo People]," mimeo, Jogyakarta, Indonesia, 1990; Agus Purnomo, "Waduk Berwawasan Lingkungan, Diskusi: Dimensi Sosial dan Kemanusiaan dalam Pelaksanaan Pembangunan: Studi Kasus Kedung Ombo," YLBHI, Jakarta, 22 March 1989.

20. In 1992 the Indonesian government abolished IGGI, some members of which were persistently raising criticisms about killings and human rights abuses in East Timor, then formed the new Consultative Group on Indonesia (CGI), which is chaired by the World Bank and does not have human rights concerns under its purview. The government hoped INGI would dissolve, but it merely changed its name in November 1993 to the International NGO Forum on Indonesian Development (INFID). It has continued to raise the problems at Kedung Ombo with the World Bank.

21. George J. Aditjondro calls this "borrowed democracy." Personal communication.

22. See Michael Cernea, *Involuntary Resettlement in Development Projects: Policy Guidelines in World Bank–Financed Projects* (Washington, D.C.: World Bank, 1988).

23. Dennis Purcell, *Back to Office Report for Thomas E. Davies, AS5AG Unit Chief* (Washington, D.C.: World Bank, April 1989).

24. Stanley, *Seputar Kedung Ombo*, p. 277.

25. See Cernea, *Involuntary Settlement.*

26. INGI, "Letter to Barber Conable, President of The World Bank," INGI Brussels Conference, Nieuwpoort, Belgium, 26 April 1989. See also INGI, "Brief Report on a Visit to Kedung Ombo," unpublished manuscript, March 1990, and INGI, "Aide Memoire: Seventh INGI Conference," Washington, D.C., April 1991.

27. See Purcell, *Back to Office Report.*

28. See William Partridge, "Lessons Learned from Kedung Ombo," office memorandum to Mr. A. Cole, World Bank Office, Jakarta, Indonesia, 1989.

29. Stanley, *Seputar Kedung Ombo.*

30. *Ibid.*

31. Officials of LBH Semarang, interview by George J. Aditjondro.

32. See Hiroko Tanaka, "How Is 'Success' Defined? An Involuntary Resettlement Case Under Japan's Development Assistance Project," M.A. thesis, Department of City and Regional Planning, Massachusetts Institute of Technology, Boston, Mass., 1994.

33. See Purcell, *Back to Office Report.*

5

The Philippines: Against the People's Wishes, the Mt. Apo Story

Antoinette G. Royo

In the early 1980s, the Philippine government began exploring options for sources of domestically produced energy that would reduce expensive oil imports. Geothermal power was one such renewable and cheap source of electric power. The key conflicts surrounding geothermal power mainly concern the social and environmental impacts associated with the location of the thermal wells and power plants. In the early 1980s, the Philippine government began exploring the North Cotabato region in the southern Philippines around Mt. Apo as the site for a geothermal plant. The Philippine National Oil Corporation tried to obtain international funding for the project from a series of aid agencies, beginning with the World Bank. Each attempt was rebuffed, due in part to the transnational campaign that emerged in opposition to the plant. The project, however, continues to this day and provides a clear example of how the impact of geothermal development has involved significant environmental and social costs.

The Mt. Apo project was very significant because it was the first project the World Bank refused to fund solely on environmental and social grounds in response to the transnational campaign that emerged around the project. For six years a transnational alliance, led by Philippine nongovernmental organizations (NGOs), successfully intervened to prevent all possible sources of international aid financing for the project.

This chapter first explores the Mt. Apo project, the emergence of the opposition, and the evolution of the World Bank as the opposition's central target. It also looks at the divisions within the World Bank that contributed to the success of the campaign, and then examines the dynamics among the local community groups, support institutions, national networks, and international NGOs to explain the success of the transnational alliance to prevent all international funding for the project. Finally, it looks at how the campaign's legacies have altered domestic politics in the Philippines.

Mt. Apo: Background

For us, the Lumad of Southern Mindanao, the land is our life; a loving gift of Magbavaya (The Creator) to our race. We will die to defend it, even to the last drop of our blood.[1]

Mt. Apo is an area that possesses a mix of features of keen national and international interest. Standing 10,311 feet high, it is the country's highest peak, a dormant volcano that straddles the territorial boundaries of Davao City and the provinces of Davao del Sur and North Cotabato in the Southern Philippines. In 1936, President Manuel L. Quezon's Proclamation 59 declared it a national park to preserve its pristine and biologically diverse environment. In 1984, the Association of Southeast Asian Nations (ASEAN) recognized Mt. Apo as one of the richest botanical reserves in the Southeast Asian region, and it was classified as an ASEAN Heritage Site.

The environmental features of Mt. Apo are not found in any other mountain or volcano in the Philippines. It is central to the ecosystem of southern Mindanao. The source of twenty-eight rivers and creeks, it is the watershed of Davao City and the provinces of Davao del Sur, North Cotabato, Bukidnon, and adjacent areas. Four national irrigation systems are fed by four of these rivers, and approximately 50,000 hectares are irrigated by these systems. In the lower reaches of the volcano are closed canopy forests from 10 to 20 meters high, covering a total of 21,477 hectares. A total of eighty-four different species of birds have been recorded in the park, including the world famous and endangered Philippine eagle. A number of animals live in the park, such as the Philippine monkey, wild pig, Philippine deer, Philippine cat, and Philippine tarsier. But much of the biological diversity in the remaining forests remains understudied.

Six tribal groups live around the 72,814 hectare mountain: the Manobo, Bagobo, Ubo, Ata, K'lagan and Kaulo. For these people, it is not only the dwelling place of their gods and of Apo Sandawa, the supreme God, but also the burial ground of their ancestors. These tribal groups are part of a broader collectivity of indigenous cultural communities in the Southern Philippines known as the Lumad. The term means "of the soil" and a census conducted by Lumad-Mindanao in 1991 estimates that they number approximately 460,000 within the entire Mt. Apo area.

The Project and the Players

Mt. Apo was first explored by the Philippine National Oil Corporation (PNOC) in 1983 when it undertook a geoscientific study to evaluate its potential as a geothermal site.[2] The government's initial objectives in pushing for the development of the Mt. Apo steam wells were to prepare the southern Philippines for increased industrial demand for electricity, a purpose that continues as part of the current administration's commitment to acclerate industrialization and make the Philippines a newly industrialized country by the year 2000.

After additional exploration in 1986, PNOC began building an access road and drilling two exploratory wells the following year. PNOC planned to build twenty-one production wells and seventeen reinjection wells, capable of producing 120 megawatts (MW) of power in the first phase of the project. It would manage the steam field development, and the state-owned National Power Corporation would manage the construction and operation of the power plants.

Both the exploration and the drilling occurred without environmental impact assessments (EIAs) as required by law. The oil company's exploration in Mt. Apo was in clear violation of a law prohibiting commercial exploitation of areas within a national park and was initiated without any environmental or social impact studies. PNOC's operations within the park also violated the Philippine commitment to preserve Mt. Apo as an ASEAN Heritage Site, although the Philippine government did not regard the commitment as binding. By early 1988, PNOC had drilled, tested, and capped two wells within the park.

Financing for various components of the project was sought from a range of sources, both domestic and international. The National Power Corporation was negotiating an energy sector loan with the World Bank, in which the construction of two power plants at Mt. Apo were listed as possible subprojects. The loan negotiations were completed but did not include the Mt. Apo project. PNOC continued, however, to try and have the Mt. Apo project incorporated as a component of the loan.

The Opposition Emerges
Opposition to the project first emerged in 1987–1988. The campaign against the Mt. Apo project began because local Lumad communities felt threatened by the geothermal drilling and construction: developments were simply too fast, too intimidating, and too complex.

In this period the Lumad too had organized in opposition to the project and on April 13, 1989, declared their opposition through the D'yandi quoted at the beginning of this chapter. The D'yandi is a peace pact, and this one was sealed with the blood of all the defenders of the sacred mountain—thus making it irrevocable, unlike a contemporary contract or memorandum of agreement. It was a statement of protest, its gravity something that perhaps only a Lumad can completely understand. All twenty-one tribal chiefs, except for those now dead, still stand by this pact to oppose the PNOC geothermal power plant. This declaration, according to one of the elders, was not against the government of the Philippines per se, but was simply a statement of their will to live within and protect their ancestral land. The D'yandi has been renewed every year by the tribal elders as a reaffirmation of their commitment to oppose the project, a commitment that has inspired other members of the campaign, both inside and outside the Philippines, and has served as the touchstone against which campaign members have evaluated the appropriateness of their own activities.

Opposition also emerged in nearby regions—namely, North Cotabato and Davao—and later among Manila-based NGOs. The core of the campaign consisted of three task forces[3] set up between 1987 and 1988 to articulate the issues, negotiate with government, and expand public environmental consciousness. The local task force was composed primarily of the Lumad people directly affected by the project. In addition, the Catholic Church in Kidapawan played a key role. The regional task force, based in Davao City, and the national task force were composed of environmentalists, consumer and religious groups, as well as lawyers and academics. Because it was based in the national capital, the Manila Task Force was more able to engage all branches of government: legislative, executive, and judicial.[4] It engaged both the Department of Environment and Natural Resources (DENR) and the legislature to push for a halt to the project and respect for the ancestral domain rights of the Lumad living in the Mt. Apo region. From 1989 to 1990, campaign allies in the legislature introduced four resolutions calling for a halt to PNOC's operations.

The Legal Rights and Natural Resources Center (LRC, Kasama sa Kalikasan–FOE Philippines) was central to the activities of the regional and national task forces and the lobbying work that targeted the multi-lateral development banks (MDBs). Composed mainly of lawyers and paralegals who practice alternative law or developmental legal assistance, LRC combined policy advocacy with grassroots legal services in the case.[5] It worked closely with major indigenous peoples' coalitions, conservation groups, national professional societies, and international support groups for indigenous peoples within and outside the task force. Although functioning as the legal/policy advisor, tactician, referee, chief policy communicator, and documenter of the facts and events related to the campaign, it always acted in only a representative capacity vis-à-vis the affected Lumad communities and facilitated, rather than led, the MDB lobbying efforts.

Numerous solidarity networks of indigenous peoples abroad supported the campaign, but the most consistent international supporters were part of the MDB lobby of the Mt. Apo campaign. With the initiative of the Friends of the Earth–U.S. (FOE-U.S.) representative in the Philippines and the Bank Information Center in Washington, D.C.,[6] the World Bank lobby began to take form in 1989. The Manila-based representative of FOE-U.S. was the focal point of international MDB campaigning, while in Washington, D.C., the Philippine Development Forum (PDF) took over the mantle of leading the campaign inside the World Bank. PDF had been founded in 1989 as a network of primarily Washington, D.C.–based environmental, development, church, and human rights activists committed to working with Philippine NGOs for just and sustainable development in the Philippines.

Within the task force in Manila, each member represented an organization that took on one major task assigned to it by the campaign. Some individuals (mostly concerned scientists) voluntarily offered their services to the task force. The MDB lobbying, even if assigned to FOE-U.S., was implemented by a team of at least four members who were mainly from legal organizations.[7]

The team's focus was the review of PNOC's compliance with Philippine environmental, forestry, administrative, and other pertinent laws and regulations. It also included contacting key people at the World Bank to whom the PDF network could feed its latest findings. Actual meetings for dialogue with Bank officials rarely occurred in the Philippines, except when the latter actually visited the sites. In these cases the Lumad community and local task force members discussed their case directly with Bank officials.

In sum, opposition to the PNOC project by the MDB team in the task force consisted of preparing legal briefing papers; documenting violations on site; gathering information on PNOC's latest moves; writing letters to the international support groups and Bank contacts; encouraging local people to continue writing petitions; engaging in dialogue with Bank officials at meetings the team arranged; and inviting other international support groups to visit the site.

The MDB work strengthened national advocacy for recognition of indigenous peoples' rights to their ancestral domain. This link assisted in pressuring the Bank to act—given its strong policy commitment to support indigenous peoples, as reiterated by Bank officials—in the Philippines in particular.

The Environmental Impact Assessment

The opposition brought the case to public attention in 1988. In the middle of that year, pressure from local communities and NGOs encouraged the DENR secretary to order PNOC to suspend its operations pending compliance with the law that required all environmentally critical projects to submit an EIA. The oil company produced a nine-volume EIA in 1990 to satisfy the EIA requirement. With very limited time, campaigners analyzed the documents and found the EIA environmentally and socially inadequate. There was no discussion of alternatives, however. Public hearings did not allow for sufficient time (as required by law) for campaigners to prepare counterproposals and for the local community to discuss its impact substantively. A joint study on the Mt. Apo project criticized PNOC's contention that Mt. Apo was an ideal production site: data from PNOC were reportedly insufficient to assure the proper installation of safeguards.[8] The EIA process was far from satisfactory, even as various government allies took extra efforts to try and bridge the gap.[9]

Two reviewers from the World Bank were asked to comment only on the quality of the EIA, not on the merits of the project. With one focusing on physical and the other on social impacts, their findings reinforced those of the project's critics. Following the review, for the first time ever, the Bank's regional environmental office chief did not clear a project. A May 1991 memorandum from the chief stated that the Mt. Apo Geothermal Project could not be cleared "since this project is located in a national park of international significance and involves extremely sensitive social issues"—the first time since the creation of its environmental guidelines in 1989 that such a position had been taken.

In January of 1992, despite the indigenous peoples' plea to reject the EIA and perhaps because of great pressure from the presidential palace, the DENR issued an environmental compliance certificate (ECC) to PNOC.[10] The certificate had over twenty conditions attached to it, violations of which could lead to its suspension and therefore a halt to PNOC operations in Mt. Apo.[11] Later that month, then-President Aquino issued a proclamation segregating a 701-hectare area in Mt. Apo National Park as a geothermal reserve.[12] PNOC had special access to the president's office because presidential Executive Secretary Catalino Macaraig, was also chair of the board of PNOC.

In January of 1992 the campaign responded to the issue of the ECC by filing a petition in the Supreme Court of the Philippines for a review on *Certiorari*, alleging grave abuse of discretion of the respondent, DENR, in issuing the ECC. This petition involved an enormous amount of work by the legal groups and in the end was unsuccessful. The Supreme Court dismissed the case in February 1992 on the technicality that ten out of twenty-four copies of the 350-page petition lacked certified true copies of the annexed DENR decision in question. Upon motion for reconsideration, the Supreme Court again denied the case on grounds that "there is no compelling reason to warrant the reversal of the questioned resolution and in any case there is no grave abuse on the part of the public respondent in rendering the questioned resolution."

With domestic legal remedies exhausted, the Bank was now the main target. Although the campaign had been pressuring the Bank since the late 1980s, the Bank became the central target after the Supreme Court decision.

The World Bank in Focus

Taking on the World Bank was not exactly new for Philippine NGOs. Struggles against Bank-sponsored projects had been a crucial component of struggles against the Marcos dictatorship. Perhaps the most famous campaign was in the late 1970s when the Kalinga people successfully defeated a World Bank–sponsored series of hydroelectric dams on the Chico River that would have inundated numerous villages and large areas of agricultural land.[13]

The external pressure from NGOs coincided with splits inside the Bank over the project. Two Bank missions to the Philippines resulted in conflicting recommendations, which raised the profile of the project even higher inside the Bank. The first mission in 1989 resulted in an official letter from the Bank to international NGOs, stating that they were "well aware of the social [and] environmental issues surrounding the proposed geothermal development ... including those related to tribal people.... [I]f undertaken at all the project has been deferred to beyond 1992 and will, therefore, not be financed under the Bank's proposed energy sector loan." The letter concluded that the World Bank intended to work closely with the Environmental Management Bureau and the Philippine National Oil Company to strengthen their capacity to carry out environmental impact assessments. Of particular significance, the letter mentioned that the Bank planned to help PNOC find an alternative site for the Mt. Apo project.

A second Bank mission in early March 1992 was for the Integrated Protected Areas (IPA) Project. Bank officials met with local people in Mt. Apo, Davao, and Kidapawan during visits and noted people's concerns. Afterwards, they met informally with the Manila campaigners. A report by the mission's local consultant reflected the opposition's views,[14] concluding that the geothermal project posed a dilemma for the integrated protected areas system and the goals of the Mt. Apo National Park.

The Bank's Energy and Environment Departments were on opposing sides, a division that bode well for the transnational campaign. Increased external pressure meant fewer possibilities for internal consensus.

Ancestral Domain

While the Energy and Environment Departments debated, a parallel process was taking place within the Bank that also had significance for the project. From 1989 to 1992, the Bank was negotiating with the Philippine DENR on the Sector Adjustment Loan for the Environment. The loan included a substantial section on sustainable forestry with an emphasis on community-based forest management. While that loan was being negotiated, forestry policy lobbyists,[15] such as LRC, maintained regular contact with Bank officials and provided regular input on recognition of ancestral domain. This recognition was anchored on a proposed policy of divestment whereby the DENR would relinquish control of the so-called "uplands,"[16] which could be proven to be ancestral territory.

The viability of the ancestral domain arguments relied mainly on the idea that properties held by tribal or indigenous peoples since time immemorial should not form part of the public domain of the state.[17] Research conducted mainly by the LRC was provided both to the Bank and to various members of the task forces, thus educating not only the indigenous peoples' organizations and NGOs, but also the government and international institutions about the legal basis of ancestral land claims.

The World Bank responded with much interest, on the basis of its own new guidelines mandating respect for indigenous cultures and ancestral land rights in its projects.[18] As a result, in its preliminary documents for the Sector Adjustment Loan for the Environment it listed the implementation of targets for divestment as a high priority. The DENR disregarded those priorities on the basis that there was no ancestral domain law as yet. In the end, the loan proceeded, with the Philippine government agreeing only to interim measures that could lead eventually to the recognition of ancestral land rights. Recognition of ancestral domain would have major implications for the Mt. Apo project because much of the region would fall under the ancestral domain provision.

The Bank Rejects the EIA

Based on their contacts with the local communities and the national task force, PDF and other Washington-based groups pursued a lobbying and letter-writing strategy with World Bank officials in Washington. PDF members met with Bank officials several times in the first six months of 1992. Using the connections and credibility that the Washington-based environmental NGOs had with Bank staff, and using information provided by LRC and other Manila Task Force members, they were able to provide Bank staff in Washington with evidence of extensive local opposition to the Mt. Apo project.

The Bank's strict application of its environmental and social guidelines to the Mt. Apo project proposal resulted in the rejection of the EIA in mid-1992. The grounds for rejection were:

It did not adequately assess alternative sites for geothermal power generation or alternative means of generating energy; Consultation with local groups was inadequate both in terms of who was consulted and the methods used; It did not address the capability of local institutions to manage the proposed social and environmental mitigation programs; It did not consider the cultural and religious importance of the Mountain to indigenous groups; It did not evaluate the impacts of sudden growth of population on the local area due to construction.[19]

The oil company then withdrew its loan request from the World Bank.

The Bank's decision to reject the EIA and PNOC's withdrawal of its request strengthened an ongoing process of consolidation by the opposition campaign's task forces.[20] The National Coordinating Committee of the campaign was established to monitor the progress of the overall campaign and encourage all the task forces to unite on targets and focus. The campaign decided on a strategic focus on ancestral land and tribal peoples' rights to self-determination, side by side with regular monitoring of PNOC activities and militarization in the area. Some members disagreed with the decision, wanting to focus more on independent environmental monitoring, but eventually the emphasis on tribal peoples' rights prevailed.

Militarization of the area increased following PNOC's withdrawal of its request for a loan from the Bank. Of the three task forces, the local one was most vulnerable to repression. PNOC's use of the military to provide security and the subsequent harassment of the community members became a crucial lobbying issue for the opposition,[21] exposing a critical weakness in PNOC's strategy due to strong international opposition to human rights violations associated with the implementation of development projects.

Once the Bank rejected the EIA and PNOC withdrew its request, PNOC explored other sources of international financing, including the U.S. Trade and Development Program, as well as U.S. and Japanese export-import banks. In late 1992, the Japanese bank, under pressure from Philippine and Japanese NGOs, denied PNOC's request for funding, and in 1993, under pressure from Philippine and U.S. NGOs, the U.S. bank followed suit. With additional pressure from European-based groups, the opposition approached the banks, briefed them on the issues, and expressed their serious concerns and reservations about PNOC's undertaking. The Asian Development Bank was also targeted with the same information and warned of impending problems if it supported any type of proposal from the PNOC to develop Mt. Apo.[22]

The Current Situation

In 1983 PNOC had projected that the power plant would be fully operational a decade later. As of late 1995, the geothermal steam fields still remained untapped, but the project has moved steadily ahead, even though the Philippine government plans to develop Mt. Apo as a national park. The government, however, has also passed a law allowing for geothermal explorations within the park.

The company's response to continued local opposition has been to try to persuade the local communities with both spiritual and material methods. In March 1992 PNOC sponsored a counterritual to the D'yandi, performed by some assimilated tribal peoples supportive of the project. In a more material vein, PNOC established the Mt. Apo Foundation in 1993, which was designed to assist socioeconomic and educational needs of the tribal peoples affected by the project. It agreed to commit one centavo per kilowatt hour of electricity generated by the plant to an environmental endowment fund under the management of the foundation, which is currently headed by a non-Lumad appointed by PNOC.[23]

Some Philippine NGOs active in the Mt. Apo campaign have become active in the NGO consortium that is managing the Integrated Protected Areas Program, which is financed by the World Bank through the Global Environment Facility (GEF). In mid-1994 the Bank gave many activists the impression that it was weakening its strict application of environmental and social standards. Many participants in the Mt. Apo campaign understood the Bank position to be that no projects that threatened the environment could be allowed in the protected areas sites. When the Philippine Congress modified the National Integrated Protected Areas Act by specifically allowing the construction of geothermal plants in parks, the Bank did not protest. As a consequence, the World Bank's requirements were circumvented. Campaigners viewed the Bank's de facto support of the watered-down provisions of the Protected Areas Act as a retreat from its initially strong opposition to the Mt. Apo project. The Bank claimed that because the project had already broken ground and the government was determined to build the project, it was hopeless to oppose the whole legislation for just one area, and that Mt. Apo's inclusion as a priority area in the Integrated Protected Areas System (IPAS) could help to mitigate any destructive impacts caused by the project.

The government, local communities, and NGOs continue to debate proposals on appropriate approaches to manage the Mt. Apo National Park. NGOs also continue to monitor the development of the geothermal site and highlight the numerous violations of PNOC's environmental management agreement with government. These violations are mainly high pollution levels and the development of areas outside the project site. To date, no sources of international financing have been tapped. Although the government claims that it can continue to draw on internal funding and foreign reserves, many people, including some within the government itself, have doubts as to whether valuable foreign reserves would actually be used, given the substantial debt burden the Philippines faces.

The Dynamics of the Transnational Alliance: Decision Making and Accountability

The transnational alliance that emerged over the Mt. Apo project was notable for at least four elements: decision-making processes, accountability mechanisms, the formation of strategic alliances, and the exploitation of contradictions within the World Bank and the Philippine government.

Decision Making
The task forces that ran the campaign in the Philippines coordinated the work with international partners. As the protest moved into full swing in 1989, the task force structure provided a venue for interested groups and individuals to share experiences and discuss strategy. The advocacy work vis-à-vis multilateral development banks was seen by many task force members as collaborating with the "enemy" and hence was viewed initially as best performed by foreigners.

The antagonism toward the MDBs in general, but the World Bank in particular, was due to the Bank's history of support for the Marcos dictatorship from 1973 to 1986.[24] The Bank was seen as a key provider of international legitimacy as well as funds for the regime, and the regime's relatively recent departure meant that those feelings still ran strong.

Task force members possessed a range of political viewpoints. The Lumad opposition was motivated fundamentally by their desire to prevent the desecration of their sacred land and to realize their ancestral domain rights as carved into the Philippine 1987 Constitution.[25] The conservationists were in the minority and also the least progressive. Their interest was to protect the endangered species on Mt. Apo. Farmer and consumer groups shared concerns of protecting the vital watersheds. The Catholic Church and human rights advocates joined the environmentalists in calling for a just and wholesome environment. Other groups engaged in legal and public policy advocacy were interested in taking advantage of the democratic space within the Aquino government to challenge its legal, socioeconomic, political, and environmental paradigm, which was very similar to that of the Marcos government. The Mt. Apo project provided a prime opportunity to engage in constitutional and legal battles with the Aquino government, as well as to challenge the economic policies pursued by both the MDBs and the Philippine government.

LRC belonged to this latter group, and its involvement in the Mt. Apo campaign started in that context. LRC's work in the campaign included not only information exchange, Bank visits, delegations, and negotiations, but also the development of policy input, the facilitation of community visits to the Bank, and the promotion of community dialogue with Bank officials. Information exchange entailed working with task force members to provide regular monitoring of PNOC's field maneuvers as well as its financial plans and options, and developing environmental and legal arguments against choosing Mt. Apo as a site.

What began as an international information exchange on the Mt. Apo project developed into a strong, ongoing lobby effort for a number of reasons. First, there existed a network of NGOs and organized local communities capable of facilitating fact-finding missions and processing vital information into news releases, information updates, appropriate executive/judicial actions, or policy agendas for potential dialogues with the government. A second factor was the full-time attention that a resident representative of an international environmental NGO put to the multilateral development bank campaign.[26] Third, the church-based opposition received support from large networks such as the Catholic Bishops Conference and the Association of Major Religious Superiors of the Philippines. Fourth, preexisting international solidarity networks of the campaign's member NGOs in Australia, Japan, the Netherlands, Switzerland, the United Kingdom, Canada, and the United States sporadically distributed numerous news releases and action alerts opposing the Mt. Apo geothermal plant and focused mainly on human rights issues. All of these factors shaped the context of the transnational alliance.

But it was members of the Philippine Development Forum in the United States—a new organization—who became the most involved of all international partners. In Washington, D.C., groups such as the Friends of the Earth–U.S., the Bank Information Center, the Columban Fathers for Justice and Peace Office, Greenpeace, and the Environmental Defense Fund—many of whom were members of the PDF—took a lead in responding to calls from the Manila Task Force. They informed Bank staffers of the problems being raised by the local peoples and the individual task forces. Bank lobbying eventually became the unifying point of the transnational alliance.

The campaign's decision-making process was sophisticated. The Manila Task Force made key decisions in its regular meetings on different strategic and tactical dimensions of the national and international campaigns, including the MDB campaign. These decisions were generally based on information coming from the Mt. Apo region and on the legal and scientific arguments developed by researchers. PDF received information directly from the Manila Task Force, especially from LRC. Any additional input PDF needed, LRC supplied directly. Position papers prepared by PDF were sent to the task force members for comments and data check before publication. Updates from PDF were disseminated through action alerts, long-distance communication, e-mail, or fax. Decisions made for a period of six to eight months were evaluated through the Solidarity Conferences attended solely by Philippine-based groups.[27] The Solidarity Conferences served as mechanisms to ensure accountability to the local communities, build consensus on demands, strategy, and tactics, and develop a coherent division of labor.

Accountability

Three separate dimensions of accountability emerged from the campaign: (1) the accountability within the Philippine Mt. Apo campaign; (2) the question of the accountability of the international partners to the Philippine campaign; and (3) the accountability of the PDF coordinator and membership to the leadership of PDF. The first two dimensions involved questions of accountability to the local communities. The challenge was how to ensure that both Manila and Washington-based NGOs engaged in lobbying the Philippine government and the MDBs would remain accountable to the demands of the Lumads in Mt. Apo.

Accountability within the Philippine Campaign   An enormous amount of energy was put into discussions of the appropriateness of engaging the government and the MDBs in a dialogue in the effort to stop the geothermal project. Members of the task forces opposed to dialogue made at least four arguments. First, they believed that this effort would compromise the communities because the MDBs could use them to legitimize the project after participating in such meetings. Second, meetings with government and MDB officials would have no real results because the communities' counterproposals would be ignored. Third, they argued that the government's laws served only the status quo, using the military presence in the area to protect PNOC as proof. Fourth, they felt that lobbying the World Bank diminished the strength of indigenous peoples' position for self-determination because it was seen as compromising the peoples' independence from the dictates of government policies.

The various members of the Mt. Apo campaign had at some point shared social and political responsibilities within their own jurisdictions. The transnational alliance recognized that they had a full mandate to proceed with the MDB advocacy, primarily vis-à-vis the World Bank. They also knew, however, that they were constrained in negotiating with the Bank: they were empowered only to oppose the funding for the project but not to bargain for continuing plant operations on behalf of the local community. The same limitations were applicable in the other levels of the campaign as well. The lawyers could not file the court cases without the full agreement of the tribal chiefs; groups and individuals in the task forces could not freely join the PNOC-organized environmental-monitoring teams or its foundation without being treated as "sellouts."

The transnational dimension of the campaign with regard to lobbying the World Bank was essentially a fait accompli. Had this dimension been discussed for the first time in the solidarity meetings, it most likely would have been discouraged. Foreign participation in the campaign succeeded because the transnational alliance membership was willing to allow local groups to scrutinize their political standpoints with regard to MDBs. Some members of the campaign believed that the banks used sustainable development rhetoric to disguise their primary mission: to lend and make money. Some, however, believed in reforming the MDBs, but most task force members did not. This difference would have been the primary difficulty in initially setting up the coalition, but there was no chance for these sentiments to surface. Because PNOC requested money from the World Bank, PDF fulfilled the urgent need to lobby the Bank. By the time the solidarity meetings were held, the lobbying effort was already underway, and by the time of Second National Solidarity Conference, the Bank had already rejected the EIA.

PDF Accountability to the Philippines   Although the PDF was distanced from the local politics of the Mt. Apo campaign, it was still accountable to the local people. Jennifer Smith, then the PDF coordinator, visited Mt. Apo and came back with a personal commitment to oppose the geothermal plant on the basis of the D'yandi declaration, the gravity of the project's environmental impact, and the World Bank's violations of its own rules. PDF's work focused on compiling, packaging, and bringing appropriate information to their Bank contacts who sympathized with the project opponents. Through Jennifer Smith and others in PDF[28] pressure was placed on the World Bank to support the Lumad position against PNOC's funding application. Their mandate was clear: Stop the Mt. Apo project. PDF was never in the position to negotiate with the Bank on conditions for continuance of the project.

Internal PDF Accountability   During this campaign, an ongoing internal debate in PDF focused on whether it should be a "campaigning network" or just serve as a "coordinator of information" on Philippine development issues. The Philippine partners of PDF[29] were more inclined to have it be a low-key support network in Washington, D.C., which meant that there could be alliances in pertinent campaigns only when solicited by Philippine groups and only when politically acceptable. PDF's positive and successful participation in the Mt. Apo transnational campaign, however, created a debate within the PDF-U.S. board about the degree to which its actions should be member led or staff led, especially given the facts that the focal point of the transnational alliance was PDF and the main spokespeople at the World Bank delegations were PDF members. At around the same time in the Philippines, the Philippine partners discussed whether PDF should just be an information conduit or also serve as an advocate. In the context of the Mt. Apo campaign, the advocacy role of PDF was made clear: information dissemination also meant that the disseminator could take sides. As an advocate, PDF was accountable to those people on whose behalf it was working—the Lumad—and opened itself up to public scrutiny for the positions it took and the policies it advocated.

Making Strategic Alliances

For any party involved in the Mt. Apo situation, dealing with the various interest groups in the Mt. Apo campaign, including the local communities, was the biggest challenge. The number of "interested parties" with financial and/or bureaucratic interests was quite large. In the Mt. Apo area, local government officials sought to build political support for future elections from the military, the revolutionary forces (National Democratic Front and the New People's Army), and other local tribal groups with suspect interests.[30] National government officials included: the National Power Corporation to manage and distribute the power produced by PNOC, the foundation set up by PNOC to collect royalties for the tribal peoples' trust fund, the presidential representative for Mindanao who was charged to oversee the project's overall economic development, the DENR, the legislature, and the judiciary. On the financial side, there was the World Bank, the Asian Development Bank, U.S. and Japanese export-import banks, and the private companies willing to sign up for build-operate-transfer (BOT) schemes.

The campaign eventually took the perspective that negotiations or dialogue with any of the above interested parties should assist in protecting and promoting the right of the indigenous people to oppose the Mt. Apo project. Campaigners had met with some or most of these interested parties at one point, officially or otherwise, to negotiate the terms for this opposition. The negotiators were, at minimum, elders of local peoples or elected representatives of the tribal peoples networks.

The transnational dimension of the alliance dealt with only the more nebulous structures outside of Philippine politics and only from the point of view of the tribal peoples. Its mandate was the statement of elders and local leaders seeking support. Inside the alliance, actions were based solely on that informal mandate and the basic understanding of the issues, acquired from direct visits and "immersions" of the advocates. The World Bank and the export-import banks were within direct reach of the transnational alliance, as were PNOC, the National Power Corporation, and the government. Alliances with any individuals, partners, consultants, or advisers of any of these institutions began and ended solely on the basis of whether or not such alliances would be beneficial at some point in the campaign. Further, such alliances were made with full knowledge that these institutions might also take advantage of the contacts made by the transnational alliance to further their interests.

Taking Advantage of Contradictions

When the NGOs identified the existence of internal conflicts within the Bank and other government departments, they exploited these openings. The opposition saw that for as long as its complaints concerned the Bank or its administrative policies, the protest was viewed as valid and even contributed to divisions within the Bank. The transnational alliance became an effective tool to tackle such divisions because the international contacts had a stronger reputation for engaging the Bank.

World Bank officials reacted positively to external pressure by showing openness to NGO letters, visits, and dialogue[31]—for three main reasons, according to Bank sources: first, information about local sentiments and the existence of a strong local lobby reached the Bank; second, the Bank's review of PNOC's EIA found it inadequate;[32] and third, a Bankwide review of its environmental principles was occurring at the same time, so Mt. Apo served as a test case for the Bank's new environmental standards.

Bank sources claim that these congenial relations were facilitated by the fact that the Washington groups (mainly the PDF) acted with much restraint and tackled specific policy issues to which the Bank could respond. The regular visits by the PDF coordinator and PDF members, together with other active campaigners, focused on technical and policy questions based on information from Philippine groups. PDF's focus on delivering one clear message—"the World Bank has to play by the rules"—made their advocacy easier.

The Impact of the Transnational Campaign

The campaign had three sets of targets against which impact could be measured: (1) the PNOC; (2) the World Bank, other MDBs, and international aid agencies; and (3) other Philippine government agencies.

As far as impacts on PNOC, it would probably not have set up a Mt. Apo Foundation without the opposition to the project. Clearly, the campaign was a force to be reckoned with and, hopefully, demobilized. The Mt. Apo project still continues, however, even without international financing, so the campaign was not successful at actually halting it.

The outcomes did coincide with the campaign's demands regarding the actions and policies of the World Bank and other international aid agencies. But it is difficult to ascertain the degree to which the Bank's decision to reject the EIA was due to the pressure created by the trans-national coalition. World Bank officials admitted that it was NGO pressure that ensured strict scrutiny of PNOC's EIA, thus leading to its rejection.

Clearly, there was a lack of consensus among the Bank's departments. Inside the World Bank, strict application of its environmental and social guidelines supported the views of individuals who made key decisions concerning the project. Also, the Philippine government's policies re-garding the recognition of rights of indigenous peoples may have also provided some persuasive argument for Bank staff who opposed the project. In the end, the PNOC project was said to have poorly packaged its EIA so that Bank evaluators did not need to work hard in order to defend their position not to support it.

The Bank was not easily influenced, however, and in many previous cases had acted against the interests of communities negatively affected by development projects. Campaigners found Bank policies to be key in the Mt. Apo case, and they challenged these policies via documentation and presentation of facts culled from their own direct field visits. Like a case study, those facts were packaged in a way that, if pitted against Bank policy, they would speak for themselves. All the campaigners had left to do was to push for some action inside the Bank.

Success with the Bank translated into pressure on the Philippine gov-ernment for legal changes to respect ancestral domain, for consultation and dialogue with NGOs, and for changes in environmental impact assessment procedures. (This triangulation of pressure—directly from below by NGOs and peoples' organizations [POs] and indirectly via an international institution—raised some dilemmas for some campaign members, which are addressed in the concluding section of this chapter.) The campaign succeeded at catching the Philippine government in a contradiction between its development goals and its constitutional obli-gations: namely, the policies it pursued to achieve newly industrialized country status clashed with its own commitment to recognize the rights of its indigenous/tribal peoples to their ancestral land. The geothermal plant project became a showcase for measuring the government's intent to carry out its support for the indigenous peoples. As a consequence, lobbying for the passage of the ancestral domain bill rose to its peak during the campaign, with government departments themselves submitting their own drafts for Senate or House sponsorships.

The internationalization of the Mt. Apo campaign brought home the message that the Philippine government needed to take positive steps toward recognizing and protecting indigenous peoples' rights and toward taking NGOs seriously. This internationalization could not be said to have pushed the Philippine government to dialogue, as much as it did the World Bank, but Bank recommendations to the Philippine government were treated like mandatory requirements ("conditionalities") in loan negotiations. Power to persuade the Bank therefore constituted power over the Philippine government.

Lessons and Legacies of the Mt. Apo Campaign

Like the campaign against the Chico Dam in the late 1970s, the Mt. Apo campaign has left many legacies that have influenced ongoing campaigns against MDB-funded projects in the Philippines—such as the Masinloc coal-fired power plant, the Integrated Protected Areas System, and the Industrial Tree Plantation lending schemes. Similar legacies have influenced Philippine government policy, government–NGO relations, and relations among NGOs.

The two main areas of Philippine government policy affected by the Mt. Apo campaign were the EIA process and the emergence of ancestral domain onto the national policy agenda. The issue of EIAs has become an important focus of discussions about assessment and reform, with the Philippine government recognizing the need for technically adequate assessments to meet donor criteria. The government has requested and received several technical assistance loans from multilateral and bilateral aid agencies for refining the process of environmental impact assessment and monitoring. Mt. Apo has become one of the classic case studies used to illustrate the shortcomings of the existing process.[33]

In addition, ancestral domain was placed firmly on the national policy agenda. The respect for and recognition of indigenous peoples' rights to their ancestral domain has increased, and it is considered politically inappropriate (or incorrect) for any government office at this point to reverse this trend. Current land-use studies take into account the type of land, the private claimants and the nature of claims, the existence of indigenous resource management systems, the feasibility of combining these with recent technology, and so on. The Department of Environment and Natural Resources has issued several Certificates of Ancestral Land Claims (CALC) and delineated areas of ancestral domain under an administrative order.[34]

Since the Mt. Apo campaign, all proponents of development projects in the Philippines mention consultation with NGOs and POs as a priority. The nature of these consultations, however, varies greatly because much depends on the experience of the NGOs, local government, and the financing institutions themselves. A related development from this campaign is highlighting the value of using models of "tripartism" that reflect the multiple constituencies affected by development projects. This "tripartism" can take many forms—such as government, NGO, and PO; or government, MDB, and NGO; or more recently, national government, local government, and PO—as a means of shaping the formulation and implementation of development projects. The Philippine government now admits that NGOs play a key role in ensuring full implementation of its projects.

The government also has new respect for NGO opposition to projects. Some government officials have accepted that NGO opposition, expressed in workable legal or administrative terms, can guide the government to study possible alternatives, work out legal remedies, and establish compromises that ultimately take into account the interests of local communities. Hence, in the case of Mt. Apo, some NGOs that continue to oppose the Mt. Apo project are now part of a panel of advocates to discuss the planning, implementation, and monitoring of the Integrated Protected Areas System. Others from the opposition are looking into biodiversity in the Mt. Apo National Park and are doing cultural research regarding its indigenous peoples with a view to getting support from other foundations.

The NGOs

The Mt. Apo campaign left two legacies regarding relations among NGOs. One has to do with the organizational innovations of the campaign itself; the other is more a dilemma faced by Philippine NGOs (and Southern NGOs more broadly) engaged in advocacy with the World Bank or other MDBs.

The organizational innovation of the campaign was the development of a "unified front"—where the national, regional, and local task forces worked together to implement a coordinated opposition and to provide international campaigners with the ammunition for their communications with the MDBs or international aid agencies. The Northern members of the alliance played the role of articulator or messenger for the Southern members, a structure that enabled the campaign to combine the flexibity and efficiency that comes with a division of labor without sacrificing accountability to the local communities.

The success of the Mt. Apo campaign reinforces an unresolved paradox for Philippine NGOs (including ones that participated in the campaign) that are politically oriented toward supporting national sovereignty. By convincing the MDBs to act in a given way, and using conditionalities to deny funds to the national government, they indirectly erode the very sovereignty they claim to support because the impact of their campaign would be to move their governments to comply with Bank conditionalities. A more obvious contradiction is the fact that by talking to the banks, Philippine NGOs are legitimating their country's dependency and foreign indebtedness and complicating the implications that go along with working with the banks in servicing external debts. Further, NGOs involved in community organizing and empowerment will often view working with the Bank as "sleeping with the enemy." The Mt. Apo campaign does not offer easy answers to these paradoxes.

Appendix

Task Force Members

Task Force Apo Sandawa (TFAS), Kidapawan Members
1. AGM-Services Network (ANI) Foundation, Inc.

2. BATUNA
3. Cotabato-Davao Sur Federation of Cooperatives (CDSFC)
4. Consortium for the Development of Southeastern Mindanao Cooperatives, Inc. (CDSMC)
5. Cotabato Peasant Forum
6. Diocese of Kidapawan
7. GKK-Kidapawan Foundation
8. Indigenous Peoples Project for Natural Resource Management (IPPNRM)
9. Kidapawan Parish
10. Kapisanan ng mga Mabubukid sa Kotabato
11. Philippine Rural Reconstruction Movement (PRRM)-Cotabato
12. Sinabadan ka Katigatunan ka Sandawa (SKS)
13. Task Force Detainees of the Philippines (TFDP)
14. Tribal Filipino Center for Development (TFCD)
15. Tribal Filipino Program, Diocese of Kidapawan
16. Women's Development Center (WDC)

Task force individuals include Cong. Gregorio Andolana, Carlos Bautista, Efren Catedrilla, Danilda Fusilero, and Atty. Solema Jubilan.

Task Force Apo Sandawa (TFAS), Davao Members[a]:

1. Alliance of Health Workers (AHW), Region XI
2. Association of Social Development Agencies in the Region (ASDAR, Region XI)
3. Assumption School of Davao
4. Basic Christian Communities–Community Organizing (BCC-CO)
5. Care for the Earth Ministry (CEM), Malabog
6. Care for the Earth Ministry (CEM), Redemptorist Parish
7. Community Organizing Davao Experience (CODE) Foundation
8. Development Center for Human Resources (COREDECENT)
9. Davao Apostolate for Youth and Community Action (DAYCA), San Pedro Parish
10. Episcopal Commission on Tribal Filipinos (ECTF- Mindanao)
11. GABRIELA, Southern Mindanao Region
12. Humanitarian Alliance against Disaster Foundation (HALAD)
13. Holy Cross of Davao City College (HCDC)
14. Health Alliance (HEAL), Mindanao
15. Education Forum (EF), Mindanao/KAMKEM
16. Kinaiyahan Foundation, Inc. (KFI)
17. KONSUMO, Dabaw
18. LRC-KsK Mindanao Branch Office
19. Lunhaw Mindanaw
20. Missionaries of the Assumption (MA) Sisters
21. Mindanao Environment Forum (MEF)
22. Mindanao Inter-Faith Peoples Conference (MIPC)

23. Mindanao-Sulu Social Action (MISSAS-CBCP) Secretariat
24. Media Mindanao News Service (MMNS)
25. Promotion for Church Peoples' Rights (PCPR), Mindanao
26. Peoples' Response against Inequality and Consumer Exploitation (PRICE)
27. Rural Missionaries of the Philippines
28. Social Action Center (SAC), Davao
29. Sisters Association of Mindanao (SAMIN)
30. Southeastern Mindanao Regional Ecumenical Council (SEMREC)
31. Socio-Pastoral Institute (SPI), Mindanao
32. Tribal Filipino Apostolate (TFA), Malabog
33. Urban Integrated Health Services (UIHS)

Task Force Sandawa (TFS), Manila Members[b]

1. Developmental Legal Aid Center (DLAC)
2. Environmental Policy Institute
3. Friends of the Earth, International (FOE)
4. Green Forum, Philippines
5. Health Action and Information Network (HAIN)
6. Haribon Foundation for the Conservation of Natural Resources (HARIBON)
7. Kabataan para sa Tribong Pilipino (KATRIBU)
8. Kayumangi Forest Development Foundation
9. Legal Rights and Natural Resources Center (LRC)
10. Lingkod Tao–Kalikasan (LTK)
11. People's Action for Cultural Ties (NCCP-PACT)
12. Philippine Ecological Network
13. Philippine Rural Reconstruction Movement (PRRM)
14. Structural Alternatives and Legal Assistance for the Grassroots
15. Scientists, Technologists, and Engineers for the People (STEP)
16. Tunay na Alyansa ng Bayan Alay sa Katutubo (TABAK)
17. Ugnayang Pang-Agham Tao (UGAT)
18. Volunteers in Scientific and Technical Assistance (VISTA)

[a] Members also include some concerned individuals.
[b] Members also include some concerned individuals.

Notes

1. See the full text of the D'yandi printed in the special issue of the *Philippine Natural Resources Law Journal* 2, no. 2 (May 1990), pp. 26–7.

2. Some of the history given in this section comes from a briefing paper prepared by R. J. de la Rosa for the Alternate Forum for Research in Mindanao Board (mimeo, undated).

3. At the local level, Task Force Sandawa(TFS), Kidapawan; at the regional level, Task Force Sandawa (TFS), Davao; and at the national level, Task Force

Sandawa (TFS), Manila. A list of the different task force members is attached as an appendix to the chapter.

4. A list of all task force members is included as an appendix to the chapter.

5. As then-counsel for the largest indigenous peoples network in the Philippines, LRC lawyers and campaigners immersed themselves in the culture (i.e., stayed with the Lumad in the area for long periods of time) and acquired a deeper insight into the problems from the local peoples' perspective. That experience greatly informed their handling of the case as alternative lawyers.

6. The initiative was taken especially by Chip Fay, the Manila-based representative of Friends of the Earth–U.S., and by Chad Bobson, the head of the Washington, D.C.–based Bank Information Center.

7. This team included LRC, SALAG, PANLIPI, and Tanggol Kalikasan.

8. See Russell Andaya et al., "Geothermal Power Generation and the Mt. Apo Environment," *Philippine Natural Resources Journal* 2 (May 1990), a journal published by Legal Rights and Natural Resources Center, LRC-Ksk, FOE-Philippines.

9. It was not difficult to refresh government officials' memories of the promises they had made to the people in 1986. A consistent few, especially in Congress, had come a long way in supporting the demand of the Philippine indigenous peoples for the recognition of their ancestral domains. The Congressional Committees on Indigenous Cultural Communities and on Natural Resources, both in the Senate and House of Representatives between 1989 and 1990, issued four separate resolutions supporting the campaign. In the executive branch, within the Department of Environment and Natural Resources (DENR), which was crucial to the EIA process, the campaign had an ally in the newly appointed head of the Environmental Management Bureau, who was an advisor to DENR secretary Fulgencio Factoran Jr.

10. Trying desperately to balance the interests of the government and the indigenous peoples as well as the important environmental issues raised by the opposition, the DENR imposed twenty-one conditions on PNOC in the ECC, conditions that reflected most of the concerns raised by the opposition. The violation of any of the conditions could result in its cancellation. The lament of the opposition, however, was that PNOC could then resume operations without serious regard to complying with conditions set. Sure enough, in early 1993, PNOC's blatant violation of the ECC (roadclearing without appropriate permit) did not result in its cancellation.

11. For a brief discussion of some of the conditions of the ECC, see Jose A. Magpantay, "The Mt. Apo Geothermal Project: An Evaluation," a study commissioned by the Alternative Forum for Research in Mindanao (mimeo, 1995).

12. President, Proclamation 583, Republic of the Philippines, Ermita, Manila, S. 1992.

13. For background on the Chico Dam case, see Anti-Slavery Society, *The Philippines: Authoritarian Government, Multinationals, and Ancestral Lands* (London:

Anti-Slavery Society, 1983), and the occasional coverage in the *Southeast Asian Chronicle* in the 1970s and early 1980s.

14. See Ponciano Bennagen and Shelton Davis, *Integrated Protected Areas System: GEF Appraisal: Mission Site Visit Report,* draft (Philippine Environment and Natural Resources SECAL, World Bank, Washington 1992).

15. These forestry policy lobbyists included the Anthropological Society of the Philippines, PANLIPI, and some supportive academics and individuals. Among them, Professor Owen Lynch Jr. played a significant role in developing the historical research for ancestral domain recognition and in the writing of the draft Ancestral Domain Bill, which in 1995 was still pending in the Philippine Congress.

16. Under the Revised Forestry Code (Presidential Decree 705), which is the primer for all forestry policies and activities of the DENR, any land that is above 18 percent slope is classified as *upland.*

17. This is the ruling of the United States Supreme Court, in *Carino v Insular Government,* 41 Phil 935 (1909), a landmark case in Philippine jurisprudence for the recognition of individual ownership of ancestral land.

18. See paragraph 4 of the *World Bank Operational Manual Statement 2.34* (Washington, D.C.: World Bank, 1982), which contains the earliest policy statement of the World Bank for indigenous cultures, entitled "Tribal Peoples in Bank-Financed Projects." See Shelton Davis, "The World Bank and Indigenous Peoples," (April 1993). See also, *World Bank Operational Directive 4.20* (Washington, D.C.: World Bank, 1989). See the chapter by Andrew Gray in this volume for further discussion of the Bank's policy regarding indigenous people.

19. As contained in the report of the Philippine Development Forum Coordinator, Jennifer Smith, to its members, (13 June 1992).

20. Several networks had proposed national conferences to assess strategies for Mt. Apo, which later came to be known as the series of four Solidarity Conferences, open to all interested campaigners from center- to left-associated groups in Philippine politics who had done work in Mt. Apo. First a preconference for the regional groups was held in Davao City on 14 April 1992; then a National Solidarity Conference was held in Kidapawan on 29–30 April 1992; third, a general meeting of all working committees set up in the previous conference was held in Davao, 4 July 1992; and fourth, another National Solidarity Conference was convened in Kidapawan, 2–3 July 1993 to assess the work of the previous year.

21. As of September 1992 the government deployed additional troops from the 603rd Brigade to join the soldiers of the 37th, 35th, and 27th Infantry Battalions of the Philippine Army in the areas of Sauta Cruz, Digos, Bansalan, Makilala, Kidapawan, and the nearby foothills of Mt. Apo. Information from the 602nd Infantry Battalion based in Davao province stated that the military learned of a New People's Army (NPA) plan to disrupt the project and that the military armed forces should thus secure the area. Armed confrontations and land mine explosions were reported in the Santa Cruz and Davao del Sur side of Mt. Apo. Since then, the military has increased its presence in the area.

In addition, tribal peoples were harassed by military forces. Two of those who complained of not getting payment for work done were killed ("Army Men Killed Tribal Opposing Apo Project," *Philippine Daily Globe,* 4 October 1992); surveillance was conducted of trips to the farms ("Geothermal Site Turning into a Hamlet," *Philippine Daily Express,* 15 September 1992); and families were displaced ("68 Families Displaced by GEO Project," *Philippine Daily Express,* 15 September 1992). Public hearings of the Philippine Senate on the issue of increased militarization reported that PNOC trained a special army as additional security for the plant, the Special Citizens Auxiliary Army. It was reported at those hearings that PNOC requested the 35th Infantry Battalion through Lt. Col. Antonio Romero to assist in the creation of an eighty-eight man Citizens Armed Forces Geographical Unit (CAFGU).

22. PNOC asked the Asian Development Bank to pick up the Mt. Apo project, but after intensive NGO lobbying ADB decided against financing, avoiding confrontation with the government by just saying it was more interested in other sites and had limited funding. It should also be noted that in one meeting with ADB staff, campaigners were told informally that after the ADB saw the New People's Army in the Mt. Apo area, which announced it would attack the project if implemented, one bank staff said, "If we were considering funding before, we surely won't step in now."

23. There are one hundred centavos in a Philippine peso. The peso's exchange rate is about twenty-five to the U.S. dollar.

24. For good discussions of the relationships between the Marcos dictatorship and the World Bank, see Walden Bello et al., *Development Debacle* (San Francisco: Food First Books and Philippine Solidarity Network, 1982), and Robin Broad, *Unequal Alliance: The World Bank, the IMF, and the Philippines* (Berkeley: University of California Press, 1988). See also Paul Mosley, Jane Harrigan, and John Toye, *Aid and Power: The World Bank and Policy-Based Lending,* vol. 2 (London and New York: Routledge, 1991.)

25. After the People Power uprising in 1986, President Aquino assembled a Constitutional Commission to write a new constitution. Professor Ponciano Bennagen, a recognized anthropologist-for-the-people, was appointed to promote and protect the interests of the indigenous peoples' network; he labored to insert pertinent provisions in the present constitution.

26. Mr. Chip Fay set up office in Manila as Asia representative of FOE–U.S./EPI upon the merger of both offices to assist the local environmental work, mainly in the area of forestry and multilateral development bank lobbying. His expertise was indigenous peoples and human rights advocacy, having worked with Survival International as Asia representative before that time.

27. The value of conducting Solidarity Conferences for the task force is discussed in the section on organizing strategies in this chapter.

28. Other PDF members active in lobbying the Bank included Richard Forrest (National Wildlife Federation), Dave Batker (Greenpeace), Chris Cobourn (Columban Fathers Justice and Peace Office), and Owen Lynch (World Resource Institute.)

29. The partners were the Freedom from Debt Coalition, Philippine Rural Reconstruction Movement, Philippine Environmental Action Network, Green Forum, Council for Peoples' Development, and Coalition of Development NGOs.

30. The local tribal groups that PNOC cites as having accepted the project are the Cotabato Tribal Consultative Council and Tribal Communities Association of the Philippines. These organizations are legally registered, and their members are composed of the educated and assimilated tribal peoples mostly living outside of the forests, many of whom proclaim themselves as chiefs, albeit unilaterally. Some of the elders found them dubious because at one point they opposed the PNOC project for having failed to sufficiently compensate local people, yet after some promises of job openings and livelihood, they retracted and in fact celebrated a Pamaas, a "counterritual" to the D'yandi—a ritual that was supposed to bless PNOC's operation in the area, but was rejected by most elders for having easily proclaimed some PNOC officials as "chiefs."

31. From the "Report of the Philippine Development Forum, U.S. Coordinator," unpublished, PDF Washington, D.C., 13 June 1992, addressed to its members. In the May 5 meeting, the attendance included: Jeffrey Balkind, country officer for the Phillipines; Darayes Mehta, staff, Industry and Energy Division, East Asia Country Department no. 1; and Thomas Weins, agricultural economist, Agricultural Division, East Asia Country Department no. 1. On May 19, Mr. Vinet Nayyar—the Bank chief for the Industry and Energy Division, East Asia Country Department—three staff from his division, and Jeffrey Balkind met with the PDF staff and some of its members—notably representatives from Greenpeace, National Wildlife Federation, and Natural Resource Defense Council.

32. This finding was contained in an office memorandum from Gloria Davis, regional environment chief, May 1991, Washington, D.C.

33. A subsequent ADB study on public participation in the EIA process included Mt. Apo as one of its four cases—the others being the Calico Coal-Thermal Power Project, the Benguet Corporation Open Pit Mining Project, and the Philippine Associated Smelting and Refining Corporation Expansion Project. The Mt. Apo campaign holds the record for the longest public opposition to a development project in the Phillipines.

34. Department of Environment and Natural Resources (DENR), *Administrative Order No. 02*, as amended, S. 1992.

6

Planafloro in Rondônia: The Limits of Leverage

Margaret E. Keck

The Rondônia Natural Resource Management Project, or Planafloro,[1] was designed to address social and ecological problems left in the wake of Polonoroeste, a development program partially financed by the World Bank in the early 1980s. Because the latter had already been a major focus of NGOs in the evolution of the worldwide multilateral development bank (MDB) campaign, nearly all of the parties concerned—officials of the World Bank, of the state of Rondônia, of the Brazilian federal government, and of NGOs—had experience with aspects of the case. Disbursements for the project began in June 1993, and its implementation is expected to take five years.

Although it involved only a small loan (the Bank's share was only $167 million), this project attracted a remarkable amount of attention both inside the Bank and out. The Polonoroeste project that preceded it generated sufficient controversy to be mentioned by name in World Bank president Barber Conable's 1987 speech to the World Resources Institute in which he announced reforms intended to improve environmental performance. It followed, therefore, that Bank officials and the nongovernmental organization (NGO) community were likely to pay an unusual amount of attention to its successor. And indeed, they did. In response to claims that the Bank wasn't paying enough attention, one Bank official would eventually state in frustration that the Planafloro was the most carefully scrutinized project in the Bank's history. NGOs, in turn, mounted around the Planafloro what appeared from the outside to be the strongest set of local-international linkages around any transnational environmental campaign. As a result of concerted NGO pressure, in 1991 the Bank convinced Rondônia state officials to permit an unusual degree of consultation and even participation in project implementation by local organizations in Rondônia. Yet, by the end of the project's first year, NGOs were already threatening to vacate the participatory roles allocated them; by the end of its second, the Rondônia NGO Forum—with support from Friends of the Earth—had brought the case before the independent review panel set up in the wake of the Morse Commission. What happened?

Understanding what happened requires that we move beyond a two-actor model in which an NGO network linking local and international organizations pressures the World Bank to change its policies—the vision of the Planafloro case that is most widely disseminated and that is fundamentally misleading. Simply put, Brazilian national- and state-level political variables are central to the outcomes of this story. The fact that key local (and foreign) NGOs initially did not fully understand the nature of the political game they were playing magnified their already notable political weakness locally; by casting these NGOs in an internationally visible but locally symbolic political role, state-level political elites rendered them largely irrelevant. In the resulting play of give-and-take between the Bank and the state government, the latter had a compelling weapon always in hand—the shadow of the worst-case scenarios that provided the counterfactual justification for the project.

In 1986, a group of technical personnel from the Rondônia state government, with consultants and with technical assistance from the UN Food and Agriculture Organization (FAO) and the World Bank, began to formulate an agro-ecological and economic zoning plan for the state that was to become a framework for future development activities and the basis for the development and conservation projects proposed in the Planafloro. Two factors motivated their initiative:

• After the 1984–85 midterm evaluation of the Polonoroeste, it was clear that what began as a Bank-supported project intended to rationalize settlement patterns in the state had instead increased both deforestation and immigration; and

• The state faced acute financial crisis as federal transfers and Polonoroeste financing (on which its government had depended since the early 1980s) ended at the same time.

Both Bank staff and proponents of the project in the state argued that situating development modules within an overall agro-ecological and economic zoning plan for the state would address these problems.

In many respects, the Planafloro directly addressed the problem of *sustainable development*. Using an ecologically sensitive planning tool, it aimed to intensify economic activity (especially agriculture) where the land could sustain it, discourage or ban settlement in fragile areas, regulate timber extraction, establish or strengthen extractive reserves,[2] protect indigenous areas and ecological reserves, and reinforce social infrastructure in key areas. This agenda was extremely ambitious (and probably unrealistic) for a project that involved only $167 million dollars from the Bank and counterpart funds amounting to approximately $30 million apiece from the state and federal governments.

The Planafloro thus differs from other Bank projects examined in this volume in two important respects:

• At least some of its proponents specifically intended it to address ecological issues.
• As the successor to a project frequently raised as an example of the World Bank's environmental irresponsibility, it was inevitably viewed through the lens of the earlier project. Because the Polonoroeste was the MDB campaign's first case, international NGOs were familiar with Rondônia's problems, and the Planafloro was under a microscope both in and outside the Bank from the beginning.

Whatever its formal aims, the Planafloro meant different things to different people. For some it was a pilot project in ecological planning; for some, a development project pure and simple; and for others, a way to get some money into the state. For NGO representatives and local movements, it offered a chance to monitor such a project from the beginning, with considerable knowledge of its antecedents; in addition, because of international networking, the World Bank might well treat their concerns more seriously than they were treated locally.

This chapter does not analyze the technical substance of the project; rather, it explores the remarkable disjunction between the diverse (and sometimes diametrically opposed) views of what the project was supposed to accomplish, a disjunction that persisted well into the life of the project. Exploring these divergent views, the political strategies their proponents developed, and the arenas in which they espoused and pursued them helps us to map the increasingly complex interpenetrations between domestic and international politics. They also challenge social scientists to develop new theoretical tools to describe and explain such interpenetrations.

Background

Rondônia is located in the Western Amazon, bordering Bolivia to the south and southwest, Mato Grosso to the east, Amazonas to the north, and just touching the state of Acre to the east. Formerly a federal territory, it became a state in 1981 in the midst of an extraordinarily rapid settlement process. Between 1970 and 1980 the population grew at a yearly rate of 16.02 percent, continuing at 12.99 percent annually between 1980 and 1985.[3] In just fifteen years, it went from 111,064 to 904,298. Stimulated by the completion in 1960 of Federal Highway BR-364 from Cuiabá to Porto Velho, colonization accelerated when that road was paved under the auspices of the Polonoroeste program, a federal initiative with substantial financing from the World Bank.[4]

For the military government that took power in 1964, occupation of the center-west of the national territory was both a security imperative and a means of diminishing the political effects of land pressures in the south and northeast. Laws stipulating federal control of 100 kilometers on either side of federal highways and a 250-kilometer strip along national boundaries brought much of Rondônia's land under direct federal control. Colonization and infrastructure programs were controlled and financed from the capital.[5]

The World Bank and Polonoroeste

Polonoroeste began early in the 1980s as an effort to rationalize the rapid and chaotic migration process to Rondônia. Besides paving the BR-364, it involved a variety of well-intentioned but often barely or badly implemented initiatives regarding rural infrastructure and agricultural extension. The project also stipulated protection of indigenous areas and ecological reserves; lack of attention to these elements of the plan led to an unprecedented temporary suspension of disbursements in 1985 until the Brazilian government came up with a revised policy acceptable to the World Bank.

Perhaps even more important in the long run was the effect of Polonoroeste on state building in Rondônia. Statehood and the Polonoroeste loan both arrived in 1981. All critiques of the program cite weakness of implementation and the institutional weakness of public organs in the state. Although true, this criticism is ironic; the Polonoroeste period was the high point in public institutional capacity in Rondônia. It attracted professionals to the state—agronomists, planners, economists, doctors, engineers. Some stayed beyond the life of their contracts, but most did not. More significantly, however, the project reinforced Rondônia's dependence on outside transfers (from the federal government, from the Bank) for its institutional development; when the transfers dried up, so did much of the base of qualified personnel. High turnover of federal government employees, World Bank consultants, and state functionaries involved in the project seemed to impede the development of an institutional memory. Inconsistent technical oversight made it easier for state politicians to channel resources to their bailiwicks for short-term political purposes. In the "get-rich-quick" atmosphere of Rondônia in the 1980s, development monies were as easy to harvest as the coffee and *cacao* they were intended to promote. For popular sectors, access to resources depended on connections to more powerful elite actors, and the democratic transition in the 1980s did not fundamentally alter this relationship.

By the end of the 1980s the state was feeling the consequences of more than a decade of rapacious mining of soil, timber, and mineral resources. Timber extraction, mining (gold and cassiterite), and agriculture were all hard hit by a combination of resource exhaustion (in the case of gold and easily extracted, high-priced timber) and market factors. Federal transfers of funding and personnel had fallen significantly.

Altogether, Rondônia was the environmentalists' object case of colonization run amok. Efforts to rationalize settlements could not keep up with the human tide that the paving of BR-364 seemed to encourage. Rapid deforestation, frequent abandonment of homesteads, widespread malaria, encroachments by colonists and timber companies on supposedly protected reserves, mercury pollution of rivers from gold mining—all became the stuff of congressional hearings and a growing campaign in the North charging the World Bank with environmental irresponsibility. In mid-1988, as satellite data documenting the extent of forest burning in the Amazon became available just when the United States was undergoing an unprecedented heat wave and drought, Brazil (overall) and Rondônia in particular became the focus of global concern about the role of tropical deforestation in climate change. A local development problem had become (for some) a global environmental problem.

State Politics in Rondônia
At first glance, Rondônia seems a prime candidate for capacity-building programs. Institutional weakness appears natural, given the state's newness and rapid colonization.[6] Excessive dependence on federal and international programs and transfers may have distorted the local state-building process. Both "explanations" of Rondônia's weakness, however, mask deeper questions about how the impact of financial transfers is mediated locally by the organization of political interests and allegiances. Institutions are weak in Rondônia also in part because power relations among the patronage groups that colonize their top layers were unstable—leading, among other things, to frequent personnel shifts as the group in power seeks to reward allies and punish opponents.

Political parties have little relevance in determining the political positions a politician in Rondônia is likely to take. More than in most states, many major players in Rondônian politics formally belong to small parties with little national standing. Even the Workers' Party (PT), normally the exception to the nonprogrammatic and nondisciplined norms of party behavior in Brazil, is more fluid in Rondônia.

Relevant political constellations in Rondônia are patronage-based "groups" or alliances of politicians that tend to include members of federal and state congressional delegations, state government, and local governments, with support from economic interests in the state—such as timber, cattle, and mining sectors, as well as the ubiquitous construction industry. The control exercised by any given group depends on the lock they have on federal and state patronage resources and major local governments. Patronage appointments go many levels down, and local politics in Rondônia is essentially pork barrel politics. Besides development monies, "pork" may include access to land and land titles, as well as regulation of natural resource use.

Pork barrel politics requires that there be something to distribute. Many statements by local politicians in Rondônia about the Planafloro make sense only as part of the jockeying for position among patronage groups. Without reference to pork barrel politics, it is difficult to understand why the state legislature voted down the zoning plan on which the Planafloro is grounded in early 1990, with assemblyman Osvaldo Piana leading the opposition, and then passed an almost identical version at the end of 1991 when Piana was governor—or why the leader of the opposition in the second vote, assemblyman William Curi, was subsequently appointed by Piana to be secretary of planning for the state and thus in charge of the overall coordination of the Planafloro.

NGO Politics and Rondônia

The MDB campaign was conceived in 1983, not in response to particular abuses, but rather because affecting Bank policy was seen as the most economical method of influencing the ecological dimension of development in the Third World. If, the reasoning went, MDBs significantly influence Third World development policy, and if the largest donors influence MDB policy (through the system of weighted voting and the process of International Development Association [IDA] replenishment), then trying to cause a shift in banks' policies could have an important global ripple effect.[7]

The campaign activists' first cases were chosen when some combination of hearsay and examination of World Bank reports suggested that a project might be problematic. Only later did the ability to reach local people who could provide information and testimony become a central element in the campaign. Most of the information they received about Polonoroeste did not come from people who lived in Rondônia. The first Brazilian witness at a U.S. congressional committee hearing on MDB lending was José Lutzenberger (later President Fernando Collor's environmental minister), a nationally known environmentalist with only sketchy knowledge of Rondônia. None of the Brazilian signers of the October 1984 NGO letter to World Bank president A. W. Clausen, protesting Polonoroeste's failure to protect ecological areas and Amerindian groups, were from Rondônia or even the Amazon region. Although the critique of Polonoroeste included a call for more participation by affected populations in decision making, it was participation in the abstract, and affected groups were identified more as victims than as potential actors.

Contact established in 1985 with Chico Mendes and the Acre rubber tappers changed that pattern. Foreign environmentalists increasingly sought contact with local movements and populations, and asked the banks to insist not only that the local peoples' needs be taken into account, but also that their voices be heard in the process.

Clearly, these environmentalist organizations were not the first to take on the MDBs. Since the 1960s, leftist critics have called them tools of imperialism, and more mainstream development economists have lamented that assistance programs were not getting the aid to those most in need of it. To these historical critics, including many in Third World NGOs with which the D.C. Bank campaigners wanted to build partnerships, the notion that the MDBs could be made to do constructive things (and empower grassroots organizations in the process) seemed naive.

Nor was Amazonian deforestation a new issue: it had claimed the attention of conservation groups since the early 1970s. Bruce Rich, one of the MDB campaign's main instigators, was then working at the Natural Resources Defense Council (NRDC), which had coordinated an international working group on the Amazon in the early 1980s. The advocacy wing of the U.S. anthropological community, particularly the Boston-based Cultural Survival and Anthropology Resource Center, were particularly strong on the Amazon; anthropologists provided key, early information about the Polonoroeste.

Finally, intensive human rights activity in Washington, D.C., in the 1970s concerning Latin America had produced more information on Latin America and skepticism in the U.S. policy community regarding the intentions of governments there than existed regarding other parts of the world.

Planafloro Chronology

To be based on an agro-ecological and economic zoning plan of Rondônia, the Planafloro aimed to improve the state's management of resources, conservation, and development. Its main beneficiaries were expected to be Amerindians, rubber tappers and other forest dwellers, fishermen and river dwellers, and smallholders (farmers) affected by the agroforestry component of the plan. The essence of the plan was to limit frontier expansion by intensifying agriculture and agroforestry in propitious already settled areas, by improving infrastructure, and by protecting ecologically sensitive and indigenous areas. The project also called for the establishment of extractive reserves.[8]

It was an ambitious plan, requiring both institutional and policy changes at state and federal levels. These changes included better coordination of state and federal land institutes, cessation of land titling based on clearing trees, and designation of new protected areas.

Important coordination problems arose among the levels of government involved (including municipal as well as state and federal governments, as they legislate on municipal land use). There were also very serious contracting problems: the plan's designers lacked authority to make decisions binding other instances of government. In fact, the planners avoided negotiating with municipal and other local leaders, fearing their opposition. Groups that might have supported the plan's goals (rubber tappers and Amerindians, for example) were too weakly organized to be viewed as necessary interlocutors. Although a few individuals associated with NGOs or with Rondônia's only university were involved in some discussions of the zoning plan, overall, little effort was made to rally political support in the state for its *substance*.[9] Nor was there strong support nationally; the financial team installed by Fernando Collor in 1989 had deep reservations about contracting World Bank loans for impecunious states.

In 1989, when the project was about to come before the World Bank's executive directors, a consortium of environmental NGOs in the United States and Europe charged that local populations affected by the plan had not been adequately consulted. Combined with doubts in Brasília about the project's financial soundness, these charges helped to remove it temporarily from the Bank's agenda. During 1990, several Brazilian and foreign NGOs tried to organize indigenous communities, rubber tappers, and rural unions to discuss the project.

Leveraging Development Policy

The strategy developed by the multilateral Bank campaigners involved gaining influence over the policies of environmentally recalcitrant states by leveraging more powerful institutions—that is, the banks. By trying to hold the banks accountable to their stated commitments (and by pushing them to make greater commitments), environmental activists could legitimately ask them to monitor or exercise greater influence over development policies affected by their loans. This strategy is characteristic of transnational advocacy networks and frequently helps to give groups that are politically weak locally access to a more receptive constituency elsewhere.[10] (The limits to this strategy are discussed in detail below.) It is worth noting, however, that *both* the technical proponents of the Planafloro *and* the NGO actors who contested some aspects of it believed for a long time that their efforts could remain insulated from the rough and tumble of Rondônia's politics.

Two parallel stories can thus be told about Planafloro. In the insulated view of "the project," the main actors are state and federal *técnicos,* World Bank officials, professionalized NGOs (local, national, and foreign), and local social movements. This story takes place at local, national, and international levels, and the actors hope that by influencing the behavior of the World Bank they might influence the project's local implementation.

Questions about political relations elicit a quite different story, however. Here the main agglomerations are patronage *grupos,*[11] and the story is about pork barrel politics. Relevant actors are more likely to include the state's business associations than indigenous people or rubber tappers, and the agendas of *all* of Rondônia's grupos are thoroughly developmentalist. Perhaps most relevant, the state-level técnicos who are key to the first story are entirely dependent upon the grupos in the second for their jobs.

The organization of activists involved in the Planafloro issue in Rondônia was largely stimulated from abroad by members of the environmental groups that were part of the MDB campaign. Environmental Defense Fund anthropologist Steve Schwartzman, attending a February 1990 rubber tappers' meeting in Acre, discovered that organizers from

Rondônia knew very little about the proposed Planafloro. Schwartzman and the EDF then protested to the World Bank that contrary to the Bank's stated policy, local people had not participated in the project's design, a charge Bank officials denied. The environmentalists requested more information from contacts in Rondônia. The Instituto de Pré-Historia, Antropologia e Ecologia (IPHAE) surveyed eighteen rubber-tapper, rural workers', and indigenous peoples' organizations, whose representatives claimed little knowledge of the project and asked to participate in discussions about it. NGO representatives also spoke to newly appointed environmental secretary José Lutzenberger, who in turn asked the Bank to suspend consideration of the project until he could convoke consultations with these groups. The Lutzenberger letter forced the Bank's hand, and the project was taken off the executive directors' agenda.[12]

The groups in question included environmentalists, rubber tappers, rural unions, and indigenous peoples. In the early stages of NGO activities, rubber tappers played an important symbolic role because, in the wake of the 1988 assassination of Chico Mendes, they occupied a privileged position in the symbolic discourse about sustainable use of the rainforest. They became a primary focus of international groups. Rural unions were by far the most institutionalized of these groups, consisting primarily of small farmers in colonization projects. Until the mid-1990s, the most important unions were affiliated with the rural wing of the CUT,[13] Brazil's most combative peak labor organization. Although small farmers continued to push the agricultural frontier into protected areas (either on their own or as stalking horses for timber or ranching interests), some rural unions were trying to promote a long-term vision of sustainable agriculture.

In general, most of the explicitly environmental groups in Rondônia have been very small and have fought to defend specific natural areas or promote environmental education. Most have not been very actively involved in Planafloro activities, although some nominally belong to the Rondônia NGO Forum (a coordinating body established in 1991, discussed below). A significant and quite visible exception has been Ecoporé (Ação Ecológica Vale de Guaporé) in Rolim de Moura, an organization that has long monitored the frequent invasions of the Guaporé Reserve and indigenous peoples' area.

Although the state's university has supplied some cadre for environmental and other organizations, it lacks scientific facilities, library materials, and even a well-trained faculty. There is no good public archival collection on the state's history. Few of the countless reports consultants have written about the state's development are available in Rondônia itself, so no institution within Rondônian civil society has access to (or is attempting to build) a collective memory that would contribute to the development of an autonomous critical capacity.

A weakly organized civil society and fragile social networks would have made mobilization an unlikely strategy for activists trying to affect development plans. Except for rural unions, none of the organizations monitoring the Planafloro had much of a mass base. Moreover, given a history of authoritarian political relations, none of them had much experience with democratic participation. The Planafloro activists can most aptly be described as an advocacy network within a broader transnational issue network, which "includes the set of relevant actors working internationally on an issue who are bound together by shared values, a common discourse, and dense exchanges of information and services."[14] Although the Rondônian activists frequently had organizational affiliations, their actions were more characteristic of political entrepreneurs than of organizational leaders (although clearly these are not mutually exclusive categories). They were linked to each other and to external actors and organizations (domestic and foreign) either through the historical experience of Polonoroeste or through rubber-tapper support networks. The individuals central to this effort were well placed to gather, evaluate, and transmit information and to recognize what kinds of symbols would resonate abroad. In part because they did not belong to mass organizations, and in part because of their own proclivities, the Planafloro activists focused a lot of energy on perfecting a *technical* critique of the Planafloro. First, they tried to influence project design, asking the Bank to make the project respond more concretely to ecological and human problems in the state. After that, they pressured the Bank to hold the state and federal governments to the stipulations of the agreements they had signed.

The first phase of the campaign, from 1990 to 1992, was highly internationalized, relying almost exclusively on contacts in Washington (especially the EDF) to help obtain leverage. The second phase, from 1992 to 1994, brought an effort to nationalize the campaign and a greater focus on holding government institutions (in particular the land institute) accountable for their acts. A third phase, beginning in 1994, reverted to an international focus through recourse to the Bank's Inspection Panel process.

Phase One: Rubber Tappers and the Acre Paradigm

Before the assassination of Chico Mendes on December 23, 1988, most of the world had never heard of rubber tappers. His murder fostered discussion of the role of extractivist populations in sustainable use of tropical forests, and his movement seemed to posit an identity between a poor peoples' movement for survival in a particular setting and a global interest in preserving forests. Concretely, the Acre rubber tappers' proposal to establish extractive reserves seemed to open new possibilities for thinking about the peaceful coexistence of human and natural worlds.

But several factors make the Acre situation very different from Rondônia. The Acre rubber tappers began to organize in the mid-1970s in response to encroachment by cattle ranchers on the lands they had traditionally used. Soon after, an organizer from the National Agricultural Workers' Confederation (CONTAG) arrived in Acre and quickly decided to build rural unionism in Acre on the basis of the rubber tappers' organizing. The same year, the national Catholic Bishops' Conference formed the Pastoral Land Commission to respond to the upsurge in land conflicts in Brazil, and its first head was Dom Moacyr Grechi of Acre. This was a period of growth for a whole range of social movements, the labor movement, and the electoral opposition to the military regime. Also, in 1979–80 new political parties were formed. The rubber tappers' struggle in Acre had links to all of these initiatives. They worked closely with the Catholic Church, formed rural unions and made contact with new labor leaders nationally, and participated in MDB growth in the state. In 1980–81, many rubber-tapper leaders became active in forming the Workers' Party. By the time the Acre rubber tappers met foreign environmentalists (around the time they were forming a National Council of

Rubber Tappers in 1985), they had developed a broad network of social and institutional linkages both in Acre and around the nation. Their extractive reserves proposal—probably the most cited initiative regarding sustainable development in the Amazon—was the outcome of more than a decade of experience.[15]

The experience of Rondônia's rubber tappers has been very different from that of their Acre counterparts. Many have already lost their land or have migrated to urban areas due to depressed rubber prices. They are also more dispersed. Unlike in Acre, where a progressive Catholic Church hierarchy promoted grassroots organization and where rural union organization was built on top of already existing rubber tappers' struggles, in Rondônia the Catholic hierarchy promoted a paternalist view of social relations, and rural unionism was built around new agricultural settlements.

Rubber-tapper organizing in the state also reflected two sets of struggles. One, among the World War II *soldados de borracha*,[16] involved a demand for promised pensions; most of those who participated in this struggle are no longer tapping rubber, either because the land on which they had done so was lost to development or because of age. The other struggle was carried out by those still engaged in rubber tapping and focused on land rights, cost and regularity of supplies, and rubber prices. Thus, the sociopolitical context of the rubber-tapper experience in Rondônia diverges substantially from the Acre case.

Although a large group of Rondônia rubber tappers attended the National Council of Rubber Tappers' founding meeting in Brasília in 1985, they participated only sporadically in the council until 1990. Few members of the leadership of the National Council had come to Rondônia, and tensions between the two groups were obvious—which some explain as a historical rivalry between rubber tappers of the two states, others as political differences grounded in the National Council leadership's links with the Workers' Party and Rondônia rubber-tapper links to traditional clientelistic politicians.[17]

When the 1990 meetings to discuss Planafloro were held, Rondônia rubber tappers had not formed a statewide organization. In addition, there was considerable conflict among NGOs (also called *assessorias*)[18] operating in the area over who could legitimately call together such an organization.

Organizational rivalries pitted the Instituto de Antropologia e Meio Ambiente (IAMA) against the Instituto de Estudos Amazonicos (IEA) and the Conselho Nacional dos Seringueiros (CNS, National Council of Rubber Tappers); all were based outside Rondônia. IAMA, based in São Paulo, was formed in the late 1980s by anthropologists who had worked in Rondônia both on academic research and as consultants for the World Bank; by 1990 it counted a small (six people) but cohesive group in Rondônia. In 1991, partly in response to pressure from international funders to affiliate with a Rondônia-based organization, this network hooked up with a fledgling Porto Velho organization called the Instituto de Pesquisa em Defesa da Identidade Amazonia (INDIA, Research Institute in Defense of the Amazonian Identity).[19]

The IEA, based in Paraná, was headed by Mary Allegretti, an anthropologist who had worked closely with Chico Mendes on the proposal for extractive reserves. The institute had a formal agreement with the National Institute for the Environment and Renewal Resources (IBAMA), the federal environmental organ, to carry out viability studies and otherwise facilitate the establishment of such reserves. In late 1990, IEA named Brent Millikan, an American geographer doing Ph.D. research there as its representative in Rondônia. Until 1992, IEA worked closely with the National Council of Rubber Tappers (CNS).

The CNS was not an *assessoria* in the way the other NGOs mentioned here are and has a kind of essentialist legitimacy that the others do not. Nonetheless, it had no organization in Rondônia and was not automatically accepted as the representative of rubber tappers there. Increasingly professionalized, the organization owed its success by 1990 to the development of international funding sources and collaboration with IBAMA in Brasília, as much as to movement organizing.

The final group worth mentioning is the Instituto de Pré-Historia, Antropologia, e Ecologia (IPHAE), an NGO based in Porto Velho and headed by an expatriate Dutch agronomist; its main role in relation to the Planafloro was assistance in coordination of meetings and distribution of information, especially via the Internet. Its focus was agroforestry and marketing of nontimber forest products, especially the cupuaçu fruit.

Meetings around the Planafloro

The 1990 meetings held to discuss the Planafloro were formally called by the National Council of Rubber Tappers, the National Indigenous Union, and the Rondônia section of CUT-Rural;[20] only the last, however, was headquartered in Rondônia. The CNS named a local representative at the November 1990 joint meeting.

To promote trust among groups that rarely worked together, the discussions took place in stages. In June and July, rubber tappers, indigenous peoples, and rural unions held separate meetings to discuss the Planafloro, and in November the three groups held a joint meeting in Porto Velho. Although it was significant that these groups had a common cause, the meeting also highlighted their uneven organization in the state. The rotating chair among the three organizations at the meeting included only one person from Rondônia; speakers on issues of concern to rubber tappers and Amerindians were as likely to be *assessores* as members of the affected groups; and conflicts among the advisory groups were evident. The final document was written by the three cochairs and *assessores*.

The day before the meeting, rubber tappers, Raimundo de Barros of CNS, and some advisors met for the first time with representatives from the Planafloro team (including the state forest institute [IEF], the federal land institute [INCRA], IBAMA, and the state land institute [ITERON]) about land conflicts and establishing extractive reserves. The state employees said that the suspension of Planafloro meant further delay in demarcation of extractive areas. NGOs claimed to be able to validate—and get federal resources for—the planners in this area. These discussions highlighted the complexity of roles that NGOs had taken on in the Rondônia case, especially the interplay among claims to representation, legitimacy, and authority by formally constituted NGOs, local movements, and governmental institutions.

After the Porto Velho meeting, two parallel processes accelerated. The CNS and IEA immediately began working more closely with rubber tappers around the Rio Ouro Preto to organize a cooperative with links to the CNS; and at a meeting organized primarily by IAMA in Guajará Mirim in December 1990, rubber tappers voted to establish an independent statewide organization called the Organização dos Seringueiros de Rondônia (OSR, Organization of Rondônia Rubber Tappers).

Conflicts among the *assessorias* continued to complicate grassroots organizing around the Planafloro and related issues until foreign funders (especially the World Wildlife Fund [WWF]) intervened in 1991 to reduce the tensions, resulting in the departure of IAMA from most of the fray. In fact, as rubber-tapper organizing continued, it became clear that conflicts in the state had very little to do with the rubber tappers themselves and much more with their advisors. The creation of INDIA in March 1991 helped provide a local affiliation for some of the individuals previously associated with IAMA. Furthermore, conflicts among local people were rarely as heated in day to day practice as they were in the presence of formal representatives from the outside.

The Rondônia NGO Forum was set up in November 1991 in part because local groups wanted more control over who spoke for Rondônia in national and international environment/development fora. Coordination problems raised by the Planafloro became even more evident with (*a*) national preparations for the 1992 United Nations Conference on Environment and Development (UNCED); (*b*) the formation in mid-1991 of the Amazon Working Group; and (*c*) responses to World Bank initiatives.

A month after the forum was established, a World Bank mission came to Rondônia to evaluate the state of the Planafloro. Part of its program was a meeting with nongovernmental organizations. To that end, IEA, The Union of Indigenous Nations (UNI), and CNS were invited; the state government paid plane fares to bring in CNS representatives from Acre, UNI's Ailton Krenak from São Paulo, and others. The Rondônia groups met and issued a document assessing the Planafloro situation and asking for a meeting. At the meeting they accused the state government of trying to make it look as though consultation was taking place when it was not. From that meeting came a written understanding (the *Protocolo de Entendimento*) between the state government and the nongovernmental organizations. Thus, formalization of the forum was specifically geared to respond to the Planafloro.[21]

By the end of 1991, local groups concluded that their ability to influence policy in the state depended on their ability to influence the World

Bank, for which there had to be a project. They therefore withdrew their opposition to the Planafloro, which put it back on the Bank's agenda. Pressured by the Bank to involve local organizations in Planafloro project planning, the state government agreed to set up a deliberative council on which NGOs and movements would have parity of seats with representatives from the state government organs charged with carrying out the project.

Professionalized NGOs play a crucial mediating role between local movements and the state government, and between local movements and national and international NGOs. They channel information, make contacts, and write reports and briefing papers. When visitors come to Rondônia, normally with only a short time to spend, these professionalized NGOs set up meetings and appointments for them. They therefore re-present, to a wider public, the local situation as they see it. The Amazon's place on the international environmental agenda heightens their importance, especially given the weakness both of state institutions, societal organization, and the social movement sector. The mediating role NGOs play involves brokerage rather than representation; in it, the necessary political resources are not those we associate with traditional forms of political mediation—ability to mobilize, attract votes, distribute patronage—but rather, access to information, the ability to *produce* information, and flexible access to a *variety* of resources, nationally and internationally. Particularly in settings with a low level of institutional development at both state and societal levels, NGOs can render local populations and/or movements visible or invisible to broader publics.

Phase Two: Building a National Campaign

In 1992, the Planafloro was back on the World Bank's agenda. The demand that local people be included was met with the establishment of a deliberative council. Project-planning councils (*conselhos normativos*) that were to discuss concrete proposals for meeting the (generally stated) strategic goals of the project also included representatives of local movements. Local activists dedicated a considerable amount of energy to meticulous analysis of Annual Action Plans (short-term project components that were, in fact, all the normative councils discussed) and to proposals for their improvement. Some of the local movements hoped initially that these new organs would give them real input into the process. In part because of their political inexperience, they believed that superior information and knowledge of laws and policies that were on the books would sway more powerful actors. With experience, however, this view changed.[22]

The INCRA Strategy

Beginning in early 1992, Millikan of IEA and increasingly others among the Rondônia activists spent a great deal of time tracking the National Colonization and Land Reform Institute (INCRA) plans for new settlement projects. INCRA was responsible for land allocation and land titling in the state. The World Bank's own evaluation of what was needed to make the project work gave top priority to changes in INCRA activities—changes, for example, such as the elimination of deforestation as a criterion for land titling, suitable regularization policies and practices, land policy following the zoning, and agreements between the state government and INCRA about these policies.[23] Yet a considerable gap existed between INCRA's commitments and its practices in Rondônia. Aiding in the ability to track INCRA's plans was the fact that local activists in Porto Velho had access to information from the INCRA offices there, especially because some INCRA employees were committed to the zoning plan.

The opportunity to confront INCRA came when Renato Simplício, its president, signed resolutions in March 1992 creating three new settlement projects (PA Agua Azul, PA Curupira, and PA Rio do Conto) in areas designated for extractivism in the zoning plan. After the NGOs denounced the plan, Choski of the Brazil division of the Bank wrote to Marcilio Marques Moreira, then economy minister, pointing out the incompatibility. INCRA president Simplício responded publicly, telling a *Jornal do Brasil* reporter that decisions on these areas predated the zoning plan and thus were not bound to respect it. The NGO Forum wrote a series of letters to President Fernando Collor de Mello and to INCRA authorities, documenting INCRA's violation of multiple statutes and agreements—essentially without success, however. In April 1993, the forum wrote a long exposé of INCRA in a letter to its new president, Osvaldo Russo de Azevedo, posting a copy on computer newsgroups and sending a copy to the Bank.[24] According to the forum's assessment, Russo was likely to be sympathetic personally, but his hands were tied because INCRA superintendencies in the north had been allocated as political rewards to Brazilian Democratic Movement Party (PMDB) politicians. At the time, the superintendent in Rondônia was a nomination of then-senator Amir Lando and had close relationships with construction companies.[25] The forum was soon able, however, to go beyond letter writing: with the help of the IEA's new Brasília office, they scheduled a hearing for the beginning of June on the Planafloro, INCRA, and zoning in the congressional commission on consumer, environmental, and minority issues.

The NGOs' INCRA strategy meant working more closely with rural unions than before. Although part of the NGO Forum and the Planafloro deliberative council, these unions had not played a central role in any of the activities described above. By mid-1992, the forum's initial focus on rubber tappers expanded to include more attention to land issues that affected small farmers; in addition, direct cooperation between rural unions and rubber tappers had increased. For smallholder organizations to oppose new settlement projects or land titling in areas zoned for other purposes required substantial prior educational efforts with their memberships, and even then it was politically risky. For rural labor leaders to go with environmentalists to meet with the president of INCRA in Brasília, after cosigning a letter that condemned settlement projects among other things, was highly unusual.

Besides the congressional hearing and overtures to the new president of INCRA, in early 1993 (before the Planafloro disbursements began) the forum also began to use the courts. They worked with the public prosecutor's office *(Ministerio Público Federal)* in Rondônia to sue INCRA, resulting in a restraining order (issued July 18, 1994) on settlement projects on land in four areas.[26] They also worked with the public prosecutor to sue several timber entrepreneurs for invasions of the Guaporé Biological Reserve[27] and eventually uncovered a major corruption scandal involving overvalued land appropriations.

The decision in the INCRA lawsuit came at a strategic moment. Forum members were frustrated by their failure to have much impact through the Planafloro deliberative and planning councils, and appealing to federal officials seemed not to affect INCRA in Rondônia. Seemingly without success, they also participated in an independent evaluation of the project (mandated by the Bank), which was drafted in February 1994 and which noted serious technical and institutional shortcomings. Thus, on June 15, the forum sent a lengthy letter to the president and executive directors of the World Bank asking for the suspension of loan disbursements unless and/or until measures were taken to address major problems.[28]

The forum's letter—followed three days later by a federal court ruling that validated many of its claims about INCRA's behavior in the state— apparently had a significant impact. In August 1994, a Bank mission went to Rondônia to negotiate a new agreement between the nongovernmental groups and the state government. In exchange for a promise of more serious monitoring and greater access to information, the organizations withdrew their demand for suspension of the project.

The August modus vivendi was to be short lived, and by the end of the year the NGO Forum was receptive to a suggestion by Roberto Smeraldi, coordinator of Friends of the Earth's Amazonia Program, that they take the Planafloro case to the Bank's newly constituted Inspection Panel. The claim was presented in mid-June 1995. In January 1996, the Bank's executive directors turned down the panel's recommendation that a full investigation be instigated. However, they stipulated that after six months there should be a major review of the project, in which the Inspection Panel was to be involved. Meanwhile, after the claim was filed, several key problems moved toward resolution: INCRA finally signed an agreement to respect the zoning plan, and the state government began demarcating new extractive reserves.

Domestic and International Leverage

None of the strategies described above involved direct engagement between Rondônia's NGOs and local group politics. The arenas in which they operated overlapped but did not coincide. Whereas NGOs and movements tried to influence the wording of the project's operational plans in the Planafloro Deliberative Council, local politicians sought to place their cronies in directorships of the Rural Credit Bank, which would administer large quantities of Planafloro resources. In August 1992, most local activists were entirely unaware of William Curi's imminent rise, although among local politicians, businesspeople, and even university faculty it was well known.[29]

Between 1990 and the present, local NGOs in Rondônia have become better organized and equipped. The NGO Forum provides for more regular communication among groups. Successive small grants from U.S. foundations provided some infrastructure (computer, fax) and operating expenses (phone bills, bus fares). Although the number of activists involved remains small, they have acquired a wealth of technical information regarding social and ecological conditions in the state. They initiated and provided supporting materials for the INCRA hearing and lawsuit, as well as a number of subsequent requests for restraining orders; they are quite effective at getting information relevant to the MDB campaign out internationally and, in the eyes of some Brasília-based organizations, are clearly the most effective set of local organizations involved in the MDB campaign.[30]

Between 1990 and 1992, Planafloro-related activities had to pass through international channels to acquire enough clout to have local repercussions. International networking amplified the voices of local groups who would otherwise not have been heard. Such networking was both helped and complicated by the number of expatriates present in the Rondônia NGO community: helped in that these expatriates were more easily able to capture the attention of foreign NGO representatives, but complicated in that they internationalized the campaign perhaps before sufficient attention was given to localizing it. After 1992, international networking was supplemented and occasionally supplanted by national networking (encouraged by a national seminar on the MDB campaign sponsored by EDF and others in Brasília on March 10–11, 1993). Moving the strategic focus from Washington to Brasília was important if the Bank campaigners were to address institutional issues in political fora as well as technical issues in supposedly "nonpolitical" ones. Nonetheless, the distance between Brasília and Porto Velho (or, even more, the interior of the state) is not appreciably smaller in symbolic terms than the distance between Porto Velho and Washington. Decisions and policies made outside the state continue not to affect political dynamics fundamentally there. Indeed, federal appointments at the state level are normally patronage appointments designed to win support from the state's political leaders—in Rondônia, as in much of the rest of Brazil. Although the MDB campaign will not change this process, it may add one more small voice to the growing demand for administrative responsibility and transparency, something of value in itself.

Yet, however impressive the NGO Forum's achievements, it seems unlikely that this configuration is politically sustainable. First, too few individuals provide the lion's share of technical and institutional research, and there is heavy reliance on expatriates. Although large representative organizations (specifically the rural union movement) become involved at some moments, they are unlikely to take on primary responsibility for monitoring Bank projects and government activities on development policy; their members, to whom they are held accountable in elections, have a much more diverse set of interests regarding the Planafloro than do the other constituencies involved in the forum. Smaller representative organizations, such as the rubber-tappers cooperative established on the Rio Ouro Preto Reserve and others in formation, are struggling for survival.

Although Planafloro activists have developed good relations with some técnicos in state agencies, they are not well connected politically. Several are active in the Workers' Party, but the party itself has not played a role in the Planafloro discussions. In spite of having won control of several municipalities in recent elections, the Workers' Party remains quite weak in the state overall. Thus, the political support for the Planafloro activists comes from outside and is based on their ability to generate accurate and useful information for the MDB campaign or for policy activists in Brasília. The activists recognize this, at least insofar as they resent the foreign environmental groups' limited attention span and the tendency for information flows to be unidirectional. They still see this foreign support as their best shot, however, and do not believe it will take a long time to build an effective local base of support.

The difficulty in building local support, coupled with intense concentration on negotiations around technical aspects of the project, helps to explain the activists' initial lack of attention to the highly relevant backroom politics taking place in the state around the Planafloro. However skeptical they may have been about their ability to influence the project through official channels, forum leaders with whom I spoke in 1992 still believed that the Deliberative Council would be the site of important decisions. Given the council's lack of representation of the most powerful interests in local society, this outcome was always unlikely. A council that allocates an equal number of seats to representatives of state and federal government agencies and to a group of twelve organizations whose active memberships, excepting rubber tappers and rural union organizations, probably add up to fewer than fifty people was more likely to serve as window dressing than decision maker. Moreover, the council initially did not include representatives from the industrial federation, agribusiness, mining, timber, or other major business groups. On what major developmental questions could it deliberate? And even after it opened up to include more stakeholders after 1994, the persistence of an authoritarian decision-making style relegated it to playing a rubber stamp role.

The MDB campaign has had an important impact on First World environmentalists and has in some cases helped local organizations to play a role that they might otherwise have been prevented from playing due to political weakness. The Planafloro case has been a fairly extreme example of a pattern of local-international linkage. Within Brazil, some of the organizations active at the national level—the Ecumenical Center for Documentation and Information (CEDI), for example—stressed the importance of bringing a national focus into the campaign and, of accumulating resources in order to have a more lasting impact on public policy at the national level. The Institute for Socioeconomic Studies (INESC) pioneered a strategy focusing on Congress, which has to approve all foreign loans. Although the Rondônia NGOs did move in this direction by focusing on INCRA, direct pressure on the World Bank continued to seem a more rewarding strategy. The Rondônia case also makes clear both the difficulty in drawing clear boundaries between what groups are "local" and what are "international," and the essentially political nature of any decision over who should be "represented" in discussions of a project and by whom. Recently, many development institutions have adopted the notion of "stakeholder" consultations— recognizing that a variety of contending interests can facilitate or impede the implementation of projects like the Planafloro. Even here, however, a tendency to generalize the discussion of "stakeholders" obscures a recognition of their very real differences in political and social power.

It is an ironic and thorny paradox that many of the interactions among international NGOs and local movements—interactions intended to empower these movements and increase their ability to participate in decisions affecting them—involve removing the site of those decisions from the ambit of local politics and potential democratic accountability. The normative issues raised by this kind of process are extremely difficult to resolve. On the one hand, shifting the decisional arena often appears to be the only way to ensure that the needs of powerless groups are addressed; on the other, it does not necessarily (although it may) build the network of local alliances that would be necessary to alter the configuration of power relations locally.

The Limits of Leverage

The Planafloro is a compelling example of the *limits* of outside leverage: the limits of the World Bank's ability to get governments to do things they were not prepared to do anyway and the limits of the NGO community's ability to support local partners who have not established a strong local base of sustenance. Precisely because it has been under a microscope from the beginning, the Planafloro is also a striking example of project inertia—a tendency, once begun, for loan disbursements to continue however evidently problematic they become.[31]

When a project is this difficult to implement politically, deciding how to measure its effectiveness is a genuine problem. However, relying on counterfactual justification—that is, the idea that without the project, things would have been so much worse—renders any evaluation moot and undermines political learning. The Bank used this kind of argument to justify the Polonoroeste before, during, and after the project in spite of its many "errors"; ironically, even before the Planafloro was approved, the task manager at the World Bank responsible for it asked me to think about the costs of its not being done.[32]

The Planafloro story illustrates the problems of trying to pursue sustainable development strategies based on counterfactual arguments. All of the major actors in the unfolding drama conceived their support for this project as a defense against something much worse, but they defined the "something" differently. For project proponents in the Bank, it represented a last-ditch chance to steer Rondônia's development policies into a more constructive direction, prevent further wastage of forested areas, and reinforce the position of state officials who genuinely sought a more rational management of the state's natural resources. Not surprisingly, that handful of state officials shared this view. Local social movements and nongovernmental organizations justified their participation in the project, insofar as they were allowed to and did participate, on the grounds that without their input the environmental and social outcomes would be much worse. By participating, they could at least monitor what was going on and at best contribute to positive outcomes. Local developmentalist and rent-seeking interests alike rationalized their support for the project by a similar logic: without green provisions, they reasoned, no money was going to come into the state at all. Once the money was coming in, they believed that controlling where it went would be difficult for anyone. Thus, few (if any) of the key actors undertook the project on the basis of a strongly *positive* commitment. However much optimists might claim that a process of "sustainable development" could speak simultaneously to those concerned with local empowerment, environmental protection, and economic development, bringing these groups together in any real sense would have required recognizing their very real distinctiveness.

According to foreign analysts, the Planafloro seemed a likely success story. Though small, the loan was politically salient for the Bank. The Bank *did* respond—and forced the Rondônia state government to respond—to pressure for greater grassroots participation: NGOs in Rondônia won an unprecedented voice (though not a decisive one) in project implementation. We should note, however, that the government's responsiveness was particularly notable after the NGO Forum complaint was filed with the Bank's Inspection Panel; the pressure of this new development stimulated the demarcation of several reserves and (finally) a formal agreement by INCRA to respect the zoning plan.

Linkages between local organizations and foreign environmental NGOs kept information flowing between Porto Velho and Washington, D.C., allowing close scrutiny of the project itself and of the relationship between the Bank's claims and the process on the ground. Indeed, the local-international linkages around this project appeared to be among the most effective—if not *the* most effective—of any in the transnational NGO networks monitoring Bank activities. Why, then, has the project accomplished so little and disillusioned so many?

How we answer that question depends on how we view the constellation of actors involved in the story. If our focus is on institutional change in the World Bank itself, as it is in this volume, then this case raises serious questions about institutional learning. The project proceeded in the first place despite considerable skepticism by many Bank personnel knowledgeable about the region; none of the personnel seemed to think they could (or should) do much to prevent its going forward. The 1992 assessment of Polonoroeste by the Bank's Operations Evaluation Department (OED) had placed among the most central problems in that loan's implementation the disbursement of highway-paving funds prior to funds for mitigating institutions and infrastructure.[33] In the overall Planafloro project, where highway paving plays a much smaller role, once again the first component for which funds were disbursed was highway paving—not surprisingly, also the component for which operational plans were most quickly drawn up. We could undoubtedly find other instances of such problems as well. Nonetheless, in the Planafloro, as in the Polonoroeste case, the counterfactual remains powerful. The same OED report notes that however ill-advised the rapid paving of BR-364 had been,

Given the fact that [it] did go ahead, there can be no doubt that most, if not all, of the environmental and Amerindian protection measures associated with the program, as well as many of its activities to support small-farm development, occurred as a result of the Bank's participation. Had the Bank not become involved and on the assumption that the Government would have proceeded sooner or later to *improve the trunk road anyway, the situation in terms of ecological damange and settler conflicts with tribal populations would undoubtedly have been even worse.* In this connection, the strong concern of Bank staff and consultants with the program's evolving human and physical environmental impacts and their efforts to prevent deforestation, encourage the use of perennial crops and protect indigenous areas should be clearly recognized even though these efforts were frequently frustrated because the Bank did not possess sufficient means to coerce POLONOROESTE's executors to take more vigorous and timely corrective actions.[34]

Statements made to me over the course of my research—by former federal officials involved with the Polonoroeste program, as well as by state government personnel and NGO activists—support the idea that what protection did occur would have been unlikely without Bank pressure. It also seems more than probable that Bank pressure was stronger because of the pressure brought to bear by NGOs and their allies in the U.S. Congress and other governmental institutions. In the Polonoroeste phase, effective protection of the environment and indigenous rights in Rondônia depended heavily on actions taken in the environs of Dupont Circle in Washington, D.C. The fact that this remained largely the case a decade later, when the Planafloro was enacted, was a significant problem given the explicit focus of the project on environmental sustainability.

Really two issues are at stake here. One issue is the question of institutional learning: Is the World Bank changing its behavior in relation to those affected by the projects it undertakes, and has its field of vision expanded to recognize their needs? Here the answer is mixed. Certainly, the Bank's rhetoric and stated intentions have changed; still, the case study presented in this chapter suggests that the distance between intention and implementation remains quite vast.

The other question is whether targeting the banks has had its desired impact in other areas. The initial decision by MDB campaign organizers to target the World Bank rested on the assumption of its influence over the development models and policies adopted by Third World states. In this view, the Bank is an intermediate target whose actions in turn influence the ultimate target—development plans and the behavior of public and private interests in relation to ecosystems and indigenous peoples.

The adoption of a proxy target can be a good tactic for MDB campaigners if its limits are well understood. Specifically, a two-actor model (NGOs and the Bank) simply does not work here. It should be clear from the foregoing that the central roles in this particular story belonged to Brazilian governmental actors—at both national and state levels—and that the political game involved was one that neither the Bank nor the NGOs affected more than tangentially. It is the *political quagmire* the Planafloro represents, and not its technical failings, that made the project so problematic.

The Bank's Operations Evaluation Department review of the Polonoroeste program recognized this problem as well. In the paragraph immediately following the one cited earlier, we have the following:

At the same time, the unconscious role of Bank support for the program in providing legitimacy to the various internal interests behind POLONOROESTE and/ or which took advantage of it in order to pursue their own agendas vis-à-vis the attainment of statehood for Rondonia, additional roadbuilding, the encroachment of reserves, the spread of rural settlement in inappropriate areas and even local and national political ambitions, should also not be overlooked. As a result, the Bank unwittingly became a party to actions which were to have unfortunate social and environmental consequences. *One of the general lessons to be derived from POLONOROESTE for similar future situations, accordingly, is the need for the Bank to more carefully sort out the various economic and political actors and interests involved and to determine how they are likely to respond to proposed program actions in terms of natural resource use and possible misuses.*[35]

But it is precisely this kind of political assessment that the Bank is most reluctant to make (or when it does, to heed).

It is worth remembering the different stories that key actors would tell about the Planafloro:

• For the Bank, it was an ameliorative effort, a chance to take some steps toward making sustainable development a less gelatinous idea. If key technical agencies of the state government could be empowered to act more effectively and if the land-use guidelines stipulated in the zoning plan could be given some reality, the Planafloro might produce demonstration effects; at worst, it could help fix some of the problems left in the wake of Polonoroeste. Rondônia could be expected to be a more economically viable state as a result. Although the project admittedly got off to a slow start and had some problems, they were not of the Bank's doing. The Bank did its best to monitor and strengthen the project.

• Foreign NGOs and their publics initially saw this project as a well-intentioned but arrogant proposal, conceived (as usual) from the top down, without participation of relevant populations in the state (especially the claimed beneficiaries, Amerindians, rubber tappers, and small farmers). It also ignored past failures to demarcate indigenous and conservation areas and presented no new proposals that might force the government to behave differently. Nonetheless, these foreign NGOs were fundamentally sympathetic to the idea that with the aforesaid local participation, international action was necessary to save the rainforest and its peoples, and Brazilians had to be brought to sustain the environment in spite of themselves. The responsibility for making sure that they did, then, rested with the Bank.

• Politicians in Rondônia for the most part saw the project as a vehicle for bringing development funds into the state—funds that could both serve local constituencies and also provide political pork. The state was bankrupt. Nothing more was likely to come from Brasília; the World Bank had been Rondônia's milk cow since the attainment of statehood. It did not much matter what the funds were coming for—any additional revenues were welcome. They saw the green components of the project as window dressing, not as central to its purpose, and did not expect them to be carried out. Although individuals within the state government did care about its environmental (and to a considerably lesser extent its indigenous) components, they depended on powerful politicians for their positions and had little independent leverage. NGO participation in the project was an inconvenience, but circumventable.

• For the central government in Brasília, Planafloro was a financially risky but relatively low-cost way of demonstrating concern about ecological aspects of frontier development and of satisfying state demands for transfers that the federal government could no longer provide; it has never been a particularly high-profile project.

• For local NGOs and fledgling movements, the Planafloro provided access to political arenas they could not otherwise attain. Because the World Bank was constrained to care about rubber tappers and indigenous peoples, the state government had to make some gestures—however symbolic—in the direction of paying attention to their concerns. Those concerns continued fundamentally the same as they had always been: gaining security of land tenure and livelihood, and protection from encroachment by the more locally powerful.

These are different stories, different political and social needs, different agendas and visions of what was and is possible. The relationships among these various elements mandated by the Planafloro mask deeper antagonisms and more lasting configurations of power.

As MDB campaign strategy developed, struggles around projects financed by the World Bank became a way to transpose domestically beleaguered poor peoples' struggles onto an international stage. Indigenous peoples being displaced by dams or trying to protect their historic space, rubber tappers trying to hold onto land use rights, all found small but strategically placed audiences abroad who valued their efforts in ways few did at home.

Just when these poor peoples' struggles began to be internationalized, however, other domestic actors were also trying to further their domestic interests in the political arena that a Bank-funded development project represented. Such projects are thus also arenas of domestic political conflict, and the goals of domestic actors must be two pronged: on the one hand, they must try to use whatever leverage they have in relation to the Bank to further their interests or goals in absolute terms; on the other, they must try to reconfigure their relations with the other actors in the arena.

The NGO campaign targeting the Planafloro has been relatively successful in demonstrating that (a) state and federal institutions were not living up to their stated commitments in relation to the project; and (b) outside of cajoling, the Bank was not doing very much to make them do so. Through the leverage NGOs gained by providing solid information about violations of agreements, by good use of symbolic capital associated with rubber tappers and indigenous people, and by activation of transnational networks, they won official positions from which to speak about the project and (after filing the Inspection Panel claim) won some concrete advances. The Rondônia NGOs have not been nearly as successful, however, on the second leg of the campaign—that is, in strengthening their position in relation to other actors in the arena—at least locally.

The persistence of local NGO weakness in Rondônia is striking given the very real national advances made among related organizations. Preparations for the 1992 UNCED conference in Rio stimulated NGO networking around a variety of environment and development concerns; although the impetus slowed after Rio, Brazilian environmentalists remained much better connected with each other and with their foreign counterparts than they had been. International NGOs such as WWF and Greenpeace established Brazilian affiliates led by prominent and dynamic Brazilian activists. Organizations not previously active on environmental questions—the Institute for Socio-Cultural Studies (INESC) and the Brazilian Institute for Socio-Economic Analysis (IBASE), for example— began to track environmental politics regularly; INESC did so with a particular focus on Congress. Beginning in 1993, a national NGO network formed with support from EDF and others to follow multilateral development bank issues: the intention, on the part of Brazilian and foreign supporters of the idea, was to "nationalize" the MDB campaign.

The Rondônia NGOs were part of the national network from the beginning, of course, and the Planafloro was one of the cases on which it intended to concentrate. Nonetheless, the national strategies being promoted by some of the national NGOs—the 1994 lawsuit, for example, that succeeded in winning an injunction against an INCRA colonization project—were not pursued with the same energy as were the international ones. Most likely they were not seen as effective by local activists, for whom the courts were likely to appear as unreliable as the other political institutions with which they were more familiar. The prominence of expatriates in the Rondônia network may also have played a role in preserving the primacy of an external strategy.

Social movement theorists are virtually unanimous in their views of the centrality of preexisting social networks in the formation of movements. By definition, however, in a frontier state like Rondônia, social networks were thin on the ground; those that underpinned the NGO Forum, although they did exist, were not very robust.

Given this fragility, should national and foreign NGO activists have chosen not to pursue the case? They could not really avoid it. Because of the centrality of the Polonoroeste project in the early phases of the MDB campaign, the NGOs—like the Bank—were compelled by the history of the problems involved to take it on. But it is precisely the difference between the Polonoroeste and Planafloro campaigns—and the changes that occurred in the meantime in the MDB campaign itself—that have made the Planafloro case so difficult.

Local NGO weakness was not an issue in the Polonoroeste campaign because it was wholly centered around providing alternative information. By the time the Planafloro campaign was launched, it was no longer enough to have good information about what was occurring locally; to legitimate their actions, foreign NGOs needed local partners for whom their intervention represented the making of local claims on an international stage. In the case of Brazil, the Bank campaigners' experience with Chico Mendes and the Acre rubber tappers was the model for this kind of alliance. Such an alliance was a highly positive move, but it created expectations about the local partners that were difficult to fulfill in Rondônia.

Although ultimately rejected, the Inspection Panel claim produced a flurry of activity. The Rondônia state government and the Brazilian federal government signed a long-delayed agreement committing INCRA, the Federal Land Institute, to respect the state's zoning plan, and reserves whose demarcation had been unaccountably delayed were suddenly demarcated. Bank personnel finally took a serious look at the project's shortcomings and proposed revisions that they hoped might overcome previous gridlock. The Bank's belated midterm review of the Planafloro in 1996 validated many of the NGO arguments about the state government's lack of commitment to environmental and indigenous aspects of the project. Nonetheless, the review team advised against canceling the project, opting instead to decentralize it and radically simplify its goals.

In the future the project is likely to be more affected by the national NGO network than it has been to date. As of January 1996, task management for the Planafloro moved from Washington, D.C., to Brasília, and began to be coordinated more extensively with the G-7 Amazon project. The same month, a new Bank-NGO liaison officer began work in Brasília, an individual whose previous extensive experience working with Brazilian NGOs has resulted in markedly improved communication between NGOs and the Bank. All of these developments seem likely to produce considerably greater transparency in the Planafloro process. They do not yet resolve, however, the issue of accountability. Insofar as information flows and access to decision making continue to be regulated by the good will of particular individuals rather than by the development of institutionalized relationships, their continuance remains fragile. Nonetheless, paradoxically, without such transitory spaces, the opportunity for politically weak social actors to gain the experience and legitimacy to "scale up" their efforts would probably not exist.

Acknowledgments

I acknowledge the research assistance of Elizabeth Umlas, and helpful comments from Jonathan Fox, Brent Millikan, Steve Schwartzman, and Kathryn Sikkink. My research had financial assistance from the following institutions: the Howard Heinz Endowment/Center for Latin American Studies, University of Pittsburgh, Research Grant on Current Latin American Issues; the Joint Committee on Latin American Studies of the Social Science Research Council and the American Council of Learned

Societies with funds provided by the Ford Foundation; and the John D. and Catherine T. MacArthur Foundation.

Notes

1. The project has a variety of names. Within Brazil it is most commonly called the Plano Agropecuário e Florestal de Rondônia—which translates as Agriculture, Livestock, and Forest Plan, from whence comes the acronym Planafloro. Note that the Brazilian name does not make the "natural resource" issue central in the same way the World Bank's name for it does.

2. Extractive reserves are a type of conservation unit proposed initially by the National Council of Rubber Tappers in the mid-1980s; the first ones were decreed under President José Sarney in 1989. Such reserves are under the auspices of the federal government, but in them are granted use rights for extractive populations (rubber tappers, brazil nut gatherers, and so forth), whose activities do not damage the forest.

3. Data from Estado de Rondônia, secretaria de Estado do Planejamento e coordenação geral, unpublished report.

4. Programa de Desenvolvimento do Noroeste do Brasil (the Northwestern Brazil Development Program).

5. On the colonization of Rondônia, see Lenita Maria Turchi Pacheco, "Colonização dirigida: Estratégia de acumulação e legitimação de um estado autoritário," (dissertação de mestrado apresentada ao Departamento de Ciências Sociais do Instituto de Ciências Humanas da Universidade de Brasília, 1979); and Brent Hayes Millikan, "The Dialectics of Devastation: Tropical Deforestation, Land Degradation, and Society in Rondônia, Brazil," master's thesis, Department of Geography, University of California at Berkeley, 1988.

6. At the time of the 1970 census, Rondônia had only two municipalities (Porto Velho and Guajará-Mirim); in 1980 there were seven; in 1991 there were twenty-three.

7. This information is based on interviews and conversations with Barbara Bramble, Bruce Rich, and Steve Schwartzman, Washington, D.C., as well as the interview with Brent Blackwelder in Steve Lerner, *Earth Summit* (Bolinas, Calif.: Common Knowledge Press, 1991), pp. 157–73.

8. World Bank, Agriculture Operations Division, Country Department I, Latin America and the Caribbean Basin, *Staff Appraisal Report No. 8073-BR: Brazil, Rondonia Natural Resource Management Project*, February (Washington, D.C.: World Bank, 1992).

9. The zoning plan was promulgated by executive decree; when it first came up for legislative scrutiny in 1990, it was defeated. Opponents charged insufficient consultation; many claim the real issue was fear that state governor Jerónimo Santana would use any ensuing project funds for electoral purposes. Information

on the legislative discussion comes from a transcript of the discussion of Projeto de Lei Complementar, number 026/90, on file at IPHAE (Instituto de Pré-Historia, Antropologia, e Meio Ambiente) in Porto Velho. The zoning plan finally passed at the end of 1991. For information on the plan's formulation, I relied on the following: group interview with Maria Emilia da Silva, coordinadora do Polonoroeste; João Trajano dos Santos, coordinator of Planafloro; Sebastião Ferreira Farias, subcoordinator of Polonoroeste; Astreia Alves Jordão, subcoordinator of Planafloro; Emanuel Casara; and Augusto Pinto da Silveira, representative of the Federal University of Rondônia (UNIR) on the Planafloro team; interview conducted jointly with Manfred Nitsch (Deptartment of Economics, Free Univ. of Berlin), Porto Velho, 8 August 1990. I also relied on interviews with Emmanuel Casara, Porto Velho, 6 August 1990; with Brent Millikan, Porto Velho, 10 August 1990; with Wim Groenveld, executive director, IPHAE, Porto Velho, 21 November 1990; with James LaFleur, consultant, Porto Velho, 12 November 1990; and with Vera Machado, Brazilian Embassy, Washington, D.C., 19 September 1990.

10. On transnational advocacy networks, see Margaret E. Keck and Kathryn Sikkink, *Activists beyond Borders: Advocacy Networks in International Politics* (Ithaca, N.Y.: Cornell University Press, 1998).

11. A *grupo* is not the same as a coalition, as it may include politicians from a variety of political contexts—federal senators and deputies, state legislators, mayors, and in some cases municipal council members. These alliances may or may not involve agreement on particular issues and are based primarily on personal loyalties. I am especially grateful to Ary Ott, at the time vice rector of the Federal University of Rondônia, for helping me to decipher some of the intricacies of Rondônian politics. Interview with Ary Ott, Porto Velho, 24 August 1992.

12. It is worth noting that although the Lutzenberger letter was apparently the precipitating cause of the postponement, the Lutzenberger-NGO pressure offered by no means the only or even the strongest set of objections to the loan; much more serious objections came from the new Economics Ministry, which was concerned about the viability of the loan and the Rondônia state government's eventual ability to repay it.

13. Central Única dos Trabalhadores, literally the Single Workers' Central; established in 1983, the CUT is one of three peak labor organizations.

14. See Margaret Keck and Kathryn Sikkink, *Activists beyond Borders*, p. 2.

15. For more detail on this process, see Margaret E. Keck, "Social Equity and Environmental Politics in Brazil: Lessons from the Acre Rubber Tappers," *Comparative Politics* 27 (July 1995).

16. Literally, rubber soldiers—Brazilians who were attracted to the Amazon during World War II, when supplies of Asian plantation rubber were closed to the Allies due to Japanese occupation; among these incentives was the promise of a pension that was to be funded by the U.S. government.

17. Ana Maria Avelar, who has worked with rubber tappers in Rondônia since 1983, illustrated her claim that the Acre group has not shown sufficient respect for the differences between the two groups by telling me a story about an incident at the CNS founding meeting in Brasília. Rondônia rubber tappers, she said, went to the Brasília meeting wearing T-shirts, on the back of which they had inscribed a poem about rubber tappers, taken from a calendar that was very popular in the region. This poem had been written by a poet from Acre. The Acre rubber tappers, who considered this poet reactionary, reportedly reacted negatively to the T-shirts, saying that wearing them reflected a lack of political consciousness on the part of the Rondônia group. Many of the latter, for whom the poem represented a moving evocation of their own experience, were deeply affronted. Interview with Ana Maria Avelar, Porto Velho, 11 May 1991.

18. The term *assessoria* (advisory organization) is widely used among those NGOs, which see their mission at least partly as one of promoting grassroots organization and participation. Representatives of these organizations refer to themselves as *assessores*, or advisors, in their relationships with grassroots groups.

19. INDIA was formed in 1991 at the initiative of a group of people from the university, several environmental activists, and one person from the state environmental organ. INDIA was formed because of a perception that "outsiders"— both from abroad and from southern Brazil—were monopolizing discourse and especially funding regarding Rondônia's developmental future. The desire to provide a local institutional base for the work being done by the members of the IAMA network, especially Ana Maria Avelar, was always one of the organizers' goals. The organization is still mainly a weakly institutionalized social network, but it nonetheless occupies a political space that was previously empty and as a result is likely to be taken more seriously than it otherwise might—both by state agencies and by foreign NGOs desperately seeking local institutions to support. I followed the process of INDIA's formation in numerous informal conversations with Francinette Perdigão, its first president, in late 1990 and in 1991, as well as with several other members of the original group (especially Sandra Kelly and Ana Amelia Boischio). Information on the pressure from foreign NGO funders (as well as from the Canadian Development Assistance Agency) for Avelar and the other IAMA members to work through INDIA comes from a conversation with John Butler, Amazon Program director of the World Wildlife Fund, Washington, D.C., 17 March 1992.

20. It is important to note that in the attempt to pull together affected communities—indigenous groups, rural workers' organizations, and rubber tappers, as well as environmentalists—expatriates living in Rondônia also played an important role, particularly Brent Millikan and staff from IPHAE. Although the meeting was called by the CNS, UNI, and CUT-Rural, most of the logistical work and some of the fund raising was coordinated by Millikan and IPHAE, and Millikan played an important role in the meeting itself. The role of expatriates, both in the formulation of development projects that affect the environment and in the organization of opposition to them, stares you in the face in Rondônia

and is an important component of the configuration of environmental actors elsewhere—particularly in Amazonia, but not only there. The relationship between expatriates—including the organizations in which they are prominent, if not the dominant, actors—and other local organizations became a political question worth examining in itself. It is obviously a very sensitive issue, and although I did not formally interview either expatriates or Brazilians living in Rondônia on this point, it came up frequently in conversation, and I came away from the time I spent there with fairly strong impressions of the issue.

21. Interview with Ana Maria Avelar, Porto Velho, 21 August 1992; interview with Brent Millikan, Porto Velho, 10 August 1990. Comments on the November 1990 Porto Velho meeting come primarily from my own observations at that meeting, as well as from its documentary record.

22. Not only experience, but also a considerable amount of discussion of strategies and new opportunities played a role in the gradual shift in emphasis of the campaign discussed in the rest of the chapter. The relative impact of each of these factors is difficult to assess. I should note here that I discussed the case extensively in June 1992 with several NGO activists (especially Millikan and Avelar). In these discussions I argued that they should be paying more attention to local politics; the need to do this was particularly notable in the failure to predict the rise of William Curi from leader of the legislative opposition to the Planafloro to the project's administrative steward (as secretary of planning, a post he accepted after turning down the offer of the secretaryship of Planafloro in June 1992).

23. World Bank, *Staff Appraisal Report No. 8073-BR*, pp. 16, 44.

24. Forum dos ONGs de Rondônia to Osvaldo Russo de Azevedo, M.D. Presidente do Instituto Nacional de Colonização e Reforma Agraria, 15 Abril 1993.

25. Brent Millikan to Margaret Keck, personal communication by e-mail, 22 May 1993.

26. These areas were Rio Branco, Nova Vida, Oriente, and Pedra do Abismo. The suit, brought as an Ação Civil Pública under law no. 7347 of 24 July 1985, was Processo no. 94.0001366-3, brought on 16 May 1994, Justiça Federal, Estado de Rondônia, 2ª Vara, Reg. Tombo N° 487. The INCRA response was registered on 29 June 1994, and the judge's decision was rendered on 18 July 1994.

27. Processo no. 94.0001654-9, registered on 7 June 1994, Justiça Federal, Seção Judiciária, Estado de Rondônia, 1ª Vara.

28. Forum of nongovernmental organizations of Rondônia to president and executive directors of the World Bank, Porto Velho, 15 June 1994. English translation posted on econet, conference rainforest.worldbank, on 5 July 1994 as topic 290. Portuguese version had previously been posted on alternex conference ax.brasil. Executive summary and Action Alert also posted 5 July.

29. Curi has a long history in development planning in Rondônia. He was the first head of CODERON, the parastatal Rondônia Development Corporation, in the early eighties and left amid a scandal in which he was charged with corruption

but not convicted. He worked as a consultant for the World Bank in the mid-1980s and is actively involved in the state's Federation of Industries (FIERO). Elected a state deputy in 1990, he became the chair of the legislative assembly's Agriculture and Ecology Commission, and the leader of the opposition to the sitting government. It was widely believed, however, that his opposition to Planafloro was motivated more by a desire to be courted more actively, so to speak, than by a principled opposition.

30. According to Fernando Allegretti, until recently running the IEA's public policy project in Brasília. Conversation in New Haven, Connecticut, October 1994.

31. For an excellent analysis of the dynamics of development assistance, see David Fairman and Michael Ross, "Old Fads, New Lessons: Learning from Economic Development Assistance," in *Institutions for Environmental Aid*, ed. Robert O. Keohane and Marc A. Levy (Cambridge, Mass.: The MIT Press, 1996): 29–52. On the current point, see especially pp. 32–34. Fairman and Ross draw heavily for this discussion from Paul Mosley, Jane Harrigan, and John Toye, *Aid and Power: The World Bank and Policy-Based Lending,* vol. 1 (London: Routledge, 1991).

32. Interview with Luiz Coriolo, the World Bank, Washington, D.C., 17 September 1990.

33. World Bank, Operations Evaluation Department, *World Bank Approaches to the Environment in Brazil: A Review of Selected Projects, volume 5: The PO-LONOROESTE, No. 10039 Program* (Washington, D.C.: World Bank, 1992). See especially paragraph 16:

In addition, since most of the Bank loan funds were allocated to POLONOR-OESTE's transport components whose implementation proceeded far more rapidly than other program interventions, these resources were largely disbursed before the full extent of the distortions in its execution became apparent. As a result, by the time the Bank became clearly aware of the program's deficient performance in terms of its agricultural objectives and cognizant of its increasingly adverse consequences on the physical environment and Amerindian populations, its leverage was smaller than it might otherwise have been" (emphasis added, p. viii).

34. *Ibid.,* p. xviii, emphasis added.

35. Ibid., emphasis added.

7

Ecuador: Structural Adjustment and Indigenous and Environmentalist Resistance

Kay Treakle

Over the years, NGOs involved in the international multilateral development bank (MDB) campaign have primarily focused on specific project cases—for example, the Narmada dams in India, roads into Brazil's Amazon jungle—or on specific policies involving information disclosure, indigenous peoples, and forestry. These projects and policies have provided focal points for nongovernmental organization (NGO) campaigns at the local, national, and international levels, and as can be seen by the case studies in this book, North/South NGO collaborations have precipitated some fundamental changes at the multilateral development banks—particularly at the World Bank.

In this case study of Ecuador, the story focuses not on one project or policy, but on a broader MDB-supported economic strategy that has had profound impacts on poor indigenous peoples and their communities, and that has contributed to environmental degradation in the country. Within the overall framework of structural adjustment, the recent MDB-supported economic strategy for Ecuador has included reforms in two key sectors of the economy: oil and agriculture. Both sectors are critical to indigenous peoples. In the case of oil, the lion's share of oil extraction takes place in the Amazon, home to some eight distinct indigenous peoples.[1] In the case of agriculture, the effects of recent sectoral adjustment reforms are most keenly felt by indigenous peoples in the Andes whose lands are threatened by a 1994 law privatizing communal property and water. The impacts of oil development and agriculture policy on them and on the environment form the basis of the alliances made between environmental NGOs and indigenous peoples around land rights in Ecuador.

This chapter discusses how structural adjustment policies and MDB lending in Ecuador have affected indigenous peoples and the environment; documents the indigenous and environmental NGO responses; and examines how NGOs in the North have provided strategic support to the indigenous leadership in Ecuador in their struggle to gain a voice in the national debate around economic development and how the World Bank and the Inter-American Development Bank (IDB) have responded.

The NGO–indigenous peoples alliances' focus on the MDBs and structural adjustment more generally grew out of experiences in the oil sector that began with the World Bank. As the groups received more information from Northern NGOs and gained exposure to MDBs during visits to Washington, their understanding of how Bank policy conditionality works—and in particular of the division of labor between the IDB and the World Bank—led to attempts to influence both banks.

It has been more difficult for NGOs to influence structural adjustment programs (SAPs) and loans than specific infrastructure loans and projects—in part, due to the complexities of macroeconomic policy, the profound lack of public information about SAPs, and a complete absence of participation of civil society in determining the conditions of economic restructuring.[2] But NGOs in the North and South have challenged structural adjustment policy and programs based on a critique that SAPs undermine social services, food producers, worker rights, the natural environment, and the viability of fragile democracies.

The efforts to influence MDB loans in Ecuador did not focus on overall structural adjustment policy, but instead on specific sectoral loans with tangible and identifiable impacts. Indeed, the linkages to broader structural adjustment were largely made as the sectoral campaigns unfolded. Moreover, these efforts emerged out of already established, widespread, grassroots movements. In the case of the oil sector, a vigorous campaign against industry abuses in Ecuador's Amazon has been waged for many years—with local, national, and international networks providing a strong, well-organized platform from which to launch an initiative aimed at the World Bank. In the case of agriculture sector reform, a solid indigenous peoples' movement in Ecuador has fought for decades to secure land rights and access to productive resources as part of their larger struggle for economic and social rights in the country. The focus of both of these indigenous/environmental advocacy efforts, then, was not on structural adjustment per se, but on the specific sector adjustments that followed from the SAP framework.

The Economic Background

The backdrop for structural adjustment in Ecuador is intimately linked to the country's economic dependence on oil. Prior to the discovery of oil by a consortium of U.S. oil companies in the 1960s, Ecuador depended on exporting traditional agricultural commodities such as bananas, coffee, and cocoa. The rapid development of an oil industry in the 1970s, however, created a boom economy that by 1980 placed Ecuador in the category of a "middle-income country" in Latin American terms.

The high price for oil in world markets allowed the government to expand vastly government involvement in the economy. Since the early 1970s, the state has also played a dominant role in the oil industry through the state-owned company, Petroecuador.[3] Because Petroecuador dominated the oil industry, high oil prices allowed the government to increase both spending and external borrowing by a great deal. During the 1970s, the government's priorities for spending included improving social programs; education and health programs reached some of Ecuador's poorest citizens.[4]

The oil-boom period was relatively short, however. During the 1980s, a series of external and internal events threw Ecuador into an economic tailspin from which it is still trying to recover, including the international debt crisis, the 1985–86 collapse in international oil prices, followed by a massive earthquake in 1987 that destroyed the main pipeline and shut down oil exports for five months. Similar to the decline of other Latin American countries during "the lost decade of development," Ecuador's economic decline during the 1980s was disastrous. Facing external debts equivalent to 37 percent of gross domestic product (GDP),[5] in 1983 the country began to implement a standard structural adjustment program prescribed by the International Monetary Fund (IMF), which has been implemented in the past twelve years with fifteen IMF standby agreements and numerous sector adjustment loans from the World Bank and IDB.

The structural adjustment policies sought to eliminate subsidies and reduce price controls while promoting exports and opening up the country to foreign investment. Meanwhile, the government devalued the sucre (Ecuador's currency), increased interest rates, and cut state spending. During this period of adjustment (from the early 1980s to early 1990s), Ecuador fell into an economic recession, and per capita income fell 32 percent, from U.S.$1,444 in 1982 to $977 in 1988.[6]

These austerity measures created widespread social problems, in particular in the peasant agriculture sector. Indigenous small farmers suffered from loss of income as the prices of basic necessities—including food and fuel—increased dramatically, farm credit opportunities dwindled, and crop prices fell. The economic conditions in agriculture consequently drove huge numbers of people from rural areas to the cities during this period, thus exacerbating urban problems.[7]

Despite the economic reform measures implemented during the 1980s, the country's economic picture did not improve, and poverty increased substantially. A recently completed *Ecuador Poverty Report* by the World Bank (November 1995) found that 35 percent of the population of the country is considered living in poverty, and another 17 percent are "highly vulnerable to poverty." According to the World Bank, during the decade prior to 1993 the government was unable to maintain real social sector expenditures, which at 4.2 percent of GDP in 1990 were low by Latin American standards.[8] Between 1965 and 1993, the poorest 20 percent of the population's share of the national income dropped from 6.3 percent to 2.0 percent.[9]

The Indigenous Movement in Ecuador

Ecuador also witnessed the emergence of a well-organized indigenous movement in the 1980s, as provincial and local indigenous organizations formed into regional confederations in the Amazon, the Sierra, and the coast. The indigenous peoples' agenda focused on land reform and the preservation of ethnic identity, including language and culture. In 1986, leadership from two regional organizations, ECUARUNARI (founded in 1972 in the Sierra) and the Confederation of Amazonian Indian Nationalities (CONFENIAE, founded in 1980), cooperated in the formation of the Confederation of Indigenous Nationalities of Ecuador (CONAIE) to become a national coordinating body representing most Indian groups. Their initial focus was on a "distinctly ethnic agenda that ranged from the vindication of cultural rights to more ambitious pragmatic demands such as the redefinition of Ecuador as a plurinational country."[10]

Through its nationwide membership, CONAIE represents approximately 70 percent of the indigenous peoples of Ecuador, who account for an estimated 40 percent of the country's total population.[11] The leadership of CONAIE is elected: every two or three years, delegates from around the country meet in a national congress to agree on strategy and select their president, vice president, and secretaries of education, health, and human rights. Its structure, process, and political agenda make CONAIE the most important social movement in the country. The emergence and strength of CONAIE has also been supported by a transnational alliance with Northern-based development and advocacy organizations such as Oxfam America, the Inter-American Foundation, and Cultural Survival, and with environmental groups such as the Rainforest Action Network.[12]

CONAIE was thus in a strategic position when Indians around the country launched the first national indigenous civic uprising in 1990. In June, beginning with a takeover of one of the oldest churches in Quito, tens of thousands of indigenous peasants shut down the country for a week by blockading its main highways, staging peaceful demonstrations, and refusing to bring their produce to market. Some government offices were also occupied by indigenous protesters. The uprising was a national expression of nonviolent resistance to deteriorating social conditions, lack of recognition by the government, worsening economic opportunities, and escalating conflicts over land in the Sierra.[13]

The 1990 uprising (or Levantamiento Indígena Nacional) was, according to one observer, "functionally equivalent to what in other Latin American countries appeared under the form of the so-called 'IMF riot': a display of popular protest induced by the profound impact of the economic slump and the adjustment policies of the 1980s."[14] These adjustment policies had favored businesses engaged in agricultural exports over production for the internal market, failed to respect indigenous culture, and ignored the economic needs of the indigenous farmers. CONAIE's demands, which were articulated in the "16 Points of the Indigenous Communities," included renewed agrarian reform, the legalization of territories of the indigenous nationalities, a freeze on prices of basic necessities, fair prices for farm products, and autonomy in marketing, among others.[15]

The government initial response was to call out the police and army to put down the nonviolent rebellion, but it soon agreed to open a dialogue on the Indians' grievances. On balance, the uprising won unprecedented political space for the indigenous movement in Ecuador related to ethnic identity and issues of autonomy; it did not, however, lead to significant changes in the government's MDB-supported economic strategy.

Ecuador's Economic Plan

In 1992, the new government, headed by President Sixto Durán Ballen, was elected on a platform that included a series of economic and sectoral reforms—supported by MDB lending—that would deepen and broaden the structural adjustment reforms begun during the 1980s. The government pledged to privatize public enterprises, cut the size and functions of state agencies, reform the financial system, open markets, reduce tariffs, and abolish nearly all nontariff barriers to imports and exports. The reform program sought to renegotiate Ecuador's debt under the Brady Plan by reforming laws and regulations of key economic sectors, including oil and agriculture, to make them more conducive to private investment. The two MDB loans that became focal points for the NGO/indigenous alliance were designed as part of this broader economic strategy. The World Bank technical assistance loan for public enterprise reform would have supported privatization of the state oil company, as well as other state industries. The IDB's agriculture sector reform loan would support changes in the agriculture development law, leading to increased private ownership of agricultural lands and privatization of water rights. Both the World Bank and IDB were key players in developing national policy, significantly influencing national laws.[16]

With 85 percent of new loans in the country coming from these MDBs, their "policy dialogues," supported by massive studies in specific economic sectors, profoundly influenced policies, projects, and in many cases legislation in the country. The following sections detail only two examples of the MDBs' influence in Ecuador, but they describe a process that is taking place in almost every sector of the country's economy. The

North/South NGO network attempted to debate the changes in national policy by directly lobbying the banks. Within this process, opening up political space for civil society participation in the economic policy debate at both government and bank levels has been enormously difficult.

The Campaign against Oil Pollution in the Amazon

Ecuadorian environmentalists and indigenous organizations have long struggled for accountability and justice in the areas of the Amazon that have for more than twenty years been opened up to oil development.[17] U.S. oil companies—such as Texaco, Conoco, Maxus, and Arco—have been responsible for causing extensive pollution, expropriation, colonization, and deforestation in the Amazon region, as well as cultural destruction among Amazonian indigenous communities.[18]

The "Amazon for Life" campaign began in 1990 with a nonviolent occupation of Texaco's Quito offices, where NGOs demanded that the company perform an environmental audit of its operations. Led by the national environmental organization, Acción Ecológica, the campaign also opposed a proposed World Bank loan for increased oil exploration.[19] The environmentalists' efforts merged with the indigenous movement in September 1992, when a meeting was called by CONAIE, CONFENIAE, FCUNAE, and the Coordinating Body of Indigenous Organizations of the Amazon Basin (COICA) to build a national coalition between the indigenous organizations and the ecological movement in Ecuador and to launch a permanent monitoring program on extractive activities in the Amazon, with a specific focus on Texaco and another American oil company, Maxus.[20]

Early on, Acción Ecológica, which wages a number of national environmental campaigns in Ecuador, recognized the importance of establishing international links.[21] These links between Southern and Northern NGOs paralleled the globalization process in Ecuador, reflected in the growing number of multinational oil companies and MDB development programs that campaigns to protect certain regional and local territories had to confront. Acción Ecológica targeted Northern companies, lending agencies, and donor governments. In the process, they engaged with Northern environmental organizations to strengthen their ability to generate enough pressure for the media and policy makers in the North to pay attention to the problems caused by international companies and policies. To exchange relevant information about American companies they needed allies in the U.S.; to generate pressure on World Bank executive directors, they needed contacts in the donor countries who could write letters to the appropriate government agencies and executive directors; to develop coordinated global strategies, they needed alliances with groups that could provide support for their efforts, including the launching of an international boycott of Texaco to pressure the company into taking responsibility for its activities in the Amazon. Acción Ecológica also needed help finding the financial resources to carry out their campaigns.[22] Since 1990, the "Amazon for Life" campaign has grown into an international alliance of hundreds of environmental and human rights organizations involved in letter-writing campaigns, demonstrations, a petition presented to the Inter-American Commission for Human Rights of the Organization of American States (OAS), and an international boycott of Texaco.[23]

The World Bank and Oil

MDBs' direct involvement in Ecuador's oil development has been limited. During 1987, the World Bank provided an $80 million loan for emergency reconstruction of the pipeline and other facilities damaged in the massive earthquakes that took place on March 5 and 6 of that year. Part of this loan was also earmarked for the development of an environmental management plan for Petroecuador, to be a guide for "environmentally responsible development of the petroleum sector."[24] The environmental management plan, prepared by a Colombian consulting firm, Ambientec, was intended to support an environmental impact assessment for a proposed World Bank oil production development project in 1990 that would have provided $100 million to develop new oil fields in the Oriente's Block 7 and to rehabilitate existing oil infrastructure. The loan also intended to induce the government to cut fuel subsidies. President Borja, however, could not accept the "shock treatment," and without it, the Bank did not want to make the loan.[25] The environmental management plan, although completed, was never implemented due to the incapacity of the Ecuadorian government agencies, and the loan itself was eventually canceled.

In early August 1993, Acción Ecológica contacted the Bank Information Center (BIC)[26] in Washington, requesting information about an upcoming World Bank loan for "public enterprise reform" that they had heard was being prepared to privatize Petroecuador. They expressed concern that the loan was being developed without public debate or consultation and that it could exacerbate the impact of oil development on the Amazonian environment and indigenous communities. BIC called the Bank's NGO liaison office to find information about the loan, which, according to the Bank's *Monthly Operational Summary,* was "to assist in developing a legal and regulatory framework permitting the restructuring of the energy sector, and promoting private sector participation in energy and other sectors of the economy."

The Bank's environmental classification of the Public Enterprise Reform Technical Assistance Project was listed as a Category C—requiring no environmental impact assessment or analysis.[27]

The Bank's NGO office responded that the Bank was not emphasizing privatization of the oil industry, nor was it Bank policy to increase exports. According to Bank staff, the public enterprise reform loan was to be primarily focused on the telecommunications industry. The Bank was, however, trying to get Ecuador to raise its oil prices to world levels to be more competitive and to have a "free trade in hydrocarbons" that *could* have an indirect effect on private investment in the oil industry and *may* lead to increases in production and exports. The Bank staff's argument was that, in any case, privatization would have a positive impact on the environment because private companies "are more accountable" than government agencies.

However, a staff person familiar with the loan told BIC that the loan was in fact aimed at privatizing the telephone, electricity, *and* oil sectors (e.g., Petroecuador) by providing assistance to develop new laws, which were then being written. Apparently, one environmental analysis in the Bank believed there

needed to be an analysis done of the possible environmental and indigenous peoples impacts, including land demarcation, so that once the laws are drafted that facilitate privatization, rigorous guidelines and standards would be developed and applied to regulate the private companies' operations.

Apparently, the Bank's environment staff had recommended that there be a sectoral environmental assessment in accordance with Bank operational directive 4.01, which would look at all of the implications of privatization of the petroleum industry—including the concession policy, regulatory framework, environmental audits of refineries and drill rigs, as well as the capacity of the government to enforce environmental laws and protect indigenous peoples' rights. There was also discussion of including a $2.5 million component to strengthen Ecuador's Environment Ministry or to form a new environment unit in the Ministry of Energy and Mines that would screen projects for environmental compliance.

BIC also found that apparently the Bank's environmental and operations staff had disagreed about the environmental categorization of the loan and whether or not there should be an environmental analysis. Because of the disagreement, ultimately the environment staff refused to sign off on the project because of its environmental implications.[28]

On August 9, BIC sent a letter to Dennis Mahar, then the environmental division chief for Latin America, urging that the Bank classify the loan as a Category A loan due to the "potential for increased pressure on indigenous territory, and for accelerated environmental destruction in the Amazon." On August 26, Danielle Berthelot, the project's task manager, responded in a letter that denied the Bank's involvement in "actual privatization," given that such a step would be "unpredictable." Instead, the Ecuadorian government asked for the Bank's assistance only in

ending the public monopoly in the energy sector and on deregulating energy prices. Once the energy sector is demonopolized, the state oil and electricity companies will have to comply with environmental regulations like any other companies and particularly like international companies whose standards of environmental protection are much higher than those currently prevailing in state companies.[29]

This response ignored the actual environmental track records of international companies such as Texaco and Maxus.[30]

An April 1990 internal Bank memorandum suggests that mitigating environmental impacts of oil drilling in the ecologically sensitive Amazon region in Ecuador would be impossible because the best available practices for environmental protection and mitigation are simply not used by either Petroecuador or the international oil companies operating in the country.

At the time the technical assistance loan was being prepared, Ecuador's government was developing a "modernization law,"[31] which included the controversial oil sector. The law would limit Petroecuador's functions to promotion, negotiation, and administration of oil exploration and exploitation contracts. Production companies and the Trans-Ecuadorian pipeline would be privatized. The law would also set prices based on international prices and subsidies on fuels would be substituted by some form of social compensation.[32]

Acción Ecológica staff were quite concerned that privatization would facilitate expansion of oil development in environmentally sensitive areas and indigenous territories, and would put strategic sectors of the national economy into the hands of transnational corporations while weakening the ability of the state to regulate their activities. Given the track record of oil companies operating in the Amazon, Acción Ecológica also feared that privatization would weaken the ability of popular organizations to influence either natural resource management or oil policy making.

Leaders of CONAIE also had concerns about the "modernization" process. Although agreeing in principal that the country needed to be modernized, they were concerned that the focus was on privatization of state assets as opposed to economic development for all sectors of society. Moreover, they believed that such a move would increase unemployment and precipitate a national social crisis. CONAIE argued that modernization was happening without participation by affected sectors of the population and that there was no regulation to set even minimum conditions for licensing and pricing systems.

In fact, four drafts of the "reform laws" were written during this period, none of which were made public, and the "Seventh Round" of oil leases,[33] which was to get underway in January 1994, was being designed based on the new laws that had not yet been passed. According to an article from the newspaper *Hoy* dated October 7, 1993, the World Bank was directly involved with writing the new law:

The new hydrocarbon law is now ready ... with the objective of opening the oil sector to private companies. The latest version from September 24 ... was delivered a week ago to oil companies in the U.S. by ex-minister of Mines and Energy, Fernando Santos, lawyer for the company Tripetrol which operates block one on the coast, and by Francisco Roldan, lawyer for Maxus company, who wrote the law with the *direct guidance of the World Bank*.

Sources connected with the petroleum industry in Houston told *Hoy* that the law was distributed independently in a forum which took place in this city, where the Minister of Energy and Mines, Francisco Acosta and the President of Petroecuador, Federico Veintimilla, held meetings on the five reforms that they are going to introduce in the law. Nevertheless, the *new interpretation* doesn't speak of any changes, and rather *takes a harder line on the part of the government and the World Bank to privatize the petroleum sector,* considered to be one of the most profitable and strategic of the state enterprises. (emphasis added)

Acción Ecológica and CONAIE claimed that there needed to be permanent, independent, participatory monitoring of oil companies operating in the country. For those oil development plans already approved for indigenous territories, they called for management and contingency plans that would be developed with and approved by the affected populations. If people are put in physical, cultural, or social risk, they argued, the projects should be suspended until technological and political guarantees are put in place. Moreover, they demanded that there be no oil drilling in "protected" areas until such time as conservation laws were enacted and enforceable, and until technology advanced to a point where risk would be minimized.[34]

Although the proposed World Bank public enterprise reform technical assistance loan was relatively small ($20 million),[35] the implications—in terms of expanding oil development in the Amazon—could be far-reaching. On November 10, 1993, Ecuadorian environmental and indigenous organizations wrote to Ismail Serageldin, the World Bank's new vice president for Environmentally Sustainable Development, to request that the environmental category of the loan be changed from a C to an A, in order to fully assess its potential environmental and social risks, especially regarding indigenous peoples' rights, pollution, and biodiversity. The letter also highlighted the absence of participation by those directly affected by petroleum policies, stating that

Fundamental decisions have been arbitrarily taken, without consultation with, or participation by, diverse sectors of Ecuadorian society. In this way ... [oil] concessions have been awarded in national parks, indigenous reserves and territories, including those already legally recognized. The consultants contracted to write the reforms to the Hydrocarbons Law come from the petroleum industry ... with direct advisory input by the World Bank. [F]ar from opening spaces for consultation which take into account the concerns and proposals of the populations that will be affected by these projects, within a period of 90 days, they edited four drafts which were never brought to the attention of social or environmental organizations, but only to protagonists in the industry.

The letter demanded that there be a process of public consultation of affected populations, their representative organizations, and other interested sectors of society in Ecuador.[36] It further requested that the Bank apply its policies, operational directives 4.01 on environmental assessment and 4.20 on indigenous peoples, and make available all documents developed in preparation for the loan, also noting that the scheduled January 1994 loan decision did not allow enough time for these issues to be addressed properly.

In order to generate international support for the Ecuadorians' demands, BIC, Rainforest Action Network, and Acción Ecológica forwarded the letter, along with an action alert, to NGOs in Europe, Japan, Canada, and Australia, many of whom wrote letters directly to their countries' executive directors at the World Bank. As a result, thirty-three members of the European Parliament wrote Ecuador's president and members of the Congress:

We'd like to assure you our friendship and our concern for your country and all its citizens, particularly in view of the fact that after the UNCED in Rio de Janeiro, no short-term-oriented thinking can further lead any of us to choose immediate advantages instead of sustainable development.... Could we appeal to you to include effective environmental and Indigenous-friendly clauses in your Hydrocarbons Law and the classification as "category A" of the loan negotiated with the World Bank.[37]

The Bank's response to the Ecuadorian groups' letter came on November 22 from the chief of the Bank's Regional Trade, Finance, and Private Sector Development Division, Paul Meo, who denied that the Bank had been involved with the new petroleum-licensing round, though he did not address the specific charge that the Bank had actually advised on writing the new hydrocarbons law.[38] He also neglected to address the groups' concerns related to drilling in protected areas and indigenous territories, and shifted the discussion from the loan to a Global Environment Facility (GEF)–supported grant, which was then being developed to support the government's new Institute of Forestry, Natural Areas, and Wildlife (INEFAN). This GEF project would focus on developing regulations and "community participation in the administration of such areas and territories." Basically, Meo's message was that INEFAN would take care of the NGOs' environmental concerns.

Echoing the previous assertion made by Task Manager Danielle Berthelot, Meo also claimed that the reform policy that the Government of Ecuador has asked the Bank ... to support ... would be quite far from actual privatization. The objective of the Bank's assistance is essentially aimed at ending the public monopoly in the hydrocarbons sector and deregulating hydrocarbons prices. The removal of regulatory and monopoly powers from Petroecuador can be expected to have a particular positive implication for the environment. Once the hydrocarbons sector is demonopolized, Petroecuador, like any other petroleum company, would have to comply with environmental regulations imposed by governmental environmental agencies.

A month later, Acción Ecológica's president, Cecilia Cherrez, responded to Meo's letter, challenging his assertions that the loan had nothing to do with the Seventh Round[39]:

> In relation to the government's plan to open a new petroleum licensing round, we keep our position that the main objective of the reforms sponsored by the World Bank is to create a legal framework for new contracts in the hydrocarbons sector.... Therefore, the World Bank is not in a position to deny its incumbance [*sic*] in terms of the Environmental Impacts that the new round will produce.... INEFAN deals only with protected areas, not indigenous territories, nor with every ecological conflict. In fact, the three Amazonian protected areas are under oil exploitation and suffer serious environmental problems. The indigenous territories do not have any sort of protection and they are threatened by further oil activities due to the hydrocarbon policies and the new round.[40]

Regarding the Bank's assumption that the reforms would have a positive environmental impact because Petroecuador would be held to the same regulations as private oil companies, Cherrez wrote, "There is no evidence of cleaner work by the transnational companies. Are you thinking of Texaco, or Maxus or perhaps Arco? We would like to discuss these points in detail, so we can share with you some documents which show the behavior of these enterprises in Ecuador."

In early January, the public enterprise reform loan was placed on the Bank's back burner. The Bank's reason for the postponement had nothing to do with the environmentalists' concerns, however, but with the delays in Ecuador in passing the hydrocarbons law. Meanwhile, the NGOs' campaign against Texaco heated up again when in January 1994 newspapers in Quito reported that the company denied any responsibility for environmental damages caused by their oil operations in the Amazon,[41] a denial that came just prior to the opening of the Seventh Round of oil leasing, which was to get underway on January 24. Fueled by fears that the round would lead to further violations of the Amazonian environment by oil companies, 150 indigenous representatives of Amazonian communities that would be affected by the Seventh Round, led by CONAIE, set up camp in a park in Quito to protest. When the Minister of Energy and Mines Francisco Acosta canceled a meeting with representatives of the group, many of them occupied his offices for five hours until he agreed to meet with them. In this meeting, the minister agreed to most of their demands.[42]

Although the government responded to the demands made by indigenous groups and NGOs, it was difficult for NGOs to link the Seventh Round and the new hydrocarbons law to the World Bank's technical assistance loan. The Bank's response to NGO concerns was to try to narrow the debate by explaining that they were just trying to rationalize oil prices and de-monopolize Petroecuador. They were not interested in discussing the broader implications of increasing private sector participation in the oil industry. The NGOs, however, argued that *any* expansion of oil projects in sensitive areas, without enforceable regulations and the indigenous peoples' consent, was unacceptable. The Bank argued that market-based prices and private company investment in oil would be good for the environment and that, in any case, the government of Ecuador would enforce stronger environmental regulations, which the Bank was supporting through various other mechanisms, including the GEF.[43] The fact of the matter, however, was that the Bank had for some time been advising the country on how to manage its state oil industry, though its preoccupation was largely economic efficiency, framed in terms of structure adjustment.

In this context, Bank operations staff saw the issues of the environment and indigenous peoples as obstacles rather than as legitimate factors in the oil policy debate. In one World Bank study, an offhand comment illustrates this point: "Bureaucratic delays along with increasing environmental and indigenous peoples' concerns have discouraged the participation of certain companies."[44] Addressing the environmental issue, one Bank document *does* acknowledge that redefining the role of the state requires some public sector involvement in environment and natural resource management. Recognizing that state institutions have been weak and have lacked the commitment to address environmental and natural resource issues, the Bank asserted, however, that "the Government recognizes that a central institutional framework is needed and [it] is aware of the risks of not having clearly defined environmental standards and requirements, particularly in the oil sector." The Bank, along with the IDB, agreed to provide support to design an institutional framework, but notes, "considerable preparatory work is needed and full implementation is likely to take several years."[45]

Meanwhile, despite the acknowledged absence of an effective environmental management regime in Ecuador, Bank officials responsible for the public enterprise reform loan continued to maintain that the work they were doing vis-à-vis Petroecuador had virtually no relationship with environmental and indigenous concerns. So BIC placed the loan on the agenda of a regular NGO meeting with Bank vice president for Environmentally Sustainable Development Ismail Serageldin on February 23, 1994, at which Task Manager Paul Meo was present to explain the Bank's position and answer questions.

The meeting started with Mr. Meo outlining the economic rationale behind the loan; in the context of Ecuador's large debt, it made no sense for petroleum prices to be so far below international prices. The new hydrocarbon law (passed in 1993) made it legal for the government to increase petroleum prices, but further changes were needed. In particular, he said, the government couldn't regulate itself as a producer, evidenced by the fact that "Petroecuador is an environmental nightmare." He said that the company should be demonopolized in order to separate producer functions from regulatory functions and that this process was part of the Bank's overall strategy for modernization.[46]

Meo also had quite a bit to say about the country's political situation and the role of NGOs. He said that in Ecuador there was "too much political debate" and called the country a "flaming democracy." He noted that the administration was committed to the modernization process, but was impeded by the Congress. Moreover, he said that NGOs in the country were "as feisty as their Congress," noting the numerous public demonstrations that had taken place around these issues. To illustrate the extent to which Ecuador was a "flaming democracy," he cited two general strikes led by CONAIE and characterized them as actions to "overthrow the government."

BIC staff pointed out that he was mistaken about CONAIE, explaining that they were not trying to overthrow the government but were expressing legitimate concerns that government policies were affecting the people's survival. Moreover, given that Ecuador is a democracy, they had the right to oppose such policies. Meo responded that they indeed were trying to overthrow the government: "What else do you call it when they demonstrate and burn police vehicles?"

It became clear that the notion of public debate and citizen participation was going to get no support from the Bank staff person in charge of privatization in Ecuador. Vice President Serageldin, on the other hand, was more open to hearing NGO concerns. BIC told him that oil was one of the most important environmental and indigenous rights issues in the country and that the new oil leases were controversial because they were being given in the absence of regulatory control and because they would greatly expand the territory open to oil exploitation into indigenous lands. The World Bank's loan needed to be considered, BIC argued, in the context of the broader issue of oil development in the Amazon, and the environmental assessment and indigenous peoples policies of the Bank should apply to ensure that the Bank's loan would not exacerbate the problems. Serageldin seemed to accept that there was at least some truth to this argument and asked Meo to see if the activities of the loan were "in accordance with the Bank's environmental and indigenous peoples policies."[47]

A few days later, two Ecuadorian indigenous leaders from the Organization of Indigenous Peoples of Pastaza (OPIP) and the Amazanga Institute of Indigenous Science and Technology were in Washington, D.C., on a solidarity tour with the Seventh Generation Fund, a U.S.-based indigenous peoples' organization. The leaders were Hector Villamíl, president of the Organization of Indigenous Peoples of Pastaza (OPIP, which is a member of CONAIE), and Leonardo Viteri, the director of the Amazanga Institute (affiliated with OPIP). BIC arranged for them to meet with Paul Meo so that they could explain their concerns about oil's impacts in their territories and the relationship between the Bank's loan and the Seventh Round of oil leases.

The indigenous leaders were concerned that Bank staff "had strange ideas about their struggle," somehow believing that CONAIE was trying to overthrow the government. They explained to Meo that their organizing activities were not intended to destabilize the government or to interrupt economic development in Ecuador, but rather to demand that the government consult with indigenous people when executing projects in their territories, given the destructive impacts of more than twenty years of oil development in the Amazon. The Seventh Round of concessions was slated to open up more than 800,000 hectares of their territories, and OPIP was demanding that no new concessions be given in Pastaza Province.[48] Arco had been operating there since 1989, but the impacts were unacceptable to the indigenous communities, and they were demanding joint participation with the company in monitoring activities, loan decisions, and impact assessments.

Villamíl also explained the difficulty of regulating private companies—noting that the laws in Ecuador were weak and there was no enforcement. Ample evidence revealed that private companies operated no more responsibly than Petroecuador, and he could not understand why the Bank believed that that they *would* act responsibly. Moreover, OPIP considered the government's lack of consultation with indigenous peoples in the new modernization and hydrocarbons laws to be a serious problem, especially considering that 40–45 percent of the population of Ecuador is indigenous and that the effects would be felt by indigenous communities. Meanwhile, OPIP and the Amazanga Institute were creating their own alternative plan for indigenous peoples' development of the Pastaza territory, which would deal comprehensively with health, production and commercialization, community infrastructure, alternative energy, communications, transportation, conservation and protection of areas of high biodiversity, and tourism. The plan would strengthen bilingual education and interethnic exchange, and support the recovery of degraded areas.

The indigenous leaders described their concerns and plans in hopes that this would clarify the reasons why they wanted the Bank to apply its environmental and indigenous peoples policies to the public enterprise reform loan. Meo remained convinced, however, that the loan was "purely a legal matter" and had nothing to do with the environment. He suggested that with the development of the new high-level Environmental Advisory Commission (Comisón Asesora Ambiental, or CAAM) supported by the World Bank and the IDB, the regulatory capability of the government would improve.

Later that year, due to government resistance to privatization of Petroecuador, the hydrocarbons component of the public enterprise technical assistance loan was dropped from the loan entirely.

Broadening the Debate

In the larger scheme of things, the Bank's public enterprise reform loan was not the battleground on which the oil issue in Ecuador would be won or lost, but it became clear, as NGOs delved further into the World Bank and IDB activities in Ecuador, that the macrolevel and sectoral policy advice and loans of the MDBs were key in the modernization process and economic reforms, which reached virtually every facet of economic life of the country. Given that MDB investment financing in Ecuador accounted for more than 85 percent of total external lending, their influence was enormous.

As their focus broadened, NGOs and indigenous groups began to explore other possible openings for CONAIE to use the World Bank's indigenous peoples policy to gain access to the decision-making process in Bank-financed projects. The first opportunity came in April 1994, when it was announced that Luis Macas, the president of CONAIE, had won the prestigious Goldman Environmental Prize for Environmental Leadership. Macas was going to be in Washington to receive the award, an opportunity to meet with World Bank staff. Cecilia Cherrez, president of Acción Ecológica, joined him so that both organizations could communicate their concerns together. The Ecuador Network organized a week-long series of meetings with the banks, the U.S. government, and Washington-based NGOs.[49]

Although the issues raised at these meetings ranged from GEF projects to the oil industry to modernization policy, Macas and Cherrez concentrated on the same themes: the severe environmental, social, and cultural problems caused by unsustainable, wrongheaded economic policies and government neglect needed to be addressed by allowing for the participation of all sectors of civil society in Ecuador. They also emphasized that this debate should be an informed one, and part of the problem was the lack of information about MDB activities: the government and Bank documents outlining the macroeconomic strategy were not publicly available. The only official Bank documents CONAIE and Acción Ecológica had access to were two-page summaries (project information documents) in English, about a few specific loans (documents they received from Washington NGOs because they could not get them in Ecuador), the short descriptions in the Banks' Monthly Operational Summaries, and the GEF project description. These various pieces of information suggested that the MDBs were in the process of implementing or developing loans for a wide variety of sectors and economic policies in Ecuador—including agriculture, modernization of the state, judicial reform, structural adjustment, social investment, environment, and public enterprise reforms, and adding up to a national economic strategy developed with virtually no public debate or input.[50] None of the loans were designated as environmental Category A (World Bank) or Category IV (IDB), which would have required public input into an environmental assessment process, and none of them were seen to impact indigenous communities directly, so the Banks' indigenous peoples policy was not applied. The MDB policy reforms that NGOs had focused on for so many years had not adequately dealt with *indirect* impacts or policy-based lending—a problem that became painfully apparent as NGOs and indigenous leaders began looking at policy loans that on the face of it did not have the same blatant effect as, for example, a large dam or a road into rainforest, but at another level may have more long-term and more serious implications.

The oil policy reform process suggested that the MDBs' influence in national policy exceeded that of its citizens or legislature. In a meeting at the World Bank on April 25, 1994, Luis Macas said that "democracy involves organized civil society including indigenous peoples. The communities need to be involved in the development process that the World Bank is funding." Macas and Cherrez agreed that changes needed to be made in the oil sector, but asserted that a public dialogue was also needed so that environmental and indigenous considerations would be taken into account. Although there was general agreement that modernization was essential, they noted that "society has not yet agreed on privatization."[51]

The meeting at the World Bank did not produce concrete results, but the visit did provoke some concern in the Ecuadorian government. Soon after Macas and Cherrez left Washington, BIC staff were invited to a meeting at the World Bank on the government of Ecuador's policy and coordination with NGOs and indigenous peoples. The meeting was with Galo Abril, the secretary general of the National Development Council of Ecuador (CONADE); Felipe Duchicela, the new head of the Secretariat for Indigenous and Ethnic Affairs, and the Ecuadorian ambassador to the United States. The invitation came from Marcela Cartagena, the Ecuadorian alternate executive director to the Bank. All of the World Bank staff involved in previous meetings were invited as well—as if the government felt that they needed to tell "their side of the story" in order to balance out whatever may have been communicated by CONAIE and Acción Ecológica.

Mr. Abril began the meeting by saying, "We have heard that the perception is that the government has not worked well on indigenous affairs"; he wanted to describe the important steps the government was taking to deal with environmental and indigenous issues. He explained that there had been a lack of contact with NGOs in the past, but the government was open to establishing a dialogue. The president's new Environmental Advisory Commission (Comisión Asesora Ambiental, CAAM), created to coordinate all environmental agencies, would be the appropriate place to address environmental questions. Supported by the IDB, the World Bank, the U.S. Agency for International Development (USAID), and World Resources Institute, CAAM was the first agency within the government of Ecuador to treat these issues seriously. Mr. Duchicela explained that the new Secretariat of Indigenous and Ethnic Affairs was created "to serve as a channel for aspirations and the needs of indigenous people. With the multilateral development institutions we want to help design and finance programs for the Indians ... with respect for their culture."[52]

The Latin America Division's senior advisor for operations, Ping-Chueng Loh, promised that the Bank would do what it could to support the work of the government and cited the new Social Investment Fund as an example of the Bank's commitment to environment and indigenous issues—most importantly in health, education, and poverty—using participatory processes with specific focus on indigenous concerns[53]: "We want to be open to our shareholders, NGOs and the people of Ecuador." In response to a question from Bank anthropologist Shelton Davis about the impacts of macroeconomic policy on indigenous cultures, Duchicela said that the "government policy is to achieve macroeconomic objectives, not leave out the social part of it. We started this week a National Awareness Campaign of the social work done during the past two years."

BIC asked what the government was doing in the case of oil development in national parks, protected areas, and indigenous territories. The Ecuadorian ambassador to the United States replied that "the Minister of Energy and Mines has a very close relationship with groups living in the Amazon Basin. In the Seventh Round [of oil leases], there has been a lot of indigenous and environmental involvement." Duchicela added that "one of the reasons for the Secretariat of Indigenous Affairs is that things need to be improved. Privatization has been a big issue, but we are tackling more important points with the Indians. Most Indian organizations cooperate, except CONAIE, which says that we are just another bureaucracy and intend to dismantle other Indian organizations." He went on to say that the situation with CONAIE was polarized and that they had not entered into a constructive dialogue: "I hope they put aside their political obstacles and stop making conditions on subjects," to which Mr. Abril added, "CONAIE is a political organization, not an umbrella group like they say. They are communistic [*sic*]; a side group trying to get into other issues."

The meeting made clear two things. First, the government was worried that CONAIE had been talking directly to the Bank, and they apparently wanted to show how government agencies were dealing with questions about the status of the environment and indigenous peoples in Ecuador, and to try to undermine CONAIE's legitimacy in the eyes of Bank officials. Second, neither the government nor the Bank accepted that macroeconomic policy issues should be debated by indigenous or environmental groups. BIC sent a report of the meeting to CONAIE and Acción Ecológica to let them know that at some level, their visit to Washington had gotten the attention of the World Bank and the government.

One of the messages that Luis Macas and Cecilia Cherrez brought to the World Bank and IDB during their Washington visit in April 1994 was the importance of participation of the indigenous peoples and other sectors of civil society in the decision making about development in the country. Regarding the issues that most directly affected these groups, Macas articulated a need to have a tripartite discussion among the actors—the government, the banks, and representative indigenous organizations such as CONAIE. Despite the absence of such space for participation (in 1994 Ecuador had only one indigenous representative in the national Congress), the fact that the MDBs had *mechanisms* for participation was viewed by CONAIE as a potential opening for establishing dialogue. This opening occurred, however, only after CONAIE led another national civic uprising.

The Agrarian Reform Law, the IDB, and the 1994 Indigenous Uprising
Like the World Bank, the Inter-American Development Bank has been involved at both the macroeconomic and sectoral level in redesigning Ecuador's economic policy. The NGO advocacy campaign, which had focused on the World Bank and its influence in the oil sector, later targeted the IDB's efforts at agriculture sector reform that would affect land policy and indigenous rights.

As in the oil sector, Ecuador's agricultural policy reform was designed without indigenous participation. The agricultural development law was passed by the Ecuadorian Congress June 3, 1994. The new law would liberalize the land market by making collectively owned indigenous land—much of it held by usufruct rather than private title—available to private buyers. It would effectively break up the indigenous land management system and reverse the agrarian reform,[54] under which less than 50 percent of the lands of traditional landholders had been titled, with more than four thousand land conflicts still unresolved at the time the new law was passed.[55] The agricultural development law would open these lands up to the free market. The law would have also privatized water resources and put public grazing lands and forest lands used by indigenous communities up for sale. Supported by business leaders and right-wing forces in Congress, the law favored agro-industry by giving a five-year, 50 percent tax break to new agro-industries. The agriculture minister noted that "the law is good because it replaces the idea of agrarian reform with that of productivity, and landowner with that of efficient businessman."[56]

Economic Liberalization, Land Conflicts, and Indigenous Mobilization
The pro-business change in the land law was driven by macroeconomic and sectoral policies set by the IMF structural adjustment loans that required economic liberalization (including conditionality in a March 1994 IMF standby agreement), World Bank support for privatization, and in particular a then-pending loan from the IDB for agriculture sector adjustment.[57]

Throughout the period of structural adjustment, the peasant agriculture sector, which is largely indigenous, suffered from loss of income, increased prices, high interest rates, and lack of access to credit. The new policies favored large agribusiness exports, thus exacerbating land conflicts with indigenous farmers.[58] It was in this context that the IDB developed an agriculture sector loan with the Duran Ballen government during 1993 and 1994. According to an analysis by the U.S. embassy in Quito, this $80 million loan required that the government pass a law to liberalize ownership of land and resources. The embassy analysis says that "As part of its structural reform program *and to comply with IDB sectoral loan requirements,* the GOE sent its draft 'Agrarian Sector Organization Law' to congress on an urgent basis in early May."[59]

Even before this loan was developed, World Bank and IDB economic strategy documents outline broad proposed changes in agricultural policy. For example, the July 1993 *IDB: Socioeconomic Report— Ecuador* recommends restructuring the agriculture sector to limit state interference in marketing and subsidies; restructuring the research system; developing a consistent natural resource conservation and management policy; and *modernizing land ownership and use regime,* coupled with a rapid privatization process of state-owned enterprises. The report further notes that

Liberalizing the sale of land would avoid the distortions caused by the restrictions on its use and transfer, make for more efficient land use, and serve as an incentive to invest in and introduce improvements on farms. There is no case for continuing to favor community forms of tenure that have failed in Ecuador and throughout the world.[60]

Pursuing its commitment to structural adjustment, the IDB encouraged the government to pass the agricultural development law in exchange for $80 million for debt servicing, despite its recognition of the possible social side effects of the law. The 1993 *Socioeconomic Report* acknowledges that structural adjustment will have social consequences:

Accordingly, unless the reform program includes measures to improve the coverage and quality of social services, as well as emergency programs to alleviate the negative impact of the economic restructuring, the intensification of social problems could result in political conflicts that jeopardize the viability of the transformation process.[61]

The IDB was certainly not the only international actor involved in the new agricultural development law. In fact, the law—one of the most controversial to come out of the modernization process—was originally written by a conservative agricultural policy research organization, the Instituto de Estrategias Agropecuarias (IDEA), supported by USAID. IDEA's goal has been to repeal the agrarian reform law of 1963. Its draft of the new law, which it prepared for the landowner association Camara de Agricultura de la Primera Zona, promoted opening corporate access to resources, increasing private sector investments, and liberalizing land policy in favor of large agro-industrial and aquaculture interests.[62]

IDEA's policy advice fit well within the structural adjustment framework and was given a great deal of weight by the IDB in its design of the agriculture sector loan.[63] Beginning in 1992, the IDB had begun a discussion with the government about the need for reforms such as eliminating price distortions, liberalizing trade, lifting import/export restrictions, and privatizing government enterprises. They also looked at legislation regarding natural resources—including the agrarian reform law, the water law, and regulations related to irrigation and colonization. The IDB saw the agrarian reform law of 1964 as paternalistic in that the government Agrarian Reform Institute (Instituto Nacional de Reforma Agraria y Colonización, IERAC) had to give permission to farmers to sell their land, and there were too many restrictions on people who wanted to deal as private owners in the land market. IDB also viewed IERAC as corrupt and ineffective and believed that land tenure laws contributed to deforestation because of IERAC's policy of requiring that to maintain possession of land, farmers would have to clear it.[64]

By 1993, several attempts had been made to revise the agrarian law. IDEA and the Agriculture Chamber had presented legislative proposals, and in June of 1993 CONAIE introduced their own agrarian reform legislation, which called for support for food production by small farmers for local and national markets, access to credit and irrigation, and assistance in implementing ecologically sound agriculture. This proposal emerged from consultations with more than twelve hundred indigenous communities and was presented to the National Congress on June 9, 1993, by a commission of indigenous leaders while a peaceful march was conducted.[65] The Congress's president called out the police, who violently attacked the commission, including CONAIE President Luis Macas.[66] CONAIE's agriculture policy proposal was never taken seriously by either the government or the IDB.

The IDB meanwhile continued to advise the government on proposed revisions to the law, as they prepared the $80 million agriculture sector loan. With the advent of another IMF standby agreement in March 1994, the pressure to move on agriculture reform no doubt accelerated the process of bringing a draft agricultural development law to the Congress, which was passed on June 3, 1994.[67]

The 1994 Uprising

Within indigenous communities, land is not simply a means of production, but central to Indian identity and culture. Pressures for land have long caused conflicts in the Amazon, where indigenous communities are threatened by the expansion of oil development, logging, mining, and cash-crop agriculture. Recent expansion of large-scale agriculture has driven community-based farmers off of their land, provoking resistance.[68] Regarding the agricultural development law, Luis Macas has noted, "The indigenous people cannot accept a law that promotes the renewed concentration of land in the same hands as always and prohibits indigenous access. Without a place to spread out, we will have to leave and die of hunger and misery in the cities."[69]

CONAIE immediately declared the law unconstitutional, illegitimate, and "against society, removing the definition of the social use of land

and water," and claimed that it would "open up avenues for the disappearance of communal territories which are the basis and sustenance of the indigenous peoples." At a large grassroots indigenous assembly in Riobamba on June 7, it was decided to launch a "Mobilization for Life" to demand that the agricultural development law be repealed.[70]

Beginning June 14, for almost two weeks hundreds of thousands of indigenous people demonstrated against the new law, closing strategic points along the Panamerican Highway and blocking the movement of agricultural products and other goods to urban areas. Indigenous farmers refused to bring their own produce to market, and many of them occupied government buildings.

After almost a week of protests, President Sixto Durán Ballen responded with a proposal to form a commission in which the government and indigenous peoples would work together to revise the law and draw up the implementing legislation. CONAIE rejected the proposal, demanding that the agricultural development law first be repealed. The president responded with an emergency decree, giving the armed forces broad powers to put down the uprising. Several radio stations run by indigenous people were taken over by military, equipment was destroyed, and news reports or messages related to the uprising were forbidden to be broadcast. In Cañar, the central offices of the UPCCC, a federation affiliated with CONAIE, were set on fire, trapping four people inside and killing one. Violent clashes broke out between protesters, motorists, and merchants, and nine of the country's twenty-one provinces were virtually paralyzed by the uprising. In the Amazon region, indigenous communities blocked access to several oil-drilling sites and shut down production for several days. Many of the indigenous leadership had to go into hiding, and at least four indigenous people were killed.

In addition to the protests, CONAIE brought a legal complaint to the Ecuadorian Tribunal of Constitutional Guarantees, arguing that the law was unconstitutional because (*a*) it was passed without the fifteen days of debate required by the Constitution, and (*b*) it put an end to agrarian reform, which is constitutionally guaranteed (Article 51). The tribunal agreed with CONAIE's arguments and on June 23 declared the law unconstitutional. The president announced on June 24 that the tribunal's decision was invalid, and the case then moved to the Supreme Court.

NGO Solidarity

The Mobilization for Life coalition had widespread support in the country—including more than forty indigenous peoples', environmental, human rights, development, and labor organizations. The Catholic Church also generated supportive public opinion when it denounced the law and would also prove to be a key actor in the negotiations between the government, CONAIE, and agribusiness interests. CONAIE also reached out to international organizations for support: environmental and human rights activists in Washington generated action alerts for letter writing, and Native American groups broadcast news updates and solidarity messages via electronic networks.

From their prior advocacy work with the World Bank, CONAIE and Acción Ecológica had also become quite aware of the relationship between the MDBs and Durán Ballen's economic liberalization plans. Thus, when the new agricultural development law was introduced to the Congress in May, it was clear to them that the fuel for implementing the law was the IDB's agriculture sector loan, which they knew about from official documents they had received from NGOs in Washington.

Building on networks created in previous campaigns against Texaco and the World Bank, CONAIE turned to the emerging Ecuador Network, which had formed an economics task force (including BIC, Development Gap (D-Gap), and Oxfam America) to help NGOs with information about MDB loans. At the beginning of the mobilization, BIC began contacting IDB officials searching for information about the agriculture sector loan and other related loans. BIC also forwarded up-to-date information about the mobilization to IDB's External Relations Department and Indigenous Peoples Unit in order to ensure that feedback about the loan's impacts would make it to the appropriate authorities at the IDB. IDB staff made it clear that the mobilization was not going unnoticed at their Washington headquarters.

Negotiation and Compromise

The mobilization proved CONAIE's political and economic clout, and gave the indigenous leadership unprecedented political space for negotiations with the Durán Ballen government. After two weeks of intense negotiations with the president, on July 15 the government, indigenous and agribusiness leaders, and the Catholic Church agreed to a revision of the law that reflected indigenous interests. Nina Pacari, CONAIE's secretary for Agrarian and Territorial Affairs, as well as its legal counsel and spokesperson, said that the final agreement was acceptable because it reinstated several important aspects of agrarian reform: it would limit the free sale of land, eliminate private control of water resources, and recognize demographic pressure and ancestral possession as valid grounds for expropriation of land. Moreover, passed in June, Article 17 of the law specified that only agricultural businesses could hold title to farm land. The new agreement broadened land ownership rights to include individuals, families, communities, and associations, among others.[71]

IDB Response to the Mobilization

In Washington, IDB officials recognized their role in the crisis in Ecuador. In fact, the president of the IDB, Enrique Iglesias, was in Ecuador when the uprising began and acknowledged the bank's obligation to try and mitigate the effects of the law. He met with the cabinet and told them he "wanted the IDB to be part of the solution, not part of the problem."[72] President Iglesias proposed that a mission from its Multilateral Investment Fund (MIF) go to Ecuador to discuss projects that could be developed directly with indigenous peoples' organizations as a response to the issues raised by CONAIE and others about the law's impacts on indigenous communities.[73]

Discussions inside the IDB also recognized the potential environmental and social impacts of the loan. Some IDB staff working on the loan apparently had not fully agreed with the contents of the law, but once the law was introduced to the Congress, the IDB ceased to have any influence over its content.[74] Moreover, the IDB seemed to have no choice within the framework of structural adjustment but to follow through with the loan. So, rather than changing any substantial elements of the loan, the IDB's efforts turned to mitigating the social impacts and preparing support for indigenous development in the future.

Key IDB staff were sympathetic to the indigenous concerns and supported the idea of developing a parallel program that would benefit indigenous peoples directly through the MIF. It was also helpful that they had quietly worked within the context of the development of the agriculture sector loan to try to buffer its impact. IDB environmental managers asked operations to modify the conditions of the loan and to create a parallel technical cooperation component that would alleviate, to some extent, the social impacts of the agriculture sector reforms that the loan supported.[75]

In September 1994, Luis Macas returned to Washington. With the help of BIC, D-GAP, and OXFAM America, he met with the IDB project team responsible for developing the agriculture sector loan, the IDB's Indigenous Peoples Unit staff, and with President Iglesias. Whereas at the beginning of 1994 indigenous leaders had come to Washington to try and convince the MDBs they weren't a fringe group trying to overthrow the government, in September Macas received red-carpet treatment at the IDB.

Macas's main message was that although the new agricultural development law created a crisis situation, subsequent negotiations moved CONAIE to the next stage of the process, which was to build a national consensus behind policy reform that would benefit the indigenous population.

In a September 21 meeting with the agriculture sector loan team, Macas said,

Ultimately, we've managed to change some points that affected indigenous peoples, especially small and medium producers. The experience has meant that people have seen the importance of participation. This will strengthen the democratic process in Ecuador, which should carry over into other modernization efforts, for example in the area of oil, where indigenous people want to reach an agreement between the government, the companies and the peoples on whose lands oil is being exploited.[76]

He told the IDB staff that the MDBs and government of Ecuador need to ensure the participation of the indigenous peoples in all aspects of development. He said,

Indigenous people suggested modernizing the state back in 1988. We believe that the state structures should be changed, and the political system is outdated. We should have a participatory democracy in terms of economic development. When government companies go into bankruptcy, it's a problem. In privatization of state industries, however, the companies go to very low bidders. This is not a modernization process. It's a privatization process and de-investment. We are in agreement that government agencies do need to modernize, but this is not the way. When you speak of modernization, it is not a holistic development that benefits all of the people, but only some. We want the MDBs to enforce participation in this process.

Macas also emphasized that the MDBs' indigenous peoples policies should be implemented. He said, "My question to you is how will the participation of indigenous people be guaranteed?"

The IDB's perspective, as explained by the task manager, Gabriel Montes, was that the agriculture laws of the past had not changed the income of the rural population. The IDB wanted to promote agriculture laws that would be less paternalistic, more flexible regarding sales of land, and more rational regarding the use of water and the management of natural resources. Their view was that the loan would help to accomplish these things. Macas responded,

We have a different point of view. Supply and demand may seem logical to you, but it doesn't reinforce the concept of reciprocity, which is traditional. To avoid impacts on the majority, it is key to have the government, the MDBs, and the affected people who benefit or lose to work together to understand each others' vision and ideas of development. Our suggestion is to be involved at all levels; not only the indigenous peoples but all of society.

It was clear from this meeting and others during Macas's visit that there was an internal debate within the IDB, and that at least some were challenging the agriculture staff's "vision" about who the law would help. The day before this meeting, on September 20, the IDB's environmental managers had discussed the sector loan with the task manager and had agreed on several points that would modify the loan and attempt to control some of the damage. For example, the Comité del Medio Ambiente (CMA) recommended that there must be monitoring to identify and measure the loan's effects on, among other things, forests, indigenous communities, rural migration flows, distribution of land, and water rights. They insisted that loan conditions specify consultations with affected people, particularly indigenous people, in the process of developing regulations for implementation and in developing a new water law. They also recommended that those vulnerable groups affected by the new law receive support through training and a parallel technical cooperation component that would be developed to benefit the indigenous peoples directly. These recommendations were critical to influencing, at least on paper, the final agriculture sector loan proposal.

The key changes had to do with the parallel technical cooperation, focusing on training, and the MIF project, which could deal with production, land titling, capacity building, or other programs that CONAIE may want to propose. These two programs were developed through the Indigenous Peoples Unit in dialogue with CONAIE. From the IDB's perspective, the challenge was to prepare a program that would involve CONAIE in a legitimate way in mitigating the acknowledged effects of the agriculture sector loan. They also needed to ensure that the government would not oppose these programs and would recognize the legitimate leadership of CONAIE as a representative of the majority of Ecuador's indigenous population.[77]

From CONAIE's perspective, its newly acquired "seat at the table" with both the government and the IDB signaled a new period in which indigenous peoples could have a greater voice in determining the policies that affect them. The challenge would be to maintain enough momentum to use these small successes to create a larger and more permanent space for the indigenous movement in decision making about development in Ecuador.

Conclusions

What was the impact of the indigenous/NGO effort to influence the MDBs in Ecuador? In the case of the World Bank, the NGO campaign probably had no direct effect on the public enterprise reform technical assistance loan specifically, although it did improve the Bank's attitude toward CONAIE and contributed to a growing awareness inside the Bank for the need to deal with oil pollution in the Amazon. The Bank eventually dropped from the loan the component relating to the hydrocarbons sector reform because, according to Bank staff, the government of Ecuador was not ready to move ahead with privatization of Petroecuador. It *is* likely, however, that the international campaign against private and state oil companies' operations in Ecuador had an effect on the government's perception of what would be politically possible.

In terms of the Bank's increased awareness of the need to deal with oil pollution, the campaign probably had an influence on subsequent Bank-financed operations. During 1995 the Bank prepared a program for the

Environmental Technical Assistance and Mitigation project (PATRA), which in part was designed to assist the government in clarifying the various ministerial roles in environmental management in the country and to provide assistance in strengthening government agencies and the regulatory framework to "inte[grate] environmental concerns into development policy and economic activities."[78] A $4.8 million component of the project *was* to have been for "mitigation of the impacts of hydrocarbon operations in the Ecuadorian Amazon," which included a program to design cleanup operations for areas in the Oriente outside the scope of programs being financed by Texaco and the European Union. During 1995, the Amazon component of the loan changed from a mitigation project to a sustainable development project that ceased to address oil pollution because, according to World Bank project staff, the government of Ecuador was not willing to finance mitigation.[79] It is not clear why the government opposed this component of the project; given the seriousness of the problem and its high public profile in Ecuador and internationally, it would seem in their interests to pursue a pilot program with the World Bank. What is clear is that some World Bank staff recognized the importance of oil pollution mitigation as fundamental to the Bank's future involvement in the oil sector in Ecuador and were actively pursuing it within a national environmental loan. In that sense, the NGO and indigenous efforts to influence the public enterprise technical assistance loan, as well as the ongoing international campaign against Texaco, did increase World Bank sensitivity to the political consequences of involvement in the oil sector in Ecuador.

In the case of the IDB, the indigenous mobilization and international networking profoundly influenced the process and, to some extent, the content of the agriculture sector loan. After the uprising that paralyzed parts of Ecuador for almost two weeks, Bank staff began to understand the problems and concerns raised by the indigenous leadership and consequently made changes in the loan that might mitigate some of the agriculture development law's impact. Moreover, CONAIE leadership gained unprecedented access to a policy process that had previously excluded them and were ultimately able to negotiate directly with IDB staff on the terms of the parallel technical assistance.

Although the technical assistance component has yet to be implemented, another, probably more important outcome was the inclusion in the agriculture sector loan of a requirement for the participation of indigenous peoples in the elaboration of the new water law, intended to promote privatization. The passage of an "acceptable" water law is a condition for the release of the second tranche of funds from the IDB's agriculture sector loan. Again, it remains to be seen whether the participation requirement will be implemented.

As to whether the efforts to influence these loans depended on a North/South collaboration, the ability of CONAIE and Acción Ecológica to work with Northern collaborators did help them to achieve greater effectiveness in influencing the banks regarding the two loans. Both organizations had strong ties to Northern organizations long before the two loans became "issues," but their MDB advocacy goals required them to have both information and strategic support in order to target their lobbying efforts more precisely.

Activists and indigenous peoples in the South see MDBs as especially formidable institutions. Staff in the country are often not willing to meet with NGOs and are not accustomed to providing information or consulting with civil society actors. World Bank information is for the most part not available in Spanish; plus, most information that would be relevant to the NGOs and indigenous peoples is not available or accessible in any language. NGOs and indigenous groups did not understand the MDB policies regarding access to information and participation until they began to grapple with these specific loans. For them to get information, analyze it, and then use it to argue effectively with the appropriate people at the institutions took a certain amount of guidance from organizations that (a) knew how to navigate the bureaucracies, and (b) could help interpret how the banks' policies on participation and indigenous peoples could apply in the specific cases.[80] Gaining access to the institutions at critical times may have not *depended* on the assistance of Washington-based groups, but it no doubt helped that there were NGOs in Washington who could utilize relationships with Bank staff to facilitate their meetings with CONAIE and Acción Ecológica. Northern organizations also furnished the financing for timely visits to Washington for representatives from both groups to meet with Bank staff.

Probably the most important element of this North/South NGO relationship, however, was that the Northern NGOs were clearly operating as support organizations in response to the leadership of the Ecuadorian indigenous and environmental groups. Although most of the concrete information about the loans came from the Washington-based groups, the impetus for analysis of the loans, the national political processes they were part of, and subsequent NGO action clearly came from CONAIE and Acción Ecológica. The role of the Washington-based NGOs was clear: to provide the Ecuadorian groups with appropriate financial and technical support to enable them to achieve their goals. Moreover, the Washington groups presented tactical options to the Ecuadorians—for example, suggesting who to meet with at the institutions or how to approach certain discussions—but did not orchestrate their meetings or negotiate on their behalf. The broader strategic agenda was driven by CONAIE and Acción Ecológica because the circumstances derived from their immediate experiences and needs in Ecuador.

The effectiveness of the North/South collaboration ultimately depended on the fact that CONAIE and Acción Ecológica are well-organized groups that provide leadership for two parallel, but distinct, movements in Ecuador. The relationship between the organizations is formalized, and their outreach to international groups intentional. CONAIE's strength largely comes from the fact that it is a representative organization that has remained in touch with the base organizations even while it has gained national and international stature.

Finally, what impact will these efforts have in how the World Bank and the IDB apply their environmental and indigenous peoples' policies in the future? Both the World Bank and the IDB have become a great deal more accessible to CONAIE and certainly more broadminded about the indigenous issues that CONAIE raised in the course of the dialogue described in this chapter.

During the spring of 1995, Felipe Duchicela, the (now former) secretary of Indigenous and Ethnic Affairs, came to Washington to seek MDB funds to support an indigenous peoples development program that the government wanted to launch. Since then, the World Bank has been working to develop a plan for "Integrated Development for Indigenous and Black People" in response to demands expressed by indigenous peoples to strengthen their organizations, to deal with land issues, and to provide training. The initial plan is being developed using a consultative approach with indigenous organizations, NGOs, and government agencies. In fact, the Bank staff, which in the previous year had referred to CONAIE as "communistic" and accused it of trying to overthrow the government, has asked it to participate in designing the proposal. In September 1995 CONAIE and several other indigenous organizations sent a letter to the World Bank agreeing to participate in the program.

Moreover, the Bank has supported the creation of a working group— with membership from the "Comité Decenio" (or "Decade Committee," which includes CONAIE and five other indigenous organizations) and from the government—that would decide on the management of the funds. According to Bank anthropologist Jorge Uquillas, "the main idea is to increase the participation of indigenous peoples and their organizations in the development of the project."[81] If such a project gets funded by the Bank and is agreed to by CONAIE, it could prove to be the first World Bank loan in Latin America targeted specifically to, and developed primarily by, indigenous peoples. [Eds. The loan was passed in Jan. 1998]

In the meantime, a number of current MDB loans to Ecuador worth hundreds of millions of dollars form a constellation of projects and policy reforms that have consequences for the indigenous population and the environment—including structural adjustment, judicial reform, privatization, and modernization of the state. In addition, within their programs both the IDB and the World Bank have loans to support conservation of the environment and natural resources management, as well as more emergency social investment funds; the World Bank also has an Indigenous Peoples Training Project. These programs are meant to strengthen much-needed governmental and nongovernmental institutions, and to mitigate at least some of the costs of structural adjustment. The net effect of these different lending trends remains to be seen. If MDB support for more indigenous participation in the policy process ends up creating political space, then these small loans could have even greater spillover effects.

Both the World Bank and the IDB have official indigenous peoples policies that are to be applied in situations where indigenous peoples will be affected, and that require their participation in project design and implementation. Though there is hardly a program in the country that does not affect the indigenous people, the MDBs have only begun to have meaningful discussions with CONAIE.

Perhaps a broader lesson that can be derived from these experiences is the profound lack of attention that both the World Bank and the IDB give to the potential negative social and environmental consequences of macroeconomic policy and specific loans that implement that policy. To fully understand such impacts, these banks must bring civil society into the decision-making process. Although the government of Ecuador has a clear responsibility to open political spaces for civil society actors, the MDBs did not use their leverage to encourage a more participatory process until after pressured by national NGOs, international NGO's and indigenous peoples, despite indigenous peoples and environmental policies that ostensibly required them to do so.

Ultimately, CONAIE is asking that all MDB-financed programs be developed with serious participation and consultation with indigenous peoples. CONAIE leaders have stated many times in meetings with MDB staff that their objective is not to block development in Ecuador but to participate, to present their own projects and alternatives, to benefit the indigenous communities and protect the environment. Participation means that there is active involvement, from the leadership to the base of indigenous communities, at the beginning of the process right through to the end, in order to give affected people the opportunity to identify potential impacts, create their own development plans, and propose alternatives. This is something that neither the Ecuadorian government nor the Banks have brought to the development process.

Without such participation, economic policy and development programs will continue to have negative impacts on the indigenous peoples and environment of Ecuador.

Acknowledgments

The following people provided invaluable contributions to the case study. Juan Aulestia of Oxfam America assisted in the development of the case study, explained the history of oil development in Ecuador and the evolution of the oil campaign, and described the creation of the

Ecuadorian and international indigenous and environmentalist alliance. He also participated in several interviews. Cecilia Cherrez and Desider Gomez of Acción Ecologica in Ecuador, and Maria Augusta Espinoza of Red Bancos, Washington, D.C., researched and reported on environmental and social affects of structural adjustment in Ecuador. Nina Pacari of CONAIE, Ecuador, provided the author with valuable details about the uprising and about CONAIE's perspective on the MDB loans described in this chapter. Finally, the case study would not have been possible without the initial research of World Bank and IDB loans carried out by Deborah McLaren of the Bank Information Center. The author would like to thank all of these people for their contributions and friendship.

Notes

1. These eight indigenous nationalities—whose population totals between 200,000 and 250,000—are the Quichua, the Shuar and Achuar, the Huaorani, the Secoya, the Siona, the Shiwiar, and the Cofan. See Melina Selverston, "The Politics of Identity Reconstruction: Indians and Democracy in Ecuador," in Douglas Chalmers et al., eds., *The New Politics of Inequality in Latin America* (New York: Oxford University Press, 1997).

2. *Structural adjustment* is the term describing a set of economic policies imposed on developing countries by the International Monetary Fund and the World Bank as a condition for receiving financial assistance. Designed to reduce these countries' debt burdens and accelerate economic growth, structural adjustment programs typically have the following characteristics: deregulation (to encourage trade and foreign investment), promotion of exports (to earn foreign exchange), cuts in state spending (to reduce the expenditures of the state), and privatization of state enterprises (to remove the state as an economic actor and to transfer state assets to the private sector). Structural adjustment programs also devalue local currencies and increase interest rates. See Development Gap, *The Other Side of the Story: The Real Impact of World Bank and IMF Structural Adjustment Programs* (Washington, D.C.: Development Gap, 1992).

3. Petroecuador (formerly the State Petroleum Company, CEPE) handles all aspects of oil development in the country, from production and refining to transportation and export of oil and oil products. Although foreign oil companies have been allowed to operate under concessions administered by Petroecuador, the authority for overseeing oil development in the country rests with the political interests involved in the state company.

4. World Bank, *Ecuador—A Social Sector Strategy for the Nineties,* report no. 8935-EC, Country Department IV, Latin America and the Caribbean Region (Washington, D.C.: World Bank, November 28, 1990).

5. World Bank, *Ecuador: Public Sector Reforms for Growth in the Era of Declining Oil Output* (Washington, D.C.: World Bank April 1991). Between 1982 and 1988, the external debt more than tripled, from 37 percent to 122 percent of GDP.

6. *Ibid.*, p. 2.

7. Urban unemployment rose from 5.7 percent in 1980 to 13 percent in 1988, and rural to urban migration accelerated. See IDB, *Ecuador: Socioeconomic Report,* report DES-13 (Washington, D.C.: IDB, July 1993), pp. 187–88.

8. World Bank, *Ecuador: Private Sector Development Project President's Memorandum* (Washington, D.C.: May 1993), p. 5, par. 11; IDB, *Ecuador: Socioeconomic Report,* p. 204, par. 4.153, and table IV-17. See also World Bank, *Ecuador: A Social Sector Strategy for the Nineties,* report no. 8935-EC (Washington, D.C.: World Bank, November, 1990), and World Bank, *Ecuador Poverty Report,* report no. 14533 (Washington, D.C.: World Bank, November 1995).

9. *International Fund for Agricultural Development: State of World Rural Poverty: A Profile of Latin America and the Caribbean 1993;* cited in a Friends of the Earth publication, *IMF Fact Sheet.*

10. One of the early successes of CONAIE was the achievement in 1988 of an agreement with the government of President Rodrigo Borja to implement a bilingual (Spanish and Quichua) education program on a par with the state's educational system, to be funded by the government and carried out in all of the indigenous areas of the country. See Leon Zamosc, "Agrarian Protest and the Indian Movement in the Ecuadorian Highlands," *Latin American Research Review* 29, no. 3 (fall 1994): pp. 37–68, and Melina H. Selverston, "The Politics of Culture: Indigenous Peoples and the State in Ecuador," in Donna Lee Van Cott, ed., *Indigenous Peoples and Democracy in Latin America* (New York: St. Martin's Press, 1994).

11. The latter figure is an Ecuador Ministry of Education estimate, cited in Selverston, "The Politics of Culture."

12. For an overview of the indigenous rights movement in Latin America, see Alison Brysk, "Acting Globally: Indian Rights and International Politics in Latin America", in Van Cott, *Indigenous Peoples and Democracy.*

13. See Melina Selverston, "Politicized Ethnicity as Social Movement: The 1990 Indigenous Uprising in Ecuador," *Paper Series* no. 32, Institute of Latin American and Iberian Studies, Columbia University, 1993.

14. Zamosc, "Agrarian Protest."

15. The 1990 uprising has also been analyzed as a mobilization based on ethnic identity, and indeed one of the demands of CONAIE during the uprising was to change Ecuador's Constitution to recognize the country as a plurinational state. See Melina Selverston, "Politicized Ethnicity."

16. The July 1993 IDB *Socioeconomic Report* analyzed the proposed reforms and made recommendations for implementing them. This document laid an analytical and strategic foundation for reforms in both the hydrocarbons and

agriculture sectors, among others, that subsequently led to legislative reforms that were supported by MDB loans but opposed by the indigenous peoples' organizations and environmental NGOs.

17. Most of the oil development in the country takes place in the Amazon region—the Oriente—in areas occupied by indigenous groups. Between 1972 and 1990, Texaco was the dominant oil company in Ecuador, controlling more than 88 percent of the total national production of crude oil. It operated under a twenty-year contract from the state oil company, Petroecuador, in nearly 400,000 hectares of four Amazonian provinces. Texaco also constructed the Transecuadorian Pipeline System from the Amazon across the Andes to the coast. During this twenty-year period, Texaco and Petroecuador were responsible for releasing more than 17 million gallons of crude oil and 30 billion gallons of toxic waste across the Amazonian watershed, including inside the Cuyabeno Reserve. Since the early 1980s Texaco and other oil companies have operated within the reserve, and in 1990 Petroecuador illegally expanded its oil operations there. See Center for Economic and Social Rights, *Rights Violations in the Ecuadorian Amazon: The Human Consequences of Oil Development,* (New York: March 1994). See also Acción Ecológica and Rainforest Action Network, "Four Years of Struggle Against Texaco's Dark Legacy in the Ecuadorian Amazon," (San Francisco, CA, 1994).

18. One indigenous group, the Cofan, has nearly been extinguished as their lands have been invaded first by Texaco—which drilled its first well in their territory in 1967—then by colonists.

19. The efforts to influence the MDBs emerged from already established North/ South NGO networks fighting Northern multinational companies. The international campaign is in fact primarily aimed at those companies operating in the Amazon region, and only secondarily at the MDBs.

20. See Joe Kane, *Savages* (New York: Alfred A. Knopf, 1995). Kane documents the controversial role of U.S. environmental organizations in the incipient movement to resist oil expansion in the Oriente, as well as the emergence of the national and international environmental/indigenous alliance.

21. It should also be noted that a broad range of actors have been involved in the struggle to curb the widespread environmental and social impacts of oil development in Ecuador—including petroleum workers (*petroleros*), small agricultural cooperatives, colonists, unions, and other sectors of civil society.

22. During 1993 and 1994, the author had numerous communications with members of Acción Ecológica and their international supporters, through conversations, letters, and electronic mail. In addition, Cecilia Cherrez and Desider Gomez contributed to the research for this chapter.

23. The OAS petition was presented by the Sierra Club Legal Defense Fund in 1990. See "Petición presentada a la Comisión Inter-Americana de Derechos Humanos, Organización de Estados Americanos, por la Confederación de Nacionalidades Indígenas de la Amazonia Ecuadoriana (CONFENIAE) a favor de la población Huaorani contra Ecuador" (San Francisco, CA, 1990).

24. World Bank, *Ecuador: Emergency Petroleum Reconstruction Project,* project completion report, loan 2803-EC (Washington, D.C.: World Bank, December 1993).

25. During this period, Moi, a Huaorani from the Amazon, met with the World Bank's vice president for Latin America and Bank staff who were preparing the loan. Accompanied by Juan Aulestia of Oxfam America, Moi presented slides of the oil pollution in the Napo region and asking that the Bank not allow this to happen in Huaorani territory.

26. The Bank Information Center is an NGO clearinghouse for information on MDB projects and policies. The author was director of the Latin America and Caribbean Program at the time.

27. The World Bank classifies all of its loans for their potential impacts on the environment. A Category C loan is "unlikely to have environmental impacts," whereas a Category A loan "is likely to have significant adverse impacts that may be sensitive, irreversible and diverse." World Bank, "Operational Directive 4.01, Environmental Assessment."

28. The preceding two paragraphs are based on a conversation between the author and a member of the World Bank staff, Washington, D.C., August 1993.

29. Danielle Berthelot, Trade, Finance, and Private Sector Development Division, LAC Region IV, letter to Kay Treakle, Bank Information Center, 26 August, 1993.

30. See Judith Kimmerling, *Amazon Crude* (New York: Natural Resources Defense Council, 1991) for documentation of the environmental impacts of oil. See also Center for Economic and Social Rights, *Rights Violations,* for a detailed critique of the lack of environmental protection, government enforcement of environmental laws, and impacts of oil pollution on public health in indigenous communities. See also Joe Kane's "Letter from the Amazon: With Spears from All Sides," *The New Yorker,* 27 September 1993, and his *Savages,* both of which describe the impacts of oil on the Huaorani.

31. In November 1994, the World Bank approved a $16 million technical assistance project for modernization of the state public sector reform, which was intended to implement at least in part the law of modernization in Ecuador. This law was also passed as part of the government's response to structural adjustment conditions imposed by the IMF.

32. These aspects of the law were also contained within the recommendations in the IDB's 1993 *Socioeconomic Report.*

33. The Seventh Round of oil leases was launched by the government on January 24, 1994. It would open thirteen oil blocks (ten in the Amazon and three on the coast) equivalent to almost 3 million hectares. It would essentially double the area in the Amazon open to oil exploration and exploitation.

34. Acción Ecológica and CONAIE also had a long-standing demand regarding Texaco's operations in Ecuador: the company should pay damages for the severe economic, social, and environmental damage it caused during its twenty years of

operations in the Oriente, and it should clean up contaminated sites and seal up areas of risk. During this period, two separate lawsuits were brought against Texaco in U.S. courts, launched by indigenous communities affected by the company and by a U.S. lawyer who had been working with indigenous and environmental groups in Ecuador. See "Ecuadorean Indians Suing Texaco," *New York Times,* 4 November 1994, and "Texaco Has Left Ecuador But Its Impact Remains," *The Christian Science Monitor,* 25 March 1994. And as noted, Ecuadorian, international, and U.S. organizations led by the Rainforest Action Network also launched an active boycott of Texaco.

35. The loan also covered telecommunications and energy, in addition to hydrocarbons.

36. The letter was signed by Luis Macas, president of CONAIE; Vicente Pólit, president of the Ecuadorian Committee in Defense of Nature and the Environment (CEDENMA); Valerio Grefa, general coordinator of COICA; Angel Zamarenda, president of the Confederation of Indigenous Nationalities of the Amazon (CONFENIAE); and Cecilia Cherrez, president of Acción Ecológica.

37. Members of the European Parliament, letter to the president of the Republic of Ecuador, Mr. Sixto Durán Ballen, and members of the Congress of Ecuador, 19 November 1993.

38. In an interview with a Bank staff person (Washington, D.C., 1994) knowledgeable about the situation in Ecuador, BIC was told that the Bank "did not write the law" but had been "very involved in discussions, giving advice, and so on" with a group of congressional members, the Ministry of Energy and Mines, and "representatives from oil companies."

39. Cecilia Cherrez, president of Acción Ecológica, letter to Paul Meo, 27 December 1993.

40. The three Amazonian protected areas are the Cuyabeno Reserve, Yasuni National Park, and the Cayambe-Coca Reserve.

41. See Judith Kimmerling, "The Environmental Audit of Texaco's Amazon Oil Fields: Environmental Justice or Business as Usual?" *Harvard Human Rights Journal* 7 (spring 1994), pp. 199–224.

42. These included a requirement that affected indigenous communities could set the terms for participation in environmental impact studies, management, and contingency plans and approve them. The indigenous communities would also reserve the right to suspend activities if they find that the projects are presenting a social, cultural, or physical risk. The minister did not agree to the demand for a fifteen-year moratorium on oil leasing, however. See "Indigenous Representatives, Environmental Activists Hold Meeting with Ecuador's Minister of Energy and Mines," summary of the meeting, prepared by Glenn Switkes, Rainforest Action Network, 24 January 1994.

43. The World Bank is supporting several efforts to strengthen environmental regulations and institutions in the country. For example, the Global Environment Facility is supporting a program for biodiversity protection; technical assistance

has been given to strengthen the CAAM; and the World Bank approved the PATRA loan in April 1996.

44. World Bank, *Ecuador: Public Expenditure Review: Changing the Role of the State,* report no. 10541-EC (Washington, D.C.: World Bank, August 1993).

45. World Bank, "Strengthening the Delivery of Basic Public Services," in *Ecuador: Private Sector Development Project,* staff appraisal, report no. 10027-EC (Washington, D.C.: World Bank, May 1995).

46. Author's notes. Meeting between NGOs and World Bank Vice President for Environmentally Sustainable Development Ismail Serageldin, 23 February 1994.

47. Meo never did follow up on this request and was eventually transferred to another area of the Bank.

48. OPIP had been awarded one of the largest areas of indigenous territory legally recognized by the Ecuadorian government after a massive march from Pastaza to Quito in 1992.

49. The Ecuador Network was an informal group of American NGOs— including the Bank Information Center, Community Action/International Alliance (CAIA), Oxfam America, Development Gap, South and Meso-American Indian Information Center (SAIIC/CONIC), Rainforest Action Network, Human Rights Watch, the Center for Economic and Social Rights, Institute for Policy Studies, the Inter-American Dialogue, Center for Democracy, and the Washington Office on Latin America. The author was present for most of the meetings referred to in the text. The information was derived from notes taken at these meetings and follow-up conversations with the participants.

50. BIC's research found that as of July 1994 more than $400 million in loans were being executed or planned for Ecuador by both the IDB and the World Bank.

51. Author's notes from a meeting between Luis Macas and Cecelia Cherrez, and Global Environmental Facility project manager Cesar Plaza, anthropologist Jorge Uquillas (LATEN), Fernando Manibog (environment), and Marcela Cartagena (alternate Executive Director for Ecuador), 24 April 1994.

52. Author's notes from a meeting on Ecuador government-civil society relations, held at the World Bank in early May, 1994.

53. The $120 million Emergency Social Investment Fund of Ecuador was hardly an example of participatory development. Luis Macas criticized the program for creating political conflicts by promoting party politics in the distribution of funds and by financing projects that weren't helpful. For example, he told Bank staff that not all communities wanted latrines; some had other priorities: "You can save a lot if the community participates." He was also concerned about the lack of participation by people marginalized by the modernization process. He told the Bank's Social Investment Fund manager that there was a real need for partnerships to be built between the communities, the government, and the Bank, and that projects should be developed by indigenous communities rather than the private sector. Above all, the fund should be transparent and accountable, but in

Ecuador, it was neither. For information about the World Bank and participation in Ecuador, see Thomas F. Carroll, "Ecuador Participation Study (PRONADER Project): Some Preliminary Findings," a World Bank working paper from the Workshop on Participatory Development, 17–20 May 1994, Washington, D.C.

54. Ecuador's first agrarian reform law was passed in 1963, and the second in 1973. Originally, land reform legislation was intended to eliminate an anachronistic labor system (*huasipungo*) whereby indigenous peasants gave free labor to haciendas in exchange for marginal land for producing subsistence crops. The 1963 law allowed for some expropriations of land by indigenous communities. The 1973 law promoted modernization of agricultural production and reinforced the state's ability to appropriate land if it were considered underproductive. Neither the 1963 nor the 1973 legislation resulted in substantial redistribution of land. See Tanya Korovkin, "Indians, Peasants, and the State: The Growth of a Community Movement in the Ecuadorian Andes," Center for Research on Latin America and the Caribbean, York University, *Occasional Paper,* no. 3; and William Waters, "The Agrarian Reform Debate and Indigenous Organization in Ecuador," paper presented at the Latin America Studies Association, 28–30 September 1995, Washington, D.C.

55. According to CONAIE, as cited in Jennifer Collins, "Indigenous Protests Sweep Ecuador," *Latinamerica Press* 26, no. 24 June (1994).

56. Quoted in the *Latin American Weekly Report,* 7 July 1994, p. 290.

57. Peter Romero, U.S. ambassador to Ecuador, conversation with the author, in a meeting in Washington, D.C., with U.S. NGOs, 15 July 1994.

58. See Zamosc, "Agrarian Protest," emphasis added.

59. U.S. State Department, unclassified cable, page 5, number 4.

60. IDB, *Socioeconomic Report,* par. 1.52–1.55 and 3.129.

61. *Ibid.,* par. 1.33: The report anticipates social problems:

> Implementation of economic reforms—with the reduction of certain public spending programs, the elimination of redundant public employment, the streamlining of management in the public enterprises, the tackling of international competition by inefficient productive enterprise, and the alignment of domestic with world prices—will probably worsen to some degree the level of well-being of broad sectors of society, at least temporarily. In other words the adjustment process will probably lead for some time to a heightening of social problems.

> Moreover, although the final loan document for the Agriculture Sector Loan (EC-0048) notes that laws have to be written to carry out the program and that "fundamental program decisions relating to land, water, pricing and marketing could meet with opposition from private interest groups and organized political groups which could hinder implementation," it also suggests that to minimize risk, the "Bank will insist on the use of mechanisms to consult with the interested parties on the design of all draft legislation contemplated in the program." This was probably written after the uprising.

62. See Waters, "The Agrarian Reform Debate." See also Douglas Southgate and Morris Whitaker, *Development and the Environment: Ecuador's Policy Crisis* (June 1992).

63. IDB staff, conversations with author, May 1995.

64. Another factor driving deforestation is that unequal land ownership has increased pressure on highland Indians to join other peasant colonizers invading the lowland Amazon rainforests. See also Tanya Korovkin, op. cit.

65. Collins, "Indigenous Protests."

66. Selverston, "Politicized Ethnicity."

67. In May, President Durán Ballen submitted the first draft of the IDEA-drafted agrarian law, which Congress initially rejected. It was then rewritten (with apparently stronger environmental protection) by the Social Christian majority party. Despite the attempts by CONAIE to introduce changes in the draft and insert its own recommendations, the law was passed June 3, 1994 and signed by the president June 13.

68. Promoted by U.S. AID and supported by a loan from the IDB, nontraditional export crops—including flowers, fruits, and vegetables—have encroached onto lands once farmed by indigenous communities. For more information about nontraditional export crops, see Lori Ann Thrupp, *Bittersweet Harvests for Global Supermarkets: Challenges in Latin America's Agricultural Export Boom* (Washington, D.C.: World Resources Institute, 1995).

69. Quoted in Collins, "Indigenous Protests."

70. Mandate of the Mobilization for Life, adopted at the Extraordinary Assembly of the Confederation of Indigenous Nationalities of Ecuador, CONAIE. (As reprinted in Sally Burch, "Indigenous Affairs," *Noticias Breves,* [April/May/June 1994].)

71. InterPress Third World News Agency, July 15, 1994. See also Waters, "The Agrarian Reform Debate."

72. As explained to Luis Macas at a meeting with President Iglesias of the IDB in September 1994 in Washington, D.C.

73. The Multilateral Investment Fund (MIF) is a separate fund administered by the IDB to provide financial assistance to improve the region's market economies. The MIF has three "windows": a Technical Cooperation Facility, a Human Resources Development Fund, and a Small Enterprise Development Facility. The MIF can directly finance NGO and community development projects. In 1994, the MIF financed twenty-nine projects worth $64 million.

74. IDB staff, informal conversations with the author, Washington, D.C., June, July 1994.

75. IDB staff, confidential discussions with the author; also from internal Bank documents.

76. Author's notes from a meeting between Luis Macas and officials at the Inter-American Development Bank, 21 September 1994.

77. There was a lot of discussion within the IDB about CONAIE's legitimacy as representatives of indigenous communities. CONAIE estimates that its membership (through all of its regional and base organizations) was about 70 percent of the indigenous population of Ecuador.

78. World Bank, "Environmental Technical Assistance and Mitigation Project," project information document, 15 March 1995.

79. World Bank Technical Annex, *Ecuador: Environmental Management Technical Assistance Project,* report no. T-6716-EC (Washington, D.C.: World Bank, March 1996).

80. Although advocacy groups often allege links between macroeconomic programs, sectoral loans, and changes in a country's legislation, in this particular case international allies have provided hard evidence for such links, thus strengthening the bargaining position of NGOs by highlighting conflicts with established MDB indigenous and environmental policies.

81. Conversation with the author.

III

Bank Policies

8

Development Policy, Development Protest: The World Bank, Indigenous Peoples, and NGOs

Andrew Gray

The World Bank's policy on indigenous peoples has undergone several revisions since it was first produced in 1982. These shifts have involved both positive and negative elements, reflecting the ongoing political ebb and flow between the Bank's internal actors' self-image and the concerns of the world outside. Discussions about indigenous policy produce tensions that arise from conflicts between the Bank's approach to promoting economic growth and poverty alleviation, on one hand, and the need to address its projects' human and ecological costs, on the other.

Within the Bank these differing perspectives take the form of disagreements between the staff responsible for moving projects along and those responsible for ensuring compliance with social and environmental mandates.[1] Meanwhile, when the Bank has to deal with external criticisms, these internal Bank tensions usually become transformed into a publicly united defense against the persistent protests of nongovernmental organizations (NGOs) and indigenous peoples. The World Bank has responded with an indigenous peoples policy whose implementation, in practice, remains uneven and largely dependent on sustained, external vigilance and advocacy.

This chapter looks at the shifts in the World Bank's policy on indigenous peoples, the dynamics of the coalitions among NGOs and indigenous peoples' organizations, and the interactions between these networks and Bank policy changes. An estimated 250 million indigenous people throughout the world are descended from the inhabitants of various regions before they became nation-states. They are often dispossessed of their lands and discriminated against culturally; their political institutions go unrecognized.

The World Bank has helped governments all over the globe to "develop" resources located on indigenous territories. Projects affecting indigenous peoples range from the extractive work of oil, mining, and logging companies, to the impact of hydroelectric dams flooding their lands, or to state-sponsored models of forest colonization where poor migrants from cities or impoverished areas are encouraged to "develop" indigenous territory. These projects are by no means always funded by multilateral development banks (MDBs), but international institutions such as the World Bank attract considerable attention because they are visible proponents of economic development and have the means to prevent the adverse impact of projects on indigenous peoples.

Policy Development and Indigenous Protest

The Bank's development of an indigenous peoples policy emerged against the background of a constant stream of protests against the harmful effects of specific projects. Such protests have occasionally caused Bank staff to rethink specific projects and policies, but more often the campaigns have provided an enabling environment for subsequent change. The contents of policy documents have been influenced to a considerable extent by close cooperation with the International Labor Organization (ILO) and in particular by the provisions for indigenous peoples' rights in ILO Convention 107 and more recently in its revised version, Convention 169.

The World Bank produced its first policy establishing standards for development in the territories of indigenous and tribal peoples in 1982, after several years of preparatory work. Since the 1960s, an indigenous movement had been growing throughout the world, spawning a myriad of local, national, and international organizations. Indigenous peoples have been demanding more influence over their lives and gaining greater attention. In 1972 a United Nations Special Study was undertaken on the "Problem of Discrimination against Indigenous Populations," and in 1977 indigenous representatives crossed the threshold of the United Nations in Geneva for the first NGO Conference on Indigenous Peoples.

The indigenous movement received its main impetus in the 1960s from Canada, the United States, Australia, and Scandinavia, spreading throughout Central and South America and the Arctic in the 1970s, and throughout Asia and the Pacific in the 1980s. In recent years indigenous peoples in Africa and Russia have become strong proponents of their rights. Among the first indigenous peoples in Asia to assert their rights were the peoples of the Cordillera in the Philippines: the indigenous resistance to a World Bank–financed hydroelectric project in the Philippines' Chico River Basin placed their movement on the international stage for the first time. This case is important because it shows how project-specific protests can raise broader concerns that in turn can influence policy—in this case, the Bank's guidelines for indigenous peoples and forced resettlement.

The Chico River Dam Project was linked to then–World Bank president Robert MacNamara's highly technocratic approach to poverty alleviation. He made support for top-down development in the Philippines a priority, lending $2.6 billion through sixty-one projects between 1973 and 1981, following martial law in 1972.[2] The Chico dams were to be financed under World Bank power sector loans, which were intended to support the production of electricity in the wake of the oil crisis. The indigenous Kalinga and Bontoc peoples did not learn of the project until a year later, in 1974, when survey teams entered their territories. The plans would have flooded 2,753 hectares of ancient rice terraces, affecting ninety thousand indigenous people. Official petitions were ignored by the government, and in May 1975 the "Bodong" Peace Pact was signed by 150 indigenous leaders, prohibiting local people from working on the dams and boycotting all activities supporting their construction.[3]

The National Power Corporation postponed dam construction while the government's indigenous affairs agency attempted to divide and repress resistance. Armed clashes between the local peoples and the government forces followed. By 1976 the whole area was militarized and the New People's Army (NPA)—the armed wing of the Communist Party—offered the Bontoc and Kalinga peoples its support. At an International Monetary Fund (IMF) conference in Manila in 1977, protesters forced Robert MacNamara to say that "no funding of projects would take place in the face of continued opposition from the people."[4] From this point on, the Bank became more cautious in its treatment of the project. Gradually, the project was reframed, discussion of large dams began to disappear, and statements from Bank officials emphasized its support for the irrigation program downstream. Local campaigners were not convinced. Relocation surveys were taking place in the area, and they also knew that for the irrigation to succeed large dams would have to be built.

For the next three years opposition to the Chico dams increased, and the NPA armed and trained local militia units to defend their territories: "By 1980, the Chico Valley had become a virtual war zone."[5] The World Bank by this time acknowledged the opposition to the project and withdrew its support from the Chico dams, transferring the funds to other areas of the power sector loan. Without the guaranteed funding and in the light of unrest and the international campaign against the dams, the government postponed the Chico River Dam Project indefinitely in 1981. As Walden Bello and his colleagues put it, "It was a silent retreat, but this did not detract from the fact that the Bontoc and Kalinga had accomplished something exceedingly rare in the Third World: the Bank's withdrawal in the face of popular resistance."[6]

The Chico protest showed some of the limits to the power of the usually invincible partnership between the World Bank and borrowing governments. This indigenous mass movement set a precedent—for the first time blocking a project before it was built and before the World Bank even had a policy on indigenous peoples.

The First World Bank Policy on Indigenous Peoples

In 1980, the World Bank was considering support for road projects in the Brazilian Amazon, in particular the Polonoroeste project, which severely threatened the indigenous peoples of Rondônia. On the basis of this concern, Bank staff drew up "a set of guidelines for the World Bank to follow in situations where projects they funded threatened to infringe the rights of residual ethnic minorities."[7] The first World Bank policy guidelines were therefore drawn up with the aim of mitigating the effects of the Bank's colonization schemes in the Amazon.[8]

The idea of guidelines had arisen from a climate within the Bank in which policies were considered to be a useful means of organizing frameworks for projects. For about six months before the policy guidelines appeared, the Bank held meetings with NGOs and experts concerned with indigenous rights, such as the Boston-based NGO Cultural Survival and the U.S. section of Survival International. The consultation took place largely with members of the Bank's in-house "sociology group," which since 1977 had held seminars on questions of social and environmental interest to the Bank.[9]

In 1981, the World Bank published the document entitled *Economic Development and Tribal Peoples: Human Ecologic Considerations*. This report, written by Robert Goodland, originally appeared with a blue cover, which meant it was approved by the board of the World Bank, and was accompanied by a press release that identified it as the "World Bank's policies for projects that may affect tribal people."[10] The color code in the World Bank is significant as an indicator of the status of documents and reports. Apparently, the color of this publication was an administrative error. A document originally launched as an outline of World Bank policies toward tribal peoples, and called *Tribal Peoples and Economic Development: Human Ecologic Considerations*, reappeared in May of the following year with a red cover—signifying a change in status of the document to an "orientation paper."[11]

These changes created uncertainty as to whether this new version of the document constituted Bank policy or not. Even though the red cover signified a drop in the document's strength, its foreward says, "By taking into account the policies put forward in this paper, the bank has reached a consensus on appropriate procedures to ensure the survival of tribal peoples and to assist with their development."[12]

In 1984, a Spanish version said that the document was unofficial, and in 1986, at the Committee of Experts on the Revision of ILO Convention 107, the World Bank representative stated that the real policy was contained in a confidential document that was not yet available to the public. Yet the following year, World Bank President Barber Conable sent a copy of the 1982 document to Amazonian indigenous representatives as an example of Bank policy toward them.[13]

The confusion is centered around the differences between the open policy report and the confidential policy, which was made public by the International Work Group for Indigenous Affairs in its 1986 *Yearbook*. This internal policy directive was issued in February 1982 (exactly midway between the two versions of the policy document). Entitled *Tribal*

People in Bank-Financed Projects and classified as Operational Manual
Statement OMS 2.34, this report differed in several ways from the public
policy statement.[14]

The 1981 document is the strongest World Bank document dealing
with indigenous peoples to date. The first chapter identifies tribal peoples
and makes the following statement, which encapsulates a tension that
runs throughout the paper:

The Bank's policy is, therefore, to assist with development projects that do not
involve unnecessary or avoidable encroachment onto territories used or occupied
by tribal groups. Similarly, the Bank will not support projects on tribal lands, or
that will affect tribal lands, unless the tribal society is in agreement with the
objectives of the project, as they affect the tribe, and unless it is assured that the
borrower has the capability of implementing effective measures to safeguard
tribal populations and their lands against any harmful side effects resulting from
the project.[15]

Although the document acknowledges indigenous peoples' right to
veto projects in their territories, it also considers "development" to be
an inevitable process that requires mitigating measures such as the protection of tribal areas, the provision of adequate services, the maintenance of cultural integrity, and the establishment of fora for indigenous
peoples to make their views known.

Tribal peoples (in later documents the term *indigenous* is used inter-
changeably) are placed on a "continuum of acculturation" ranging from
isolated peoples to peoples fully integrated into the state.[16] In chapter 2,
the policy paper advocates an intermediate position between these alternatives to ensure the survival of indigenous peoples. The resolution
comes through a notion of self-determination that juxtaposes two conflicting ideas:

Such a policy of self-determination emphasizes the choice of tribal groups to their
own way of life and seeks, therefore, to minimize the imposition of different social or economic systems until such time as the tribal society is sufficiently robust
and resilient to tolerate the effects of change.[17]

Although the document advocates self-determination, it also assumes
that integration of indigenous peoples into the wider society is inevitable.
Policy should therefore aim to mitigate the effects of development, rather
than provide alternatives. The policy document suggests that tribal
peoples' rights to land, ethnic identity, and cultural autonomy should

be protected and that interim safeguards be provided to enable tribes to deal with outside influences and to shield them from competition for resources on their own lands.

The policy document was criticized on various counts. John Bodley's critique questions four assumptions:

1. The view that tribal cultures will either acculturate or disappear. Bodley argues that incorporation into national economies is the result of "the expansionist policies of industrial states, not an inevitable process initiated by tribal cultures."

2. The view that the problems with development are ultimately technical and can therefore be mitigated with sound planning. Indigenous peoples are too often impoverished by development, however.

3. The Bank's weak endorsement of "cultural autonomy" and "local political sovereignty," which allow real control over territories and access to resources.

4. The assumption that indigenous peoples should become ethnic minorities, rejecting political sovereignty and control over economy and culture.[18]

Bodley's criticism of the published policy looks at the contradictions within the document: control over development, territorial rights, and self-determination versus inevitable acculturation, mitigation of development encroachment, and survival through cultural autonomy. He asserts that "the World Bank tribal policy is clearly to accommodate tribal peoples to national development goals while minimizing deleterious side effects."[19]

Bodley's concern over the implications of the World Bank's tribal peoples policy was justified by the subsequent revelation that the weaker internal OMS 2.34 was in fact the real policy. OMS 2.34 makes no reference to self-determination or the right of indigenous peoples to veto projects on their territories.[20] It assumes that development projects are inevitable and advocates mitigating measures (land recognition, social services, cultural integrity, and participation through fora) to delay integration and acculturation until indigenous peoples are ready for it—an approach broadly similar to the integrationist orientation of ILO's Convention 107 on tribal and indigenous populations.[21] This connection between the ILO and the Bank appears at several points in the history of the development of the indigenous peoples policy.

The policy can be interpreted in two ways. Some Bank specialists argue that the policy arose from a genuine internal concern that its development programs were causing indigenous peoples harm and that their rights and well-being should be respected. However, from Bodley's perspective, the policy can also be interpreted as providing guidelines to enable development projects to encroach on indigenous territories and resources with as little trouble as possible.

Ultimately, the Bank's policy on indigenous peoples did not lessen the criticism of projects. In the early 1980s, opposition to the World Bank gradually spread throughout the world. Several different strands of protest converged. Support organizations—such as Cultural Survival, the International Work Group for Indigenous Affairs (IWGIA), and Survival International—were promoting indigenous rights while grassroots indigenous organizations and NGOs in the Philippines, Brazil, and India were protesting directly against World Bank–funded encroachment on their lives.

Coupled with this discontent was new attention from the large environmental lobby, which closely scrutinized the World Bank's minuscule Office of Environmental and Scientific Affairs and found it wanting. After strong lobbying by the National Wildlife Federation, the Natural Resources Defense Council, and the Environmental Policy Institute in June, 1983, the Congressional House Subcommittee on International Development Institutions and Finance held two days of hearings on social and environmental problems resulting from World Bank projects.[22]

At this hearing, criticisms of the World Bank's published policy on tribal peoples were presented, as well as substantial evidence of World Bank encroachment from environmental NGOs and indigenous organizations.

Campaigning against the Bank

The World Bank is a formidable opponent. Inevitably, campaigners encounter difficulties in dealing with such a large institution, even though it may be sensitive to criticism and bad publicity.[23] The World Bank has an army of technical experts at its command and access to masses of information. Unless a campaign is based on solid research and firm grounds of complaint, the Bank will speedily demolish it. For this reason, campaigns depend on the quality of their information.

No Northern advocacy NGO can operate credibly without detailed information from the affected area and a clear understanding of the implications of a project for local peoples, which means that an alliance between North and South is essential. On the other hand, Southern NGOs and indigenous organizations lack access to World Bank information about the project, which is where Northern NGOs can help because of their ongoing Bank monitoring.

Washington-based environmental NGOs have demonstrated a talent for gathering information and, by gaining access to Bank material, have thus been able to argue against projects on the Bank's own terms. However, dealing with Bank reports is not always easy. The Bank, governments, and business concerns assert their power through a blend of secrecy and the preparation of a plethora of prolix documentation arguing policy positions that revolve around an issue rather than directly address it. As Cheryl Payer says in her critical analysis of the World Bank, "A great deal of what the Bank publishes is self-serving propaganda, much of the best is research that has little or nothing to do with the actual projects that absorb several billion dollars yearly."[24]

Project documents are often either devoid of material or have a welter of overwhelming technical detail that is difficult for NGOs to analyze, particularly those with little knowledge of international bureaucracy. In this context, specialized NGOs such as the Bank Information Center (BIC) play a key role in broadening not only access to, but also understanding of official project information.

After years of NGO lobbying for making information more accessible, the Bank's information disclosure policy was established in January 1994, which put many formerly confidential project documents into the public domain. Much crucial information remains confidential, however (see also Udall's chapter in this volume).

Once project information is in hand, campaigners have to demonstrate that their argument is the most reasonable. The Bank at all times tries to hold the monopoly over reasonable argument when confronting outsiders and uses several tactics to achieve this. In the first place, published World Bank studies strive to present their findings as a reasonable person's balanced position.[25] A "balanced" position, however, can be a methodological sleight of hand if "extreme" positions (held by critics of the project) are created to contrast with the desired "balanced" conclusion.

A second tactic consists of the Bank's argument that "worst examples" should be balanced by "best examples." Most Bank reports foreground "successful" aspects of its work, downplaying the negative outcomes. For campaigners, this tactic is frustrating because each Bank report needs a skeptical reading to interpret what is actually taking place. The result is that campaigners who are trying to ensure accountability to those who are affected detrimentally by Bank projects need to emphasize the worst examples in order to gain more information, secure reform within the Bank, and change or cancel problematic projects.

The Bank's third tactic stresses the counterfactual—"what would have happened had development program not been followed?"[26] Clearly, this conjectural history does not prove anything, as one could just as easily argue that local peoples would have been better off without the interference. The Bank frequently acts as a catalyst for development and is not simply a passive purveyor of loans. When it approves a project, it not only provides money but policy guidance, political legitimacy, technical support, access to private funding, and other benefits.

The counterfactual argument often depends on the assumption that governments would necessarily go ahead with projects regardless of whether the Bank was involved. Three of the project campaigns reviewed below—the Narmada Dam, the Indonesian transmigration program, and Amazon roads—were all continued by their national governments after the Bank withdrew—which might seem to support the Bank's counterfactual assumption. One could argue, however, that these national governments were able to proceed with such controversial projects in part because the original World Bank loans permitted them to raise sufficient capital to continue the projects even after the Bank had been pressured to disengage.

There is no definitively "reasoned" position when assessing the relationship between the Bank and indigenous peoples because the power relationship is so imbalanced. A "balanced perspective" in the Bank's sense is therefore bound to be different from the views of indigenous peoples. Notions of "best practices" and "worst practices" are contested by both sides, yet the power relationship is sharply unequal.

Assessing campaign impact is difficult because many dimensions of Bank decision making remain unknown. For example, little or nothing is known of projects that are canceled before information about them becomes public domain. When projects are changed as a result of complaints, it is also easy for the Bank to point to internal factors as causes for the changes in order to limit the credit gained by NGOs and popular organizations. Nevertheless, several extensive advocacy campaigns over the last fifteen years have had an effect on the Bank, providing external pressures that have influenced the parallel process of internal policy formation and institutional change.

Environmental NGOs in Alliance: Polonoroeste

In 1983 the U.S. Congress heard an account of the experience of an anthropologist who, as a consultant for the World Bank in 1980, had warned that Brazil's huge road-paving and forest colonization project known as Polonoroeste would threaten the lives and welfare of the indigenous Nambiquara and other peoples living in the affected area. His warnings, however, went unheeded.[27] Nevertheless, as was noted earlier, the concern over Polonoroeste at its initial stages was sufficient to have a direct influence on the formation of the first indigenous peoples policy in the Bank.

In 1982 the World Bank provided $457 million as one-third of the cost of paving the Cuiaba–Porto Velho road, as well as support for colonization, health, and land regulation projects.[28] Ignoring the loan conditions regarding the protection of the rights of the indigenous inhabitants, the government quickly paved the road. Previous unpaved roads had endangered the Indians and the environment, but they were used only in the dry season. The paved BR-364 doubled the population of the region in less than a decade,[29] a growth that had a devastating effect on the environment when 24 percent of the surface area was lost to colonization. The indigenous peoples of the area were directly affected because the government and its agency for indigenous peoples, the

National Indian Foundation (FUNAI), were neither willing nor able to implement the required mitigating measures.

The highway cut Nambiquara lands in half, and the promised demarcation of their lands was blocked by big agricultural companies. The indigenous peoples of the Aripuana Park benefited from the titling of their lands during the initial period, but the process was poorly carried out, so invasions continued.[30] The Lourdes area of the Gavino and Arara was so badly demarcated that by 1984 six hundred settlers were occupying about one-third of the indigenous area.[31] Colonization around the highway was particularly threatening to the Uru-Eu-Wau-Wau because small, neighboring nonindigenous settlements turned into boom towns as a result of the colonization scheme. In their 1987 report, C. Junquera and B. Mindlin comment on the absence of any real support for indigenous peoples:

According to a contractual clause, the Brazilian government ought to give 26 million dollars over five years to the Indians of the area affected by the highway. Up until 1986, about 13 million dollars had been spent but the Indians had felt little effect. It is quite apparent that the meager resources which did arrive in the communities were wasted and did not stimulate indigenous autonomy.[32]

Polonoroeste's disastrous effects on the environment and indigenous peoples provided the impetus for a major campaign against the World Bank. Environmental NGOs gathered information from researchers who had been in Brazil and from Brazilian NGOs such as the Ecumenical Documentation and Information Center (CEDI) and the Indigenist Missionary Council (CIMI). They lobbied successfully for a hearing before the U.S. Congressional Subcommittee on Agricultural Research and the Environment, which was attended by the Brazilian environmentalist Jose Lutzenburger and representatives of Brazilian NGOs that supported indigenous peoples' rights. This alliance of environmental and pro-indigenous organizations from the North and indigenous peoples' organizations from Brazil continued to work together throughout the 1980s.

In 1984, the World Bank's response to the NGO campaign emphasized how carefully the project had been planned and stated that the Brazilian government would take effective action. In 1985, with support from Senator Robert Kasten, who chaired the Senate Appropriations

Subcommittee on Foreign Operations, the Washington environmentalists met with World Bank president Tom Clausen. And in March of that year the World Bank suspended disbursements of the Polonoroeste loans.

It was an extraordinary double precedent: for the first time, the Bank was forced to account to outside NGOs and a legislator from a member country for the environmental and social impacts of a lending program; also for the first time, a public international financial institution had halted disbursements on a loan for environmental reasons.[33]

The campaign had been a success from the NGOs' point of view. They had achieved a breakthrough in influencing the World Bank and in coordinating their protests internationally. However, the suspension of the project proved only temporary. Before it was officially completed in 1987, World Bank documents reported corruption in FUNAI, land invasions, and severe health problems in indigenous areas.[34]

The alliance among the NGOs of the North brought together different political streams and strategies of protest. However, two elements were missing from the campaign in Brazil: (1) the presence of indigenous organizations themselves and (2) any interest in or understanding of indigenous peoples' rights on the part of the state.

The indigenous rights NGOs followed two distinct approaches. In the early 1980s, some considered that devastating projects could be ameliorated or prevented by placing knowledgeable and committed people within the Bank—an approach taken by Cultural Survival, which had been a member of the Bank's informal "sociology group" for several years and had a positive relationship with Bank staff.[35] In their view, World Bank projects were inevitable, and NGOs could ameliorate the dangers as much as possible by ensuring that subprojects addressed indigenous needs.[36]

Survival International and IWGIA, in contrast, took a skeptical view of the World Bank's desire and capacity to implement its policy guidelines. Furthermore, projects such as Polonoroeste were so appalling that the campaign's aim, according to IWGIA, was to try to stop them completely. Nevertheless, during the period of the campaign, both approaches coexisted within the alliance.[37] Bodley compares these positions in his definition of organizations that want to "place greater emphasis on integration and 'accommodation' rather than political liberation."[38]

A second disagreement arose between the environmental and pro-indigenous organizations, which was not particularly antagonistic, but rested primarily on how the concerns of indigenous peoples were to be presented. Indigenous rights NGOs were sensitive to those rights being subordinated to environmental concerns. For them, fundamental indigenous rights to self-determination, control over territories and resources, and the power to veto development projects on indigenous lands were the starting point. Environmental organizations, however, wished to avoid the appearance of taking ideological positions when their concerns focused on natural resource conservation.

Because these different approaches coexisted, they did not threaten the campaigns against World Bank projects in the 1980s. The differences have sharpened since then, however. The division between Cultural Survival and Survival International, for example, was brought to a head during the debate on "harvesting the rainforest," which drew out political and ideological distinctions obvious to many throughout the previous ten years.[39]

Polonoroeste was important because it was the first big Washington-based campaign and in its early days had acted as a catalyst for the World Bank's indigenous-tribal peoples policy. However, the project cannot serve as an indicator of the extent to which the OMS 2.34 was implemented because it was already under way by 1982. Assessing the role of social or environmental policies in changing or canceling projects is difficult because many other factors could be involved. The Bank will sometimes, in retrospect, attribute changes in a project to the influence of a policy. However, direct evidence connecting policy implementation with projects on the ground is exiguous.

Circumstantial evidence from the comparison of Brazil's Polonoroeste and Gran Carajas projects suggests partial policy compliance. The Gran Carajas project in eastern Amazon was organized around an iron mine near Maraba and administered by the Vale do Rio Doce state enterprise. The World Bank provided a loan of $304.5 million in 1982 to fund extracting iron ore from the world's biggest deposit and transporting it on a 900 kilometer railway to the coast; it also provided $13.6 million to protect indigenous rights. By 1987, when the program should have been completed, twenty-three out of twenty-seven indigenous territories had not been demarcated satisfactorily, and two years later only twelve such areas had been completed.[40]

By 1992, 50 percent of the titling was completed in the Carajas project, which took place under the policy.[41] Only 33 per cent of the Polonoroeste project, which predated the policy, had its titling completed. The 50 percent appears to be a positive figure in comparison to the figures for previous projects, but it remains low. Indigenous people continue to be unprotected. The Awa Guaja, for example, were awarded a land demarcation in May 1988, only to have it quickly reduced by 60 percent at the insistence of cattle ranchers. Although campaigners were still trying to secure land demarcation in 1994, the World Bank considered the work completed and was assessing a further loan to the Carajas project.[42]

Human Rights NGOs in Alliance: Transmigration

Another World Bank project that threatened indigenous peoples during the 1980s was the Indonesian government's "transmigration program," a state-sponsored relocation of poor people from the overcrowded inner islands to the outer islands. Transmigration is a legacy of a Dutch colonial policy to provide plantation labor. About 250,000 transmigrants moved under Dutch rule, whereas 400,000 moved in the years from independence to 1965. President Suharto increased transmigration, utilizing five-year plans to cause 3.5 million people to transmigrate officially and another 3.5 million to transmigrate "spontaneously" over twenty-five years.[43]

The disastrous effects of the transmigration program have been amply documented. Criticisms have been made from several different angles: transmigration took no account of the rights of the indigenous peoples living in the outer islands;[44] transmigrants encountered serious food shortages from poor soils and the imposition of inappropriate wet rice cultivation;[45] many new sites were in rainforest, resulting in the loss of about 3.3 million hectares;[46] finally, the program was a geopolitical exercise, creating "security belts of strategic importance."[47] In 1985 serious conflicts arose on Sulawesi and Kalimantan between indigenous peoples and transmigrants; yet these were small scale in comparison with the armed struggle taking place in East Timor and West Papua, which were prime targets for transmigration.[48] Indeed, in West Papua alone, some twenty-four transmigration sites had been established on 700,000 hectares of indigenous land by 1984.[49]

The World Bank provided Indonesia with a total of $500 million between 1976 and 1986 to support the transmigration program. The loans went to support transmigrants, establish "nucleus estate" plantations, and cover planning and site selection for two million people. According to Bruce Rich, "In total the World Bank can take the credit for assisting in the 'official' resettlement of 2.3 million people and for catalyzing the resettlement of at least 2 million more 'spontaneous' migrants."[50]

The NGO alliance built to challenge the Brazilian Polonoroeste project also opposed the Indonesian transmigration program. The World Bank became the target of the campaign because its loans provided the means for the increased scale of transmigration—ten times the previous total.

In 1986, the special issue of *The Ecologist* devoted to transmigration was launched in Washington and a letter of protest presented to the president of the World Bank from representatives of thirty-five human rights and environmental organizations from all over the world.[51] The Bank response depended on the counterfactual: if they, with their policies on resettlement, environment, and indigenous peoples did not support transmigration, other funders without such policies *would* support the program, which would have even more devastating results.

The Washington NGO alliance was a significant mobilization force, but European organizations such as *The Ecologist,* Survival International, the human rights group TAPOL, the Dutch Komitee Indonesia, and IWGIA ensured a shift in emphasis and balance from primarily environmental argumentation to a stress on human rights abuses. The transmigration campaign was difficult to organize because of limited access to information from the areas affected. In Brazil, local NGO researchers could publicly confirm eyewitness accounts and participate in the lobbying; in Indonesia, however, much information had to come from close monitoring of the Indonesian press and World Bank internal documents. Local Indonesian organizations had to be extremely careful in their complaints against the government. Nevertheless, stalwart local environmental organizations, the insistent complaints of indigenous peoples, and the accounts of undercover travelers provided much-needed direct information.[52]

In spite of the World Bank's 1985 approval of the transmigration loan for resettlement, the NGO campaign did affect both the Bank and the Indonesian government. By 1988, the World Bank shifted the focus of its loans from new colonization to improvement of existing sites. The government was extremely defensive. Concerned about the possible loss of international support, the Bank scaled down its official sponsorship of resettlement.[53]

The World Bank's 1987 Five-Year Indigenous Policy Review

Within the climate of the Brazil and transmigration campaigns and the growing concern over the Narmada Dam in India, the OMS 2.34— "Tribal People in Bank-Financed Projects"—was given a five-year review by the Office of Environmental and Scientific Affairs between 1986 and 1987. The review looked at thirty-three Bank-financed projects that had been implemented since the introduction of OMS 2.34 in 1982. The four conditions cited in the OMS for establishing safeguards (land rights, social services, cultural integrity, and local participation) were used as benchmarks to evaluate the extent to which the Bank had been following its own guidelines.

Whereas the review noted an increase in the identification of the presence of indigenous peoples, it also recognized "a general tendency among Bank staff to underestimate the unique social, cultural and environmental problems that both tribal and indigenous or semi-tribal populations face in the process of development."[54] This tendency was even more apparent when the figures were analyzed. Out of the thirty-three projects, only fifteen observed the policy, and of these, only two included measures that covered all four conditions: six for land rights, two for health, two to support cultural integrity, and three for local indigenous participation.

The review reinforced the NGO conclusions that the World Bank was not taking OMS 2.34 sufficiently into account in the implementation of projects and was exercising limited influence on borrowing governments to defend the fundamental rights of indigenous peoples. However, only a month prior to the release of the review's report in May 1987, President

Conable announced that the Bank would undertake environmental reforms, which were to include increasing staff to deal with environmental matters, financing some ecologically sound projects, doubling forestry lending, and advocating more collaboration with NGOs.[55]

Indigenous Peoples Take Control: The Brazilian Dams

The environmental shift within the World Bank did not occur overnight; the changes were the climactic result of campaigns that had been taking place since 1983. The increasing strength of NGO protest became apparent with the campaign over Brazil's power sector loans. In June 1986, the national Brazilian state power company, Electrobras, negotiated a loan from the World Bank for $500 million as the first stage in a program known as the 2010 Plan.[56] Part of the money was used to complete the Balbina and Itaparica dams. Balbina threatened to displace two thousand Waimiri-Atroari people from their homelands north of Manaus, yet no plan for resettlement was made until after the dam began to be filled.[57] The Itaparica relocation affected forty thousand people, many of whom were sent to inferior lands and deprived of the financial support due to them. The 190 Tuxa Indian families were separated into two groups and as a result suffered social disintegration.[58]

Plan 2010 aimed to construct 136 dams over twenty years, with seventy-eight in the rainforest.[59] During 1987 Brazil tried to negotiate a second loan of $500 million for building six dams that would flood 18,000 square kilometers and affect four thousand members of twelve indigenous groups. The Kararao and Babaquara dams on Kayapo lands caused much attention because the Kayapo fiercely opposed their construction. Two Kayapo leaders traveled with ethnobotanist Darrell Posey to the United States to protest the dams and to meet with both NGO campaigners and World Bank officials. When they returned to Brazil, they were arrested and accused of degrading the nation's image; however, the campaign and their trip to Washington provided the context in which the Bank began to back away from supporting the loan. At the same time, an environmental assessment carried out by the Bank in the Babaquara area backed up the concerns of the Kayapo. As a result of the campaign and its own internal recognition of the negative effects of hydroenergy projects on indigenous peoples, the World Bank suspended the second power sector loan for 1987.

In an alliance with Brazilian environmental and pro-indigenous NGOs, the Kayapo called an international meeting at Altamira in 1987. Representatives of indigenous peoples from all over Brazil met with international NGOs to discuss the proposed dams. The meeting symbolized the growing public visibility of the campaign. The world's media was present, as well as international celebrities such as the rock star Sting. The campaign base was shifting from a Northern initiative in collaboration with Southern NGOs to an initiative from the indigenous peoples affected by Bank projects with support from Northern and Southern NGOs.

After the meeting, the Brazilian government decided not to go ahead with the dam. Several factors were involved in the decision. Government experts reconsidered the possibility of shifting from hydro power to nuclear power as a more cost-effective form of energy. At the same time, however, the government was uncomfortable with the enormous international attention being given to the Brazilian rainforest, which clearly influenced their decision. The Altamira meeting displayed the expanding alliance between indigenous peoples and both Northern and Southern NGOs, an alliance made possible by the prior waves of protest against the World Bank.

In addition, two other features occurred in the Altamira protest that were not present earlier in the Polonoroeste campaign. In the first place, Brazil was now a democratic country and Amazonian issues were prominent topics in government circles; in the second place, the indigenous peoples themselves—not just indigenous rights NGOs—were directly and vociferously leading the campaign, utilizing the international mass media to its full extent. These features became very prominent in the Indian campaign over the Narmada Dam, which occurred at the same time.

The NGO Alliance in Full Strength: Narmada

More than two million tribal people in India risk being displaced by hydropower schemes, the largest of which has been taking place in the

Narmada Valley states of Gujarat, Madhya Pradesh, and Maharashtra.[60] The Sardar Sarovar Dam, the largest in the Narmada scheme, is 155 meters high with a 210 kilometer reservoir, and will flood about 230 communities of about one hundred thousand people, of whom nearly sixty thousand are indigenous peoples. In 1985, the World Bank contributed $450 million to the overall cost of $6 billion. The international campaign began in 1984 when an internal World Bank document acknowledged that the tribal people would be financially worse off after removal.[61] In May 1985, the Bank said that a comprehensive program for resettlement and rehabilitation had been incorporated into their loan agreements, but this statement turned out not to be true.

The indigenous peoples of the Narmada area were divided by the project into "landed" and "landless," but with no recognition of traditional land ownership. In October 1985 the International Federation of Plantation, Agricultural, and Allied Workers (IFPAAW) raised the Narmada issue, on the basis of an initiative from Survival International, to the International Labor Organization. Though the ILO said in 1986 that Convention 107 was being breached by the project, the World Bank refused to revise the loan agreement.[62]

Meanwhile, as the dam was being built, local protest grew substantially.[63] By 1989 a broad local alliance led the campaign: the Narmada Bachao Andolan (the Save Narmada Movement) under the inspiration of two leaders, Medha Patkar and Baba Amte. Thousands of local people pledged that they would rather die than be flooded. In September 1989, a massive rally of fifty thousand local people and activists at the town of Harsud demonstrated that many local people were opposed to the dam. Local and NGO protests began to be backed up by U.S. legislators and a resolution of the European Parliament. The following year a mass march drew together thousands of local people and their supporters, who walked the length of the Narmada Valley. In 1991, as protests continued, Bank president Barber Conable established the "Morse Commission" to conduct an independent review of the Sardar Sarovar Dam.[64]

After a year of intensive investigation, the commissioners' report indicted the World Bank's handling of the project and recommended that the Bank take a "step back" to consider how to improve the situation. The new president of the World Bank, Lewis Preston, decided not to stop the loan, however, but the Bank finally withdrew in June 1993, nominally at the request of the government of India.

The Narmada protest has been one of the longest and most vociferous of all the campaigns against the World Bank and illustrates the evolving alliances between NGOs and grassroots groups. Whereas in the early years human rights organizations, who were in contact with local movements, environmental organizations, and research institutions, led the protests, gradually over the decade these NGOs became increasingly more coordinated with and responsive to organized local movements. By the end of the decade, the alliance included mass grassroots protest from a combination of indigenous peoples, local NGOs, environmental groups, and human rights organizations. This local base permitted coordination between protests at the dam site with letter campaigns and legislative hearings abroad.

The Morse Commission came at an opportune moment for encouraging reforms in the Bank. It coincided with the 1992 Wapenhans Report, which looked at the overall standard of project work in the Bank and concluded that the Morse findings were not unique. Indeed, more than a third of all Bank projects were considered failures by its own criteria.[65]

1991 Operational Directive on Indigenous Peoples

The World Bank produced a new Operational Directive on Indigenous Peoples (OD 4.20) in September 1991 to update the 1982 policy, following internal discussions concerning the 1987 review. The definition of indigenous peoples was broadened to cover tribal and cultural groups, and its criteria include attachment to ancestral territories, self-identification, distinct language, and subsistence production activities.[66]

The policy mandates not only protecting indigenous peoples from the harmful effects of projects, but also providing them with the opportunity to participate in the development process, which would involve a recognition of their own distinctive needs. The emphases on participation and direct consultation are innovations to the old policy: the new OD requires indigenous peoples' presence to be recognized in many aspects of Bank activities. Special Indigenous Peoples Development Plans (IPDPs) are to be prepared for all projects that affect indigenous peoples—taking into account their legal status, providing sociocultural and geographical baseline data, and ensuring their participation.

This revision took place during the same period as the ILO's revision of Convention 107 into Convention 169, which appears to have influenced the Bank's expansion of its definition of the term *indigenous* and its incorporation of the notions of participation and consultation. OD 4.20 is an improvement, although the stronger aspects of the published 1982 policy paper are still not yet apparent in 1991 Bank policy—for example, the references to self-determination and the right of indigenous peoples to veto unwanted projects in their territories.[67] Even though OD 4.20 is in effect, as of mid-1997 it was being revised without any attempt to consult with indigenous peoples.

Implementation of OD 4.20: The Protest Continues

Most NGO campaigns against the World Bank have concentrated more on projects than on general policies, but the relationship of policy to practice appears in protests and campaigning strategies because the policies provide guidelines and benchmarks as to what the Bank itself considers to be appropriate practice. The protests that currently take place frequently use the implementation of OD 4.20 as a basis for ascertaining unacceptable treatment of indigenous peoples.[68]

By themselves, the policy changes do not seem to have produced a seismic shift in Bank practices. However, the increased size of the Environment Department has meant that one now sees more attention directed to questions concerning indigenous peoples, and the presence of the policy does provide leverage to press the Bank to implement it. Several current projects offer opportunities to implement the policy, but it is too soon to tell whether the implementation has been effective.

For example, Bank specialists claim that one project in which OD 4.20 has been implemented is in the Pacific Coast Choco region of Colombia. The Colombia Natural Resource Management Program was a $39 million loan, originally intended as the National Forestry Action Plan of Colombia, which combined environmental monitoring in black and indigenous communities with developing forest management. Before the loan went through, however, the Bank's Environment Department lobbied the Colombian government to include the black rural farmers of the Pacific Coast in its constitutional provisions recognizing ancestral land rights. Furthermore, when a new project task manager realized a major component would increase logging, it was removed.[69] This example does suggest that Bank staff can implement the directive when they are aware of a problem. In this case, though, the implementation of the policy took place only after a local Embera representative traveled to Washington and complained about potential consequences. This case appears to be one of the first World Bank projects in Latin America to address OD 4.20 comprehensively, but the outcome depended on the combination of a sympathetic task manager and informed prior local representations.[70] Because of these two atypical features, the case does not permit generalizations about the broader issue of Bank capacity to nip potentially devastating projects in the bud. Controversial extractive projects on indigenous peoples' land continue to be proposed, as in the cases of Bank loans for oil in Western Siberia and Chad.

Brazil's Planafloro project also provides insight into recent trends. Because it was designed to mitigate environmental and social costs associated with the Polonoroeste project, full implementation of OD 4.20 might clearly have been expected. Instead, as suggested by Keck's chapter in this volume and the Rondônia NGO Forum's formal complaint to the World Bank Inspection Panel, Planafloro ended up reproducing many of the basic problems associated with its predecessor. Though several Bank staff are trying to ensure that the operational directive on indigenous peoples is implemented, NGOs' work against destructive development projects will remain necessary well into the future.

Contrasting Perspectives on Changes in the World Bank

Significant changes in both the World Bank itself and in the nature of the NGO and indigenous peoples' campaigns against projects have taken place since the late 1970s. Over the past fifteen years the World Bank has increased its environmental and social staff, which has led to broader internal debate within the Bank; it also has more staff who actively try to mitigate harmful effects of Bank projects. Behind the Bank's facade of unity lie internal diversity and disagreement, which create opportunities for NGO campaigns. Yet those Bank staff who recognize the problems facing indigenous peoples remain a small minority within the institutional structure and are often located in the less influential technical (advisory) departments rather than actually having primary responsibility for loans. If a project task manager is not concerned with these issues, the influence of the advisory staff will be limited, especially when borrower governments oppose the Bank's indigenous peoples policy.

Assessments of changes in Bank policy vary widely. One insider anthropologist notes that the World Bank is the only MDB with any elaborated policy on indigenous peoples (indeed, a first attempt by the Asian Development Bank to make an indigenous peoples policy was rejected, and the process had to be reinitiated).[71] Bank staff will argue that there has been a fundamental change in the way in which the Bank approaches questions concerning indigenous peoples. Moving from an attempt merely to mitigate the problems caused by projects, the Bank has broadened its definition of indigenous peoples, encouraged their participation, and recognized their social and economic rights.

Because of the lack of systematic independent assessments of the implementation of OD 4.20 across the entire World Bank portfolio, it is difficult to tell whether the Bank's 1991 policy will be any more successful than its earlier policy concerning tribal peoples in projects that it finances. A recent official review of thirteen projects begun in the last ten years in Latin America shows that provisions for indigenous peoples' land rights were included in some projects and that the land tenure questions were partially met, although in no case were targets met completely. Shelton Davis finds that the main obstacles to policy implementation were administrative and procedural problems in the borrower governments.[72]

Some Bank critics believe, however, that the changes in the Bank do not address the fundamental problem: the approval of projects that harm indigenous peoples. Bruce Rich and others argue that ameliorating actions are a result of NGO campaigning and that only public pressure will prevent projects that wreak social and environmental havoc.

The differences between Davis and Rich's views are in fact not as marked as they seem because they deal with different aspects of the issue. Davis focuses on positive changes in World Bank policy and its increased capacity to deal with environmental questions. Rich, in contrast, emphasizes the gap between present practice and promised outcomes. The rights of indigenous peoples are still not considered at the initial stages of the projects, and some Bank projects continue to cause havoc throughout the indigenous world. Davis and Rich epitomize the differences between those who seek to mitigate problems that Bank projects cause and those who say such projects should not be started in the first place.

This chapter has drawn attention to the disjuncture between the World Bank's policy on indigenous peoples and its implementation. The World Bank sees indigenous peoples' rights as only one of several factors to consider in a project, and so it answers criticisms with reference to the positive features of policy and to a handful of potentially successful projects. The critics, on the other hand, see the rights enshrined in policy as providing the framework within which the project should take place. In this view, projects that threaten indigenous peoples should be moved or canceled, and lack of compliance with ameliorating agreements raises questions about the legitimacy of both the specific project and the broader policy.

The result is often a frustrating conflict of perspectives: the Bank staff consider the critics unreasonable because they focus on negatives, and the critics complain that the Bank incessantly misses the point and merely repeats positive statistics and rhetoric that have no meaning for the victims. The difference in perspective arises from how projects are conceptualized. According to the Bank, development projects are framed as partnerships with governments, and indigenous peoples are a subsidiary factor whose rights should be respected wherever possible. NGOs working with indigenous people, however, believe that local communities should be drawn into partnership from the beginning.

The World Bank's policy on indigenous peoples is weaker than the United Nations' draft "Declaration on Indigenous Peoples." However, many indigenous peoples' problems could be ameliorated simply by ensuring the implementation of existing Bank policies. Implementation problems are to a large extent due to the fact that the Bank remains unaccountable to those affected by its programs and projects—the so-called project beneficiaries.[73]

NGO Campaign Evolution

The evolution of Bank policy and organization over the past decade and a half has been paralleled by shifts in relationships between NGOs and indigenous organizations. Differences in political perspective and subject matter have existed among the Northern NGOs: pragmatists who do not advocate major system changes (a position held by many pro-reform Bank staff) disagree with those who hold the principle that Bank projects should not begin without the consent of the people who are affected. These positions anchor a spectrum of opinions among human rights and environmental NGOs. During the first years (1980–1985) of the general campaign that targeted the World Bank, these differences were submerged under the solidarity of opposition to Bank projects.

Over time, however, NGOs in the North have become more responsive and aware of the needs of Southern NGOs. Notably, the precedent-setting Chico River campaign in the Philippines was initiated by grass-roots resistance to a dam project with relatively little international support. The initiative remained in the Philippines, and the protest focused primarily on the national government. International institutions were targeted only when they visited the Philippines at the IMF meeting in Manila in 1978.

Northern NGOs played leading roles in the early 1980s, targeted the World Bank at its Washington headquarters, and used as their leverage the U.S. government foreign aid budgets. Their campaigns pressed legislative committees to attach human rights and environmental conditions to U.S. funding and board votes on project loans. The campaigns against Polonoroeste and transmigration lobbied the World Bank to influence national governments, and Southern NGOs helped provide key information from direct research (Polonoroeste) or from the local and national media (transmigration).

In general, over the last fifteen years campaigns have developed more coordination between Northern and Southern NGOs. The campaigns against World Bank projects involved local, national, and international mobilization and strategies that have become more coordinated and more broad based. L. David Brown makes a useful contrast between NGOs who work with the aim of resolving problems by reform from above in contrast to the more activist-type NGOs that seek to remove problems by structural transformation from below.[74] Reform and transformation-oriented NGOs may be either Northern or Southern. Within the South, grassroots organizations—in particular indigenous peoples' organizations—have gained greater influence within NGO coalitions, asserting their right to self-determination and calling for NGO accountability.

Although the campaigns discussed here have been remarkably coordinated and successful examples of international solidarity, it is impossible to avoid differences of interest, ideology, and perspective among advocacy coalition members. Nevertheless, the discussion between environmental and pro–indigenous rights organizations, as well as between indigenous peoples from both North and South, has established many sustained alliances. Shared political perspectives have drawn together new networks that bridge North and South, such as the Malaysia-based World Rainforest Movement (WRM), which in turn facilitated the creation of the International Alliance of the Indigenous-Tribal Peoples of the Tropical Forests.[75]

National politics also greatly influence campaign outcomes. Campaigns have the greatest impact when national governments have either political will or a specific vulnerability to international leverage (see in this volume Keck's discussion of the limits to leverage). The key roles played by national governments in mediating the impact of NGO advocacy efforts should not give the impression, however, that campaigns in democratic countries are necessarily more successful or easier than in oppressive regimes. In the 1980s both Brazil and Indonesia had highly repressive governments. Local NGOs had to be extremely careful about speaking out and had to encourage Northern NGOs to act as a conduit for local and national opposition (which is usually interpreted by national governments as foreign intervention). In contrast, the mass protests that took place later in the decade in Brazil, by then more democratic, and in India, where the tradition of protest is strong, provide a completely different political context for protest. Yet differences of regime alone do not account for advocacy impact. For example, the Philippines was under Marcos's repressive regime when local people successfully stopped the Chico dams, and in spite of India's democratic traditions, protesters at the Narmada sites are regularly attacked by the security forces.

When dealing with campaigners, the World Bank frequently complains that "problem projects" are the responsibility of borrower governments, yet the Bank appears to have a double standard here because it is much more prone to impose stringent conditions for structural adjustment loans than it is to enforce environmental, social, or pro-participation conditionality. The reason given by the Bank for this tendency is that it should avoid political interference in a country's internal affairs.

In this sense, a discussion of Bank involvement rests on a play with the word *political*. Clearly, when the Bank wants to intervene, the policy issue is deemed economic, whereas when it does not want to intervene the policy issue becomes "political." Bank legal expert Ibrahim Shihata argues against the Bank becoming involved in politics in the same way Brian H. Smith argues that governments should be cautious about involvement with NGOs that are "too political."[76] All who work with grassroots organizations know that it is impossible not to be political in dealing with the Bank because *apolitical* in this context means supporting the bank's definition of the status quo—itself a political position.

Bank-NGO relationships increasingly focus on differing interpretations of the words *consultation* and *participation,* interpretations that are significant because they indicate key elements in future relations between the Bank and NGOs. When the Bank uses the term *consultation* in an environmental impact assessment, for example, it refers to the possibility of affected peoples commenting on the project terms of reference at the early stages of planning. Consultation does not, however, oblige those initiating the project to accept the local perspective. From a local viewpoint, in contrast, *consultation* implies that local actors have the right to stop a project that they do not want.

Similarly, some definitions of *participation* imply that local people can contribute to the implementation of a project after the planning and decision making have taken place. However, when local people talk of *participation,* they mean that they should be involved in the initiation and running of the project and should be able to ensure that they have control over what is taking place on their lands. The Bank's recent adoption of a definition of *participation* that emphasizes "influence and shared control" over decisions that affect participants may reduce the potential for confusion and misunderstanding that has occurred in the past.

The campaigns against Bank-funded projects have met with mixed success. The Chico and Altamira campaigns stopped projects before they were built; the Polonoroeste, transmigration, and Narmada campaigns changed Bank activities, but intransigent borrower governments continued the projects, causing considerable problems for the local indigenous peoples, but encouraging policy debate within the Bank.

Successful campaigns involve several factors. First, the local NGOs and indigenous groups need to know enough about the project in order to take a stand in their own defense. Alliances between indigenous and nonindigenous NGOs in the South and with solidarity organizations in the North are all part of this mobilization. The second factor is the presence of Bank staff who take the critique seriously and are prepared to argue internally on Bank terms for compliance with policy. The third factor is that the borrowing government should also be susceptible to influence about the project it considers. The successful Altamira and Chico dam cases involved all three elements. The Narmada and transmigration campaigns were only partially successful because the borrower governments of Indonesia and India remained committed to project implementation.

The future of the Bank's implementation of an effective policy on indigenous peoples still holds many questions and many potential problems. It remains to be seen whether the Bank's environmental staff, its new stress on participation and NGO consultation, and the prospect of small projects for indigenous peoples will constitute more than cosmetic changes. A further revision of the Bank's policy on indigenous peoples is ongoing. Vigilance in watching World Bank activities is as important as ever so that at any moment campaigns can be mobilized. With the increasing influence of indigenous organizations, these campaigns can continue the long process of trying to hold the Bank accountable to indigenous peoples throughout the world.

The final word should be left to indigenous peoples themselves. The charter of the International Alliance of the Indigenous-Tribal Peoples of the Tropical Forests contains several guiding principles for evaluating World Bank projects that affect indigenous peoples:

[Article 20] Control of our territories and the resources that we depend on: all development in our areas should only go ahead with the free and informed consent of the indigenous people involved or affected....

[Articles 23 and 24] All major development initiatives should be preceded by social, cultural and environmental impact assessments, after consultation with local communities and indigenous peoples. All such studies and projects should be open to public scrutiny and debate especially the indigenous peoples affected. National or international agencies considering funding development projects which may affect us, must set up tripartite commissions—including the funding agency, government representatives and our own communities as represented through our representative organizations—to carry through the planning, implementation, monitoring and evaluation of the projects....

A halt to all imposed programs aimed at resettling our peoples away from their homelands.... Our policy of development is based, first, on guaranteeing our self-sufficiency and material welfare, as well as that of our neighbors: a full social and cultural development based on the values of equity, justice, solidarity and reciprocity, and a balance with nature. Thereafter, the generation of a surplus for the market must come from a rational and creative use of natural resources, developing our own traditional technologies and selecting appropriate new ones.[77]

No Bank project will really benefit indigenous peoples unless their rights to self-determination, control over their territories, access to their resources, recognition of their own political institutions, and respect for their cultural integrity become a reality. As first recognized in 1982, they need the right to veto any development project taking place on their lands. Without the informed consent of indigenous peoples, any project is an intrusion and encroachment on their lands and lives.

Notes

1. See Bruce Rich, "The Cuckoo in the Nest: Fifty Years of Political Meddling by the World Bank," *The Ecologist* 19, no. 2 (1994), p. 13.

2. See Walden Bello, David Kinley, and Elaine Elinson, *Development Debacle: The World Bank in the Phillipines* (San Francisco: Institute for Food and Development Policy/Phillipine Solidarity Network, 1982), p. 24.

3. See Anti-Slavery Society, *The Philippines: Authoritarian Government, Multinationals and Ancesteral Lands*, report no. 1 of the Indigenous Peoples and Development Series (London: Anti-Slavery Society 1983), p. 103; and C. Drucker, "Dam the Chico: Hydropower Development and Tribal Resistance," in John

Bodley, ed., *Tribal Peoples and Development Issues: A Global Overview* (Mountain View, Calif.: Mayfield Publishing Company, 1988), p. 154.

4. See Anti-Slavery Society, *The Philippines*, p. 109.

5. See Drucker, "Dam the Chico," p. 158.

6. See Bello, Kinley, and Elinson, *Development Debacle*, p. 57.

7. See David Price, *Before the Bulldozer: The Nambiquara Indians and the World Bank* (Cabin John, Md. and Washington, D.C.: Seven Locks Press, 1989), p. 27.

8. The effect of the Chico experience on the World Bank drew attention to the plight of indigenous peoples. According to Drucker, "The Chico conflict led to a World Bank re-assessment of its policy regarding tribal minorities threatened by development programs" ("Dam the Chico," p. 163). Although an extended reference to the Chico case appears in the context of land rights in the Bank's first policy document (sometimes cited under R. Goodland), *Tribal Peoples and Economic Development* (Washington, D.C.: World Bank, 1982a), p. 18, Bank staff have informed me that it did not have a direct influence on policy formation in the way Polonoroeste did.

9. See Nuket Kardam, "Development Approaches and the Role of Policy Advocacy: The Case of the World Bank," *World Development* 21, no. 11 (1993), p. 1780.

10. See Survival International, "World Bank Renounces Tribal Policy", *IWGIA Yearbook 1986* (Copenhagen: IWGIA, 1987), p. 148.

11. See Goodland, *Tribal Peoples*.

12. *Ibid.*, p. 3.

13. See Marcus Colchester, "Changing World Bank Policies on Indigenous Peoples", *Third World Network Features* (Penang) 1093/93 (1993), p. 3.

14. See World Bank, *Tribal Peoples in Bank-Financed Projects*, operation manual statement (OMS) 2.34 (Washington: World Bank, 1982).

15. See Goodland, *Tribal Peoples*, p. 3.

16. See Shelton Davis, "The World Bank and Indigenous Peoples," unpublished ms. (1993), pp. 5ff.

17. See Goodland, *Tribal Peoples*, p. 27.

18. See John Bodley, "The World Bank Tribal Policy: Criticisms and Recommendations," in John Bodley, ed., *Tribal Peoples*.

19. *Ibid.*, p. 411.

20. Colchester, "Changing World Bank Policies," p. 3.

21. Davis, "The World Bank," p. 6.

22. See Barbara Bramble and Gareth Porter, "Non-government Organizations and the Making of U.S. International Environmental Policy," in Andrew Hurrell and Benedict Kingsbury, eds., *The International Politics of the Environment*,

(Oxford: Clarendon, 1992), pp. 313–53; and Bruce Rich, *Mortgaging the Earth: The World Bank, Environmental Impoverishment and the Crisis of Development* (London: Earthscan, 1994), p. 111.

23. The Bank's response to Bruce Rich's *Mortgaging the Earth* is a clear example; its External Affairs Department published twelve newsheets called "Setting the Record Straight: 'Mortgaging the Earth'—A Critical Overview" (World Bank, manuscript, 1994). Clearly, the Bank found a critique from an experienced campaigner such as Rich of great concern and applied considerable effort to refute the book. However, it is difficult for people who have been following the internal reports of projects such as Polonoroeste, transmigration, or the Narmada Dam to take these refutations seriously because they concentrate on a few positive interpretations of projects and sweep away the enormous suffering and devastation that affected the indigenous peoples. As Cheryl Payer has observed, such defenses gloss over projects' real problems. See *The World Bank: A Critical Analysis* (New York: Monthly Review Press, 1982).

24. *Ibid.,* p. 7.

25. See World Bank, *Tribal People,* pp. 27–8, and World Bank, *Resettlement and Development: The Bankwide Review of Projects Involving Involuntary Resettlement 1986–1993* (Washington, D.C.: World Bank Environment Department, 1994), p. iv.

26. World Bank, "Setting the Record Straight."

27. Price, *Before the Bulldozer.*

28. See M. Leonel, B. Mindlin, and C. Junquera, "The Joint Responsibility of the International Community in the Indigenous and Environmental Issues of the Brazilian Amazon," *IWGIA Newsletter* (July–September 1992), p. 13.

29. *Ibid.,* p. 53.

30. See C. Junquera and B. Mindlin, *The Aripuana Park and the Polonoroeste Programme* (Copenhagen: IWGIA Document no. 59, 1987), p. 59.

31. Leonel, Mindlin, and Junquera, "The Joint Responsibility," p. 57.

32. Junquera and Mindlin, *The Aripuana Park,* p. 7.

33. Rich, *Mortgaging the Earth,* p. 127.

34. *Ibid.,* p. 28.

35. Price, *Before the Bulldozer,* pp. 31–2.

36. Indeed, members of Cultural Survival were acknowledged in the published World Bank policy for providing Robert Goodland with help in its preparation.

37. During the 1980s I was an executive director of the International Work Group for Indigenous Affairs (IWGIA) in Copenhagen. IWGIA participated in the campaigns against the World Bank through letter writing and publishing documents and articles on the main projects that affected indigenous peoples.

38. See John Bodley, *Victims of Progress* (Menlo Park, Calif.: Benjamin/Cummings, 1982), pp. 197–200.

39. Cultural Survival's position is that the sustainable marketing of nontimber products in forest areas can provide cash for indigenous peoples and support their integration into the market economy while saving the rainforest (see Jason Clay, "Marketing and Human Rights: Lessons from the Cultural Survival Marketing Program," *IWGIA Newsletter*, 3 (1993), pp. 4–6). Survival International, on the other hand, argues that regardless of the intent, the profits do not return to the producers, that local peoples become dependent on the demands of the international economy, which rise with the increase in value of rainforest products, and that the approach avoids local marketing strategies over which indigenous peoples could gain more control (see Stephen Corry, *"Harvest Moonshine" Taking You for a Ride: A Critique of the "Rainforest Harvest"—Its Theory and Practice* [London: Survival International, 1993]).

40. See David Treece, *Bound in Misery and Iron: The Impact of the Grande Carajas Programme on the Indians of Brazil* (London: Survival International, 1987), pp. 29–30; and Bruce Rich, "The 'Greening' of the Development Banks: Rhetoric and Reality," *The Ecologist* 19, no. 2 (1989), p. 45.

41. See A. Wali and Shelton Davis, *Protecting Amerindian Lands: A Review of World Bank Experience with Indigenous Land Regularization Programs in Lowland South America*, Latin America and the Caribbean Regional Studies Program report no. 19 (Washington, D.C.: World Bank, 1992).

42. See Survival International, letter to Robert Torricelli, Subcommittee on Western Hemisphere Affairs, 25 May 1994, London.

43. Cf. M. Otten, *Transmigrasi: Indonesian Resettlement Policy 1965–1985* (Copenhagen: IWGIA Document 57, 1986), pp. 16, 19.

44. See Marcus Colchester, "Banking on Disaster: International Support for Transmigration," *The Ecologist* 16, nos. 2/3 (1986), pp. 61–70.

45. See M. Otten, "Transmigrasi: From Poverty to Bare Subsistence," *The Ecologist* 16, nos. 2/3 (1986), pp. 71–6.

46. See C. Secrett, "The Environmental Impact of Transmigration," *The Ecologist* 16, nos. 2/3 (1986), pp. 77–88.

47. See C. Budiardjo, "Politics of Transmigration," *The Ecologist* 16, nos. 2/3 (1986), pp. 111–16.

48. See Otten, *Transmigrasi*, p. 58.

49. See Anti-Slavery Society, *West Papua: Plunder in Paradise*, Report of the Indigenous Peoples and Development Series (London, Anti-Slavery Society, 1990), p. 64.

50. Rich, *Mortgaging the Earth*, p. 36.

51. Otten, *Transmigrasi*.

52. See G. Monbiot, *Poisoned Arrows* (London: Michael Joseph, 1989).

53. See TAPOL, "Transmigration Programme under Attack," *TAPOL Bulletin* (London) (July 1986), pp. 19–20; and TAPOL, "The Privatisation of Transmigration," *TAPOL Bulletin* (London), 81 (June 1987), pp. 10–12. A World

Bank 1994 evaluation of five transmigration projects (Operations Evaluation Department, World Bank, *Indonesia Transmigration Program: A Review of Five Bank-Supported Projects* [Washington, D.C.: World Bank, April 1994, p. vii) showed that many transmigrants survive only by supplementing their subsistence work with wage labor. The report severely criticizes the project's environmental and social impact, especially on the indigenous Kubu people of Sumatra and the Dayaks of Kalimantan.

54. See Office of Environmental and Scientific Affairs, Project Policy Department, World Bank, *Tribal Peoples and Economic Development: A Five Year Implementation Review of OMS 2.34 (1982–1986) and a Tribal Peoples' Action Plan* (Washington, D.C.: World Bank, June 1987), p. 13.

55. Rich, "The 'Greening' of the Development Banks," p. 44.

56. See Nicholas Hildyard, "Adios Amazonia? A Report from the Altamira Gathering," *The Ecologist* 19, no. 2 (1989), p. 55.

57. *Ibid.*

58. Rich, *Mortgaging the Earth,* p. 159.

59. Leonel, Mindlin, and Junquera, "The Joint Responsibility," p. 16.

60. See Marcus Colchester, "The Tribal People of the Narmada Valley: Damned by the World Bank," in Sahabat Alam Malaysia, ed., *Forest Resources Crisis in the Third World* (Penang, 1987), pp. 286–87.

61. See International Work Group for Indigenous Affairs, "Dams in Central India Threaten Over One Million Adivasi," *IWGIA Newsletter*, 46 (July 1986), p. 71.

62. Colchester, "The Tribal People of Narmada Valley," p. 293.

63. See Gustavo Esteva and M. Prakesh, "Grassroots Resistance to Sustainable Development: Lessons from the Banks for the Narmada," *The Ecologist* 22, no. 2 (1992), p. 46.

64. See Bradford Morse et al., *Sardar Sarovar: The Report of the Independent Review* (Ottawa: Resources Futures International, 1992).

65. See Willi Wapenhans et al., *Report of the Portfolio Management Task Force* (Washington, D.C.: World Bank, 1992).

66. Davis, "The World Bank," p. 21.

67. Colchester, "Changing World Bank Policies," p. 4.

68. NGOs such as the Bank Information Center promoted public awareness of the new indigenous policy. See Cindy Buhl, *A Citizen's Guide to the Multilateral Development Banks and Indigenous Peoples: The World Bank* (Washington, D.C.: Bank Information Center, 1994), also published in Spanish, Russian, and Indonesian.

69. See Charles Roberts, "Increasing Accountability of the World Bank: The Role of Institutional Change and New Operational Directives in the Colombia Natural Resource Management Program," Georgetown University Law School, unpublished ms., 1994.

70. *Ibid.*

71. Davis, "The World Bank," p. 28.

72. *Ibid.*, p. 29, and Wali and Davis, *Protecting Amerindian Lands.*

73. Colchester, "Changing World Bank Policies," p. 4.

74. See L. David Brown, "Social Change through Collective Reflection with Asian Non-Governmental Development Organizations," *Human Relations* 46, no. 2 (1993), pp. 249–73.

75. See Marcus Colchester, "Global Forest Peoples' Alliance Gets into Action," *Third World Network Features* (Penang) 1052/93, (1993).

76. See Brian H. Smith, *More Than Altruism: The Politics of Private Foreign Aid* (Princeton: Princeton University Press, 1990), p. 280.

77. The International Alliance of Indigenous-Tribal Peoples of the Tropical Forests was established in Penang, Malaysia, on 15 February 1992.

9

When Does Reform Policy Influence Practice? Lessons from the Bankwide Resettlement Review

Jonathan A. Fox

How consistently does the World Bank implement its social and environmental policy reforms, and how do we know? Advocacy non-governmental organizations (NGOs) continue to document specific project cases that fall short of the Bank's own minimum social and environmental standards. Bank staff recognize problems with specific projects, though in other cases debates over the "facts" persist. Critics and defenders of the Bank differ sharply over whether individual "problem projects" are the exception or the rule.

Because of the vast number, scale, diversity, and complexity of Bank-funded investments, it is extremely difficult to document the precise scope of Bankwide compliance versus noncompliance with its own reforms. Most external critiques of Bankwide performance in entire sectors, countries, or policy areas are based on information generated by the Bank itself. However, the Bank's own information on reform-policy compliance turns out to be based more on official intentions than on independently verified, field-based information about implementation, as this volume's concluding chapter shows.

In contrast, the World Bank's 1993–1994 resettlement review set a still-unmatched precedent in terms of rigor, comprehensiveness, transparency, and self-criticism. Insider reformers strategically used public transparency as a tool for increased institutional accountability (defined here as increased compliance with official reform commitments). The 1994 report, *Resettlement and Development*, was produced by the Environment Department's Resettlement Review Task Force, with strong support from the vice president for Environmentally Sustainable

Development, environmental staff in the regional operational departments, and elements within senior management and the board.[1]

Problems with "involuntary resettlement" have provoked some of the most intense controversy between the World Bank, advocacy NGOs, and grassroots movements. As many of the other studies in this volume suggest, the *impact* of external criticism of the Bank often depends on the uneven presence and leverage of pro-reform factions within the institution itself. In order to explain this interactive process, this chapter focuses on the internal dynamics of policy reform. *Reformists* are defined here as insiders who promote compliance with or strengthening of the Bank's social and environmental policy reforms—though not all are comfortable with that label, because it implies a recognition of internal conflict.

The resettlement policy is the Bank's first social/environmental reform policy, dating from 1980, and is now highly institutionalized, including explicit operational standards and benchmarks for staff to assess resettlement issues in project design and implementation. The policy encourages project managers to minimize involuntary resettlement in the first place and then details how to "rehabilitate" those who are resettled in order to leave them at least as well off as they were before relocation. Technical guidelines were developed based on lessons from repeated social disasters over previous decades. Nevertheless, the 1994 review documented high levels of Bank and borrowing government noncompliance.

One might have expected that the external scrutiny of the 1980s would have created at least public relations incentives for avoiding overt resettlement policy violations. Mass evictions provided critics with dramatic photo opportunities, as when Indonesia's Kedung Ombo Dam forced villagers to cling to their homes while flood waters rose around them (see Rumansara chapter, this volume). Forced resettlement attracted widespread condemnation by human rights activists and environmentalists around the world, creating rapid response, media savvy networks that were ready to highlight violations of resettlement policy as examples of much deeper problems with the World Bank's approach to development.[2] Though projects involving displacement accounted for 15 percent of the Bank's portfolio in dollar terms, the controversies they provoked undermined international support for the institution as a whole. Nevertheless, policy compliance did not improve markedly until the early 1990s—and then more for future than for ongoing projects.

This study addresses two questions about the interaction between external pressure and institutional reform. First, the *Resettlement and Development* report found that Bank compliance with the resettlement policy increased in projects approved after 1992. Moreover, NGOs have found significantly fewer *new* resettlement disasters in recent years. What explains this partial progress toward institutional reform? Second, what made the Bankwide review itself possible?

In response to the first question, the timing of the upturn in compliance coincides with the height of India's Narmada Dam conflict. Because the reform policy had been largely ignored by much of the Bank's operational apparatus for more than a decade, the timing of the increased compliance strongly suggests that external political pressure was a critical factor. Internal Bank attention to resettlement issues also increased during this period. The record of internal debates over resettlement at the time reveals a pervasive staff fear of being caught with "another Narmada." To explain the second question—the origins of the Bankwide review process itself—internal factors were more important. Indeed, NGOs did not call for such a review. Key participants concur that the assessment was the direct result of a strategic staff initiative, the culmination of years of internal education, research, and debate over how to increase policy compliance. Nevertheless, it would be difficult to explain high-level management's *support* for the review process outside of the context of the Narmada conflict. The record suggests that Bank senior management decided to allow a serious preemptive search for other "potential Narmadas."

This chapter briefly discusses two alternative conceptual frameworks for explaining institutional change: *external pressure* versus *institutional learning*. A third approach—synthesizing elements of both—is proposed to explain the World Bank's experience with resettlement policy implementation. The policy itself is then described, followed by a discussion of the Bankwide review's origins and key findings. The chapter then analyzes the tensions between the task force and the more recalcitrant elements of the operational apparatus. The India portfolio is examined in some detail because it accounted for much of Bank-documented policy noncompliance, affecting many hundreds of thousands of people.

Contending Explanations: External or Internal Causes?

Among many conceptual frameworks for explaining change in large public organizations, two contrasting approaches stand out as plausible alternatives to explain the degree to which World Bank operations actually follow resettlement policy in practice. An *external pressure approach* suggests that, by themselves, large bureaucracies are not predisposed to be self-critical and reflective, and are therefore unlikely either to recognize or to learn from their mistakes in the absence of external pressures. Some would go further, arguing that effective self-evaluation in a bureaucracy is inherently contradictory because critical self-evaluation is likely to threaten entrenched interests and will therefore be treated as a threat.[3] An *institutional-learning approach,* in contrast, recognizes that merely adaptive-reactive behavior is certainly the most frequent pattern, but suggests that some bureaucracies are able to learn internally from their mistakes, become self-critical, and effectively change *without* the pressure of external sanctions.[4] The institutional-learning view plays an important role within Bank discourse, which often frames changing means and ends in terms of "learning lessons."[5] The first view would argue that genuine institutional learning leads to self-criticism that affects entrenched interests and is therefore usually ignored or crushed. The empirical discussion of the Bank's resettlement policy shows that this was the dominant pattern for more than a decade, but began to change in the early 1990s.

Both approaches make implicit predictions about the context and timing of policy change: an externally driven explanation would predict that change should only follow dramatic increases in external pressures, whereas an internal learning framework would predict that institutional encounters with new data or new institutional efforts to acquire or analyze data would lead to changed behavior. An internal-learning approach might also suggest a pattern of active intellectual search within the organization, such as an institutionalized learning process lodged in a distinct unit. An external pressure approach would suggest that such a unit would be created in response to external pressure, and even then may or may not influence the rest of the institution. An institutional-learning approach would also suggest that one should find evidence of a "learning culture" throughout the organization, where leaders encourage staff to engage in active learning and adapting. An external pressure approach, on the other hand, would focus on a dominant culture mostly concerned with deflecting outside threats, especially from actors that have leverage over the organization (such as advocacy groups with influence over foreign aid allocations). The analytical challenge that underlies the empirical issue of compliance/noncompliance with resettlement policy involves disentangling two interactive processes: (1) the ebb and flow of external pressures (international NGO advocacy and protest, mediated by the World Bank's executive directors representing donor governments that control foreign aid flows), and (2) the gradual and uneven internal advances of reformist ideas and institutional leverage.

This study focuses on the shifting balance of power between insider reformists and staff within the operational apparatus who fail to comply with reform commitments. Reformists' concerns may not be based on new learning, in the sense that they have long known the social costs of ignoring the resettlement policy, whereas much of the operational apparatus may know about the policy but disregard it in favor of other priorities. Whereas some operational staff and managers learn, others simply adapt—paying attention only when external political costs of ignoring resettlement issues go up.

Organizational learning is not a strictly intellectual process. Because interests are affected, conflict may result. Therefore change agents need to learn how to shift the balance of power within the organization. World Bank insider reformers are often allowed to do research, to write reports, and to make recommendations, but they are often ignored by those who actually manage projects. Many Bank-funded disasters were predicted by insiders, who were not heeded. Sometimes, however, insiders do manage to veto or mitigate socially or environmentally destructive projects, or even to propose beneficial ones. Explanations of World Bank reform must account for when, why, and to what degree reformists have influence in a context where "more of the same" behavior often dominates. This study has found that external pressure empowered those within the institution who had already "learned," whereas it weakened those within the apparatus who paid little attention to the already learned lessons of the past. In a process of reciprocal interaction, external critics and insider reformists each legitimate and reinforce the other.[6]

Resettlement Policy and the Origins of the Bankwide Review

In 1980, the World Bank was the first international development agency to adopt a formal policy to mitigate the social costs inherent in involuntary resettlement. Bank discourse began to reject the then-conventional wisdom that the immiseration of those evicted in the name of development was unavoidable and necessary.[7] Decades of project-driven evictions in a wide range of urban and rural settings from the United States to Africa had produced a major body of social science research, but these findings had not managed to influence the policies of international development agencies.[8] Bank social scientists developed the 1980 policy partly in response to the mass resistance to projects that forced evictions in authoritarian Brazil and the Philippines.[9] Amended in 1986 and 1990, the policy was made public in 1988; until then, such Bank policies were confidential. According to sociologist Michael Cernea, the principal internal advocate of resettlement policy reform, "translation and wide distribution of the 1988 paper was intended to increase the accountability of both the Bank and borrowing governments."[10] In its own words, the basic elements of the policy include:

• "Involuntary resettlement should be avoided or minimized whenever feasible
...
• Where displacement is unavoidable, the *objective* of Bank policy is to assist displaced persons in their efforts to improve, or at least restore, former living conditions and earning capacity. The *means* to achieve this objective consist of the preparation and execution by the Borrower of resettlement plans as development programs ...
• Displaced persons should be: (i) compensated for their losses at replacement cost, (ii) given opportunities to share in project benefits, and (iii) assisted in the transfer and in the transition period ...
• Indigenous people ... and other groups that have customary rights to the land or other resources taken for the project must be provided with adequate land, infrastructure and other compensation. (p. 5)

The policy also mandates Bank staff to set benchmarks that detail the government's commitments in advance, as an integral part of a proposed investment project. The compensation provisions are more comprehensive than almost any borrowing government provided until that time (such as their recognition of customary land rights). The Bank first reviewed compliance with resettlement policy in a 1985 portfolio review of hydro and agriculture projects approved between 1979 and 1985. This review found some improvement compared to the time before the 1980 policy, but alternatives to displacement were rarely considered, and therefore evictions were not minimized. More generally, according to the policy document, "the 'consistency curve' between projects and policy oscillated, however, running higher in projects appraised in 1980 to 1982, shortly after the policy was issued, than in projects appraised during 1983–1984, when attention lapsed and the curve declined" (p. 84).

The 1985 review led to a 1986 policy revision, which included some remedial actions, recommendations for more staff, and more explicit policy guidelines requiring that "resettlers" be offered an alternative productive base. After a brief period of improvement, widespread noncompliance persisted. The 1994 review shows that the 1986 policy revisions led to few improvements until after 1992, "when awareness of the issue increased due to the Narmada debacle" (according to one Bank resettlement expert interviewed). By themselves, "lessons learned," internal education, and better policy guidelines did not significantly improve compliance.

The Morse Commission's Independent Review

The international human rights and environmental campaign against the Narmada Dam (better known in India as Sardar Sarovar) led the World Bank board of directors to commission an independent review in 1991. As one Bank executive director put it, "When I hear what NGOs say about this project and then what Operations staff say, it sounds as if they are talking about two different projects."[11] Led by Bradford Morse, former director of the United Nations Development Program (UNDP), the review team was given unprecedented independence, access, and resources to produce a field-based assessment. The Morse Commission's main conclusion echoes many of the key issues raised by the main NGO/ grassroots critique:

We think the Sardar Sarovar Projects as they stand are flawed, that resettlement and rehabilitation of all those displaced by the Projects is not possible under prevailing circumstances, and that the environmental impacts ... have not been properly considered.... Moreover, we believe that the Bank shares responsibility with the borrower for the situation.[12]

The Morse Commission found that the Bank and the government signed the loan agreements in 1985 with "no basis for designing, implementing and assessing resettlement and rehabilitation.... The numbers of people to be affected were not known.... [T]here was no adequate resettlement plan, with the result that human costs could not be included as part of the equation."[13] The review focused not only on the estimated one hundred thousand villagers living in the submergence area, but also drew attention to the estimated one hundred forty thousand farmers likely to be affected by the proposed canal system, whose displacement was not taken into account in the Narmada project.[14] The Morse Commission highlighted the project's "non-compliance with Bank resettlement and environmental requirements," concluding that its "incremental strategy" signaled to the Indian government that resettlement and the environment were "of only secondary importance" and therefore was even "counter-productive."[15] Finally, the commission concluded that the Bank should "step back" from the project. The initial response of the World Bank's India Division to a draft was, according to Michael Cernea, "how to find fault with the facts, but they couldn't find a major fault. [They called for] small corrections, but the overall factual picture was not disputed."[16]

The Morse Commission report was a shock to Bank management. Their response was twofold. Narmada-specific damage control came first. The Bank's executive directors split over how to proceed, narrowly defeating immediate cancellation in favor of creating a procedure through which the Indian government could save face by canceling later. The Bank's second response was to assess resettlement problems throughout its portfolio. In contrast to the Morse report, which was a direct response to external demands, the resettlement review proposal was an internal initiative from the Bank's senior resettlement specialist, sociologist Michael Cernea. According to Cernea, the Bank's managing director, Ernst Stern, "agreed to a new review because it was part of the formal report to the Board and the public answer to the Morse Commission." As he put it, management wanted to know, "Are there other Narmadas hidden in the portfolio?" Indeed, Cernea had gone on record with internal warnings of Narmada's resettlement risks even *before* the project was approved—suggesting that had management listened to him then, a major problem might have been avoided. The aftermath of the Morse report positioned Cernea to seize the moment.[17]

Reviewing the Bank Portfolio

The Bankwide resettlement review was intended not only to take stock of the state of resettlement in the Bank's portfolio, but to improve institutional performance in the process. The *Resettlement and Development* report emphasizes that *"The main product of this comprehensive review is not simply its final report, but the process that the review triggered throughout 1993 across the Bank and on the ground"* (p. 2, emphasis in original). Unlike most Bank portfolio reviews, the resettlement review received the funds and political support from management needed to carry out field-based assessments of projects on the ground, which permitted independent verification of the "official story" as reflected in project documents.

A task force was created at the "center" of the Bank's structure, and assessment reports were commissioned from each of the "regions"—the operating divisions primarily responsible for project design and implementation. The regional environmental staff were the key link between the task force and the project managers.

According to Cernea, "Even after Narmada, many thought it would blow over, and put the Review on the back burner—we realized the Task Force couldn't operate [because] the Regions were not taking it seriously." The vice president for Environmentally Sustainable Development, Ismail Serageldin, requested support from the Bank's top operational

official, Managing Director Ernest Stern. Though Stern was widely seen
as unsympathetic to social and environmental policy reform, resettlement
problems had severely undermined political support for International
Development Association (IDA) contributions in several donor countries.
Moreover, policy noncompliance raised questions about the principle
of management authority over staff. Stern's pivotal December 28, 1992,
internal memo to key operational vice presidents signaled both the resettlement review's priority and the central importance of external pressures to encourage reform compliance:

As you know, the Bank has a commitment to review the status of all involuntary
resettlement components of existing projects.... *You are familiar with the widespread concern attendant on the India-Narmada projects* and their resettlement
components. The importance of getting a professional assessment ... hardly
needs emphasizing in this context. I am sure that you share this *sense of urgency*
to get the work done, and I would appreciate it if you would so advise your
managers.... We've lost much time in getting started on this exercise. *The Bank
cannot afford to fail* in compiling expeditiously a status report on its resettlement
projects. (emphasis added)

The task force designed the review to involve operational staff directly,
inviting project task managers to key meetings and briefing regional vice
presidents on work in progress. These briefings were designed to encourage them to invest their own political capital in improving project
performance before the report was finished, giving them a chance for
their projects to look better in the final review: "[we wanted them] to
coopt [us] for a good cause," according to Cernea. The goal was not to
approve bad work, "but to get the facts and trigger improvement."

In Cernea's view, "facts go a long way in Bank culture.... [F]acts can
change the culture." For the task force, the "key battle [was] over objective assessment of the facts on the ground." NGO critics had gained
leverage with a similar "fact-based" approach, documenting illustrative
case studies of noncompliance as a key advocacy tool (see Wirth chapter,
this volume). The resettlement review task force, in contrast, was able
to transcend the case study–based critique, however, to assess the whole
portfolio's "consistency with policy and outcomes" (*Report*, p. 2). Their
findings could not be dismissed as exceptions to rule because their mission was precisely to document the general pattern.

Documenting the "Resettlement Portfolio"

The *Resettlement and Development* report found that from 1986 to 1993 involuntary resettlement was involved in 192 projects, displacing an estimated total of 2.5 million people over the life of those projects (p. 88).[18] From 1986 to 1993, forty-six projects involving half a million displaced people were officially "closed" out of the Bank's portfolio, redefining the scope of study to 146 projects considered "active," representing 8 percent of all projects and 15 percent of total Bank lending.[19] More than half of all resettlement was concentrated in eleven large projects in only four countries: India, China, Indonesia, and Brazil. Projects in East and South Asia accounted for 82 percent of people to be displaced by Bank-funded projects, with 974,000 in India and 483,000 in China (p. 88). Large dams, mainly for hydropower and irrigation, accounted for 63 percent of displaced people; transportation corridors a rapidly growing sector, accounted for 23 percent (p. 92). Five large agriculture-related projects in India alone account for a full 41 percent of total Bank-funded displacement (p. 93).[20]

The task force found significantly more resettlement in Bank-funded projects than they expected. Even the lower number of 146 projects still considered "active" in 1993 was *50 percent higher* than estimated before the review. The numbers of people to be displaced by each year's projects turned out to grow over time, rising by 125 percent between 1986 and 1993, an increase attributed in part to "better identification." Perhaps the most dramatic finding was that of the

almost 2 million people in various stages of resettlement under the current active portfolio ... [t]he number of people to be resettled is 47% higher, or an additional 625,000 people, than the estimate made at the time of [project] appraisal.... Data supplied by many Borrowers at [project] preparation and appraisal have commonly understated the number of people affected. The real number became apparent only part way through the project. (p. 88, emphasis added)

This finding is critical because without knowing the number of people affected, no agency can do even minimal planning and budgeting for their resettlement and rehabilitation. In terms of who gets displaced, the report notes that the majority of the displaced are rural and poor because new projects are brought to the most under-developed, poorest areas, where infrastructure is lacking and where land and political costs are lowest.... The remote locations of many dam sites are often inhabited by indigenous peoples, ethnic minorities and pastoral peoples, which explains why ... cultural differences are so prominent in resettlement. (p. 93)

In other words, there is a direct association between large projects involving displacement and the lack of political representation of displaced peoples.

Resettlement and Development argues that humane resettlement can work if Bank policies are followed systematically. The task force recognized that some critics reject all resettlement, so it instead sided with those critics who accept that resettlement is sometimes unavoidable because of the need for infrastructure but should be minimized and carried out in a legal and humane fashion.[21] *Resettlement and Development* lists the key factors that account for resettlement successes:

a. Political commitment by the Borrower, expressed in law, official policies and resource allocations;
b. Systematic implementation by the Borrower and the Bank of established guidelines and procedures;
c. Sound social analysis ...
d. Accurate cost assessments and commensurate financing ...
e. Effective executing organizations ...
f. Public participation in setting resettlement objectives, identifying reestablishment solutions and implementing them. (p. 7)

These factors are intervening rather than independent variables, however. They all reflect political will, which in turn requires further explanation. More to the point, the report recognizes that "Resettlement works when governments want it to work.... Similarly, when the Bank itself does not consistently adhere to its policy ... project performance is weakened" (p. 8). The analytical question that follows is, in those instances where the policy was followed, why did government and Bank officials have the political commitment and resources to do so, given that noncompliance turned out to be so widespread and persistent in most situations?

"Narrowing the 'Development Gap'"

As a framework for its documentation of "consistency" between policy and operations, the report suggests that as new standards are created, a "development gap" emerges that cannot be closed overnight. It stresses that "changing entrenched bad practices takes time" (p. 97) but focuses on three main areas of partial movement toward greater "consistency" with policy.

The first area focuses on Bank efforts to change country-level policies. The review found the greatest progress in those countries and sectors that develop comprehensive, rather than project-specific resettlement guidelines. By focusing on the Bank's "unused potential" in this area (p. 98), the report implies that past problems were partly due to the Bank's failure to invest its political capital in this policy reform issue (in contrast to, for example, its high political resource investment in pro-market policy changes, such as privatization or deregulation).

In their report, the review task force conclude that the absence of country-level legal frameworks can lead to "violent displacement procedures, without due recognition and protection of the basic rights of those uprooted" (p. 101).[22] The key areas of nationwide policy reform were in the electric power sector in Brazil, Colombia, and India, urban Philippines, and across sectors in China and Turkey.[23] China accounts for the vast number of displaced people in Bank-funded projects covered by national-level resettlement policy reforms. The review claimed that China's reforms, dating mainly to the early and mid-1980s, were quite consistent with the Bank's core policy principle of "resettlement with development" for those displaced. China's reforms are explicitly attributed to past dam projects that "resulted in the disastrous impoverishment of many people and in serious social and political instability" (p. 102).[24] In addition to its stress on countrywide reforms, the report notes that many other international agencies have raised their resettlement standards in recent years as well—including the Inter-American Development Bank (1990), the Asian Development Bank (1992), the Organization for Economic Cooperation and Development (OECD) and British and Japanese bilateral aid agencies (pp. 102–3).

The report also recognizes that "the Bank has . . . encountered serious difficulties in dialogues with some Borrowers about adopting domestic resettlement regulations" (p. 103). South Asia saw little progress on this front:[25]

In India, where many resettlement projects in both non-Bank and Bank-assisted projects have failed to rehabilitate a proportion of the displaced people, no federal legislation or policy statement defines the country's general resettlement norms; resettlement is regarded as a state, not a federal matter. In turn, however, most Indian states still lack state-level resettlement policies. . . . Dialogue between the Bank and borrowing state governments, with some notable exceptions [Gujarat] has still to yield significant results. (p. 103)

The report concludes that the Bank's political strategy for dealing with resettlement problems in India—the "incremental" approach—had failed—reinforcing the findings of the Morse Commission (p. 103).[26] It notes that where progress in the "policy environment surrounding development-caused resettlement" has been achieved, it was driven by "the Bank's policy influence, as well *as a consequence of public opinion demands, of resistance to displacement by affected people, and of strong advocacy by many NGOs*" (p. 99, emphasis in original).

The second main area of performance reviewed involves the reduction of displacement by encouraging redesign of projects. Only ten such projects were mentioned, which suggests that this key policy was applied to only a small part of the Bank's portfolio (pp. 105–6). Indeed, the task force recognizes that

many engineering consulting firms, responsible for the technical design of major infrastructure projects worldwide, routinely display obliviousness to the adverse social implications of the designs they propose, sheltered by the absence of policy or legal demands in the client countries. . . . The studies prepared by such firms tend to end up with misleading budgets whenever the real, full costs of displacement and resettlement are omitted. (p. 104)

Most of the report, however, focuses on how displaced people were treated without questioning the more fundamental justification of the projects. Task force members tended to agree that many of the projects were indeed necessary because power, irrigation, drinking water, sanitation, and transportation can potentially benefit large numbers of people in comparison to those displaced. Nevertheless, they present little evidence that project planners considered alternative means to these ends.

One chapter does address the causes of displacement by criticizing energy and water subsidies that distort use and endorses demand-side management (p. 107). These issues are more directly addressed by the Bank's environmental assessment policy, which specifically requires consideration of alternatives, as well as its energy policy, which mandates support for both greater supply-and-demand efficiency. As noted in this volume's concluding chapter, however, compliance with these two mandates was still the exception rather than the rule at the Bank in the mid-1990s.

"Restoring income and livelihood" proved the weakest area of resettlement performance. The task force accepted the challenge that "the ultimate test of consistency between resettlement operations and policy is the degree to which the Bank's basic goal—reestablishing resettlers at an improved or at least the same level of living—is achieved" (p. 109). Documenting positive outcomes proved difficult, however, partly because baseline data was largely unavailable (the result of noncompliance with a basic policy norm). The report cites only one project where "incomes for all households rose after resettlement," Khao Laem in Thailand (p. 112). Even in China, which is held up throughout the report as the main success story, "projects in the poorest regions, particularly those with indigenous minorities, face difficulties and have a less satisfactory record" (p. 114). India again had systematic problems, though the report strikes an optimistic tone: "projects in India approved during the last three to four years have started out on a much better footing and are expected to yield better resettlement and rehabilitation results" (p. 114). Indonesia had a mixed record, combining some success with "serious failures," especially in urban and transportation projects (p. 115). Overall, however, "unsatisfactory performance [in restoring incomes] ... still persists on a wide and unacceptable scale" (p. 110).

Project Preparation

The Bank itself bears special responsibility for resettlement issues in the preparation and appraisal of projects because this period before signing loans is when the Bank has maximum involvement and leverage. Bank resettlement policy subjects project preparation to four basic requirements:

1. Baseline planning surveys of affected populations
2. Resettlement timetables coordinated with civil works construction
3. Resettlement plans to restore lost incomes
4. A resettlement budget (p. 129)

The report dates "significant improvement" in these four areas since 1992. but notes that "despite recent improvements, recurrent failures in project preparation and appraisal remain the root cause of much problematic resettlement" (p. 129).

The review's key findings include:

• *Baseline population surveys* are necessary, though far from sufficient, for any other mitigating measures. For the 1986–1993 period, only 44 percent of projects with resettlement included baseline population surveys. During the first five years of the resettlement policy, only 21 percent of projects were approved with baseline population information. Since 1991, 72 percent of new projects had surveys, and the rate reached 100 percent of new projects during the review year of 1993 (p. 129).

• Projects with *resettlement plans* at the time they were approved rose to 92 percent in 1993 and 100 percent in 1994. Only 50 percent of 1986–1991 projects had such plans—a decade after the policy went into effect (p. 129). For the period as a whole, less than 30 percent of resettlement plans mentioned economic rehabilitation (support for alternative livelihoods), even though cash payments have been repeatedly shown to fail as a compensation mechanism. Very few projects included resettlement timetables at the time of approval, leading to major disruptions.

• Bank loans contributed to *resettlement and rehabilitation expenses* in less than 15 percent of projects involved (p. 147). This omission reduced the bargaining power of Bank resettlement specialists and sent the implicit signal to governments that the issue was not a priority.

The Limits of Project Supervision

Even if projects have the appropriate plans on paper, government project implementation often falls far short of promises. The Bank considers direct project supervision to be its most powerful tool for assessing the progress of project implementation, including resettlement performance. In this context, the report distinguishes between the responsibilities of the Bank's operational apparatus and the borrowers:

Effective supervision depends on Country Departments' ability to allocate resources commensurate with the complexity and specific needs of individual projects, and their willingness to act promptly on the findings. . . . Project performance, on the other hand, depends largely on Borrowers' commitment to project objectives or "ownership," and their institutional and other capacities to execute the project. (p. 153)

In other words, the resources and specialized skills devoted to project supervision can indicate whether or not resettlement implementation was a Bank priority (in a process where governments clearly bear primary responsibility).

The task force found that only 56 percent of 1986–1993 project supervision missions reported on resettlement components at all, and less than 25 percent included resettlement specialists. Specialists tended to be brought in mainly when resettlement had already "become a major problem, either because it delays implementation or triggers public criticism" (p. 156). These numbers include the bolstered supervision in 1993 that resulted from the review process itself, so the 1986–1992 supervision performance was much worse. During the year of the portfolio review, in contrast, all major resettlement projects were monitored in the field, leading to a series of "retrofitting" operations that attempted to improve problem projects.

India received the majority of all specialist supervision missions in the context of the review. A 1993 Bank review of project supervision credits the Narmada debacle as being pivotal to this concentration of resources: "Sixty percent of all specialist supervision missions were in connection with ten projects in the India portfolio, the major reason being the attention the country attracted in the wake of the Narmada Sardar Sarovar Project controversy." More generally, this frank internal review concluded that

[The Country Department] response to resettlement issues [was] mainly influenced, not by specific project requirements, but by the pulls and pressures of the moment—crisis in implementation, public controversy, queries from the Board, lending compulsions, etc. . . . The main conclusion [is]: In spite of the significant progress made in the last five years, supervision of resettlement . . . has not become a routine and integral feature of project supervision.[27]

Some Bank staff stress the lack of resources as a major constraint on the capacity to implement the resettlement policy, but the internal study of supervision also stresses a sense of fear of exposing resettlement to the scrutiny of "outsiders." The feeling that greater attention and firmer actions on resettlement will affect their relationship with the borrower country is not uncommon among [task managers]. The common incentive structure in the Bank (premium given to speedy processing of the project, quick disbursement, smooth completion, etc.) also affects the quality of supervision.... [A] low supervision coefficient is often equated with better management.[28]

In other words, the dominant system of career incentives discourages task managers from risking conflict with their counterparts in borrowing governments over resettlement issues. The issue is not *whether* task managers should use bargaining power when their priorities differ from their government counterparts, but whether such pressure is worth using for the particular purpose of improving resettlement and rehabilitation performance. The report notes that the limited impact of supervision on resettlement performance does not mean that it cannot work, but instead suggests that the Bank "fail[ed] to utilize the full potential of its involvement"—in other words, Bank managers were reluctant to invest political capital in dealing with poor resettlement processes.

The Politics of Information Extraction

The task force's combined goals of gathering information, targeting noncompliance with official policy, and producing improved performance were necessarily going to provoke discomfort, if not conflict, among those operational staff members who had failed to comply with the policy. As Cernea recalled, it was like using "forceps to extract the information—people weren't happy to see the numbers aggregated." By pressing the staff of Bank regions to "see" (that is, accept) the facts, "we focused responsibility for the situation." By emphasizing this sense of internal accountability, the task force attempted to "mainstream" greater concern for compliance with the Bank's resettlement policy. As Cernea put it, the point of the exercise was not just "to spring the report on an unsuspecting audience—we would have had a superficial impact. [But] we sent back the regional reports if they were not good enough."[29] The task force's assessment of what was "not good enough" was based on its own network of experts, both on the team itself and on the ground—among NGOs, academics, and government officials outside the Bank.[30]

Levels of compliance varied greatly across the Bank's operations, and these differences proved critical to resettlement specialists in their efforts to put the "old guard" on the defensive. Resettlement specialists were able to point to an operational counterpart and say, "If X goes along with the operational directive 4.30, why can't you?" Uneven performance provided the reformers a wedge with which to isolate those project managers who were reluctant to admit to resettlement problems.

The Latin America regional staff, for example, had relatively little to fear from the review because most of their still-active projects involved small-scale resettlement (by Bank standards). Most importantly, two decades of grassroots anti-eviction protests throughout the region had led to tangible, though uneven, improvements in the ways most governments —many newly democratic by the 1990s—dealt with displacement.[31] Few new Bank-funded projects in Latin America were as large and disruptive as those of previous decades. In Latin America's more recent projects, the task force found relatively few glaring contradictions between the Bank's "official story" and the record on the ground, insofar as they were able to determine.[32]

In South and Southeast Asia, in contrast, many government officials rejected the notion that they should be somehow accountable to the often poor, low-caste, or tribal populations most often threatened with large-scale, forced evictions. According to dominant national developmentalist ideologies, the benefits of large-scale infrastructure are more important than the losses of those who are considered to be the (relatively) few. Moreover, the Bank's India Department, still smarting from the Morse Commission's public condemnation, also included staff who were veterans of the Polonoroeste Amazon rainforest road conflict—leading, according to one resettlement specialist interviewed, to an especially "paranoid" attitude. This attitude was quite understandable, however, in the sense that their pattern of noncompliance was so systematic that internal scrutiny, much less public scrutiny, could not be in their immediate interest (though it was not clear until the end of the Bankwide review process that the document would be made public).

Social and environmental staff inside the Bank had long known India's resettlement record to be devastating, but until the Morse Commission and the resettlement review, they did not have the power needed to address the issue systematically. As part of the Bankwide review process and its aftermath, however, several major problem loans in India were suspended or canceled, in some cases with little direct external pressure.[33]

Although resettlement records in other South Asia countries were not notable for their compliance with Bank policy, India dominated the internal debate because of the scale of its eviction problems (related to population size and density). The review challenged prevailing operating patterns and sought to weaken the Country Department's monopoly on information about social and environmental impact, creating more room for the regional environmental and social staff with greater expertise in (and usually commitments to) such concerns to maneuver. The review process thus increased conflict not only between the task force at the Bank's center and the operational regions, but also exacerbated tensions between the task managers in charge of projects and the regional technical staff in charge of monitoring the projects' social and environmental impact.

The internal conflict over the data on the Bank's India resettlement portfolio highlights the difficulties of establishing internal accountability. A conflictual bargaining process emerged over the inclusion and exclusion of "negative" findings in the final report—including debates over the quality of resettlement operations, the accuracy of data related to the numbers of people displaced, and the cooperation (or noncooperation) among various actors involved in the process of extracting, reporting, and organizing information.

The concern with South Asia in general, and India in particular, was highlighted by Bank senior management's request that a special separate report be written on India's resettlement portfolio.[34] Bank directors requested detailed information about India on the heels of the Morse Commission report, which concludes:[35]

Comparative analysis shows recurrent flaws in how the Bank approaches resettlement in India. They include the chronic failings in the Bank's appraisal of resettlement components. Projects are appraised and negotiated despite the absence of resettlement plans, budgets and timetables to meet the Bank's resettlement policy. All too often, decisions affecting the lives of thousands or even hundreds of thousands of peasant farmers and tribals are based on seriously deficient or flawed information and approved without requiring major conditions and actions for improvement despite a well documented record of the impoverishment caused by other resettlement operations in the same area....

An equally serious generic problem is that even when the Bank has been aware of major resettlement problems in its India projects, it has failed to act firmly to address them. Violations of legal covenants are flagged and then forgotten, conditions are relaxed or their deadlines postponed. Our review of the documentation as well as our many interviews with government officials support the view that the result of this failure is a widespread belief in India that the Bank is more concerned to accommodate the pressures emanating from its borrowers than to guarantee implementation of its policies.[36]

Some Bank project managers shared the Indian government's willingness to incur significant social costs in the name of development. Note, in particular, one report to an internal Bank focus group discussion, which states,

All governments care for the poor, but the question in the end is, where will the tradeoff be, who will get the priority? This varies from government to government. In India, there is tremendous concern for the poor—there is a democratic environment and the poor have a vote. But if there is a tradeoff between resettlement of two million people and a dam, and the government does not have the resources, what do you do? ... In the end the government for the benefit of *all* will perhaps vote for the dam and make the two million people worse off.[37]

Recalcitrant staff were not limited to South Asia. Several projects in Africa significantly underestimated the numbers displaced, such as the Tana Plain project in Madagascar and an urban project in Tunisia. According to one Bank expert, Indonesia's treatment of displaced people was "in many ways more deplorable than India," even after the Narmada controversy. But the India Department had by far the biggest problem because India alone accounted for almost half of the people to be evicted by Bank-funded projects worldwide. Moreover, *the review discovered almost half a million Indian "oustees" who were not officially acknowledged to exist when the projects that would displace them were signed* (see table 9.1). *India represented 81 percent of the worldwide number of acknowledged oustees who were excluded from Bank estimates at the time of approval of 1986–1993 projects.* Because India accounts for such a large share of the Bank's own noncompliance problem, the Bank debate over the India portfolio is a key indicator of the internal balance of power between pro- and anti-reform factions.

Table 9.1
People displaced in India portfolio: Increasing estimates

Project name	FY/sector	SAR data at appraisal	SA2 data 6/9/93	Asia regional report 12/93	SA2 data 4/1/94
Upper Indravati	83/IEN	20,000	26,500	26,500	16,078
Dudhichua Coal	84/IEN	1000	310	310	1415
Farakka II Thermal	84/IEN	0	53,500	53,500	53,500
Gujarat Medium II	84/AG	90,000	128,000	128,000	140,352
Chandrapur Thermal	85/IEN	0	2500	3800	4566
Jharia Coal	85/IEN	0	3600	3600	3502
Narmada SSP	85/AG	67,340	100,000	100,000	127,446
Andra Pradesh Irrigation II	86/AG	63,370	125,000	125,000	150,000
MCIP III Irrigation	86/AG	18,500	126,800	180,500	168,000
Coal/Gevra Sonepur	87/IEN	11,800	24,000	24,000	13,863
Karnataka Power I and II	87/IEN	2000	4000	4000	4000
Talcher Thermal	87/IEN	9600	5200	5200	14,106
UP Power	88/IEN	325		360	
Maharashtra Power I	89/IEN		1600	2600	2202
Nathpa Jhakri Hydro	89/IEN	345	400	400	1709
Upper Krishna II	89/AG	195,975	200,000	200,000	220,536
Punjab Irrig.	90/AG	825	825	825	3198
Hyderabad Water	90/TWU	35,140	51,000	51,000	42,126
2d Nat. Highways	92/TWU	3575	2500	2500	4000
NTPC Power	93/IEN	930		896	1685
Renewable Res.	93/IEN	430	430	430	612
Totals		521,240	856,300	913,556	972,998

Sectors: IEN, Industry and Energy; AG, Agriculture; TWU, Transportation and Urban.
FY: First fiscal year of project.
SAR: Staff Appraisal Report (basic project document).
SA2: India Department, South Asia Division.
Source: Internal World Bank memo, Environment Department, 4 April 1994.

Who Counts?

Counting those to be displaced matters. Few oustees in India receive even minimal compensation, but if they are not acknowledged to exist in the first place, then they are at great risk of being driven into complete destitution. For example, a Bank's Operations Evaluation Department study concluded that "Bank guidelines were seldom applied in India.... [T]his is the country with the largest number of resettlement projects, which alone would warrant special attention. In India, the overall record is poor to the extent of being unacceptable."[38] In the process of implementing a project, oustees' meager resources are expropriated and sacrificed for others who benefit from the projects, representing losses that never enter into a project cost-benefit analysis. Compensation laws based on titled property and individual male "heads of household" discriminate directly against women, the landless, and tribals, who are more dependent on common-property resources that are rarely replaced.

Who counts those who are to be negatively affected by projects? The task manager has the responsibility to make sure the job gets done as part of the project preparation. The task force insisted on confirming the accuracy of the task managers' estimates of project-affected people, but it operated at a disadvantage insofar as it was organizationally distant from the country departments, which produced and controlled the information they needed. Under the regional vice presidencies, the country departments control project funds, with day to day project-related responsibilities located under the jurisdiction of their task managers. Each region also has its own social and environmental staff, located in technical units, and many of them share the central Environment Department's concern for improving compliance with Bank reform policies. But because these social and environmental staff members are located under the regional vice presidencies, they are also structurally located under the authority of the same operational apparatus responsible for the projects themselves. Social and environmental staff from the regions can travel on mission to assess the state of resettlement in Bank projects, but only at the request of the country departments. Moreover, country department contracts are one of their key sources of financing. When input regarding resettlement questions is necessary, country departments can choose to contract outside consultants instead of experts either within their regions or in the Bank center. As a result, the staff who normally assess operational compliance with social and environmental mandates are not fully independent of those they are evaluating.[39] It was in this context of conflicting demands that the regional technical units were charged with requesting data from the country departments in order to prepare the regional reports, which were the central inputs for the Bankwide assessment. These Technical Department staff were the task force's most important allies in their effort to deal with entrenched anti-reform project managers.

Project managers were the key actors in following the policy procedures and reporting data being aggregated by the task force. A not-for-attribution focus-group discussion found that "Task Managers understand and are committed to Bank resettlement policy, but eloquently describe the lack of structural integration of resettlement ideals into Bank practices and procedures." The survey found great diversity in the attitudes of task managers, ranging from firm commitment to the resettlement policy to outright rejection, with many in between. Some resented oversight by social policy specialists:

Just dumping a directive on the Task Manager will not solve the problem. The resettlement gurus of the Bank should be in the field and work with the implementing agencies and the government. Right now they are perceived as academic individuals who are more of a hindrance than a help in project processing. We have enough NGOs and others to cope with, and we do not need Bank paid staff to add to the problem.

Others questioned key policy procedures, such as baseline data about affected populations: "People think that the affected people don't want to be resettled (but it's not true) so there is an invasion of the potentially affected areas of *more* people receiving the benefits of being resettled.... Making lists of people doesn't work."[40]

In contrast, resettlement experts concur that baseline lists of affected people are nevertheless one of the most important tools for resettlement policy because they define the nature and scope of the problem, and set a benchmark for assessing the effectiveness of "rehabilitation."

Tension between the Task Force and the South

The work of the Asia/India review committee peaked at three points. The first occurred when a Regional Highlights document was submitted in June 1993; the second involved the completion of the India review—a separate document commissioned to the India Department by the Bank's board; and the third involved the bargaining over what information was to be included and deleted from the final Bankwide review. The official estimates of oustees grew at each stage, as table 9.1 shows.

The importance of the discrepancies in data on displaced people becomes clearer if one situates diverse projects in the context of the broader pattern. Table 9.1 compares official World Bank estimates of the number of oustees in the largest India projects and shows how these numbers changed over time. The first column lists the official estimate in each loan's staff appraisal report—the official document presented to justify the project at the time of its approval. According to Bank policy since 1980, baseline surveys and resettlement plans are already supposed to be in place at this point. The second column lists the data presented by the South Asia Region to the task force as of the June 1993 Asia Highlights report. The third column shows the oustee figures as of the ostensibly final Asia Regional Report to the task force at the end of 1993. The righthand column includes the eleventh-hour revisions, which were included in the final Bankwide review.

What is the explanation for the sharp discrepancy between estimates at the time of project appraisal and the figures presented at the end of the Bankwide review? One possible interpretation of the data in table 9.1 is that the populations to be affected grew during the period between project approval and the Bankwide review for normal demographic reasons. Some officials might also wonder whether some people moved to the affected areas to be able to claim compensation benefits. In the first case, the rate of growth indicated by the data is far too high to be explained by demographic factors. In the second case, the Indian government's track record in terms of providing compensation to project-affected people would hardly encourage outsiders to try to join in—especially because, in almost all states, only those with legal property titles are entitled to even promises of compensation. One Bank resettlement specialist hypothesized that India's projects called for such massive involuntary resettlement that many of those projects would have been economically unviable if the full resettlement and rehabilitation costs had been taken into account, leading to powerful incentives for both Bank staff and government project authorities to undercount "project-affected people."

When the original sources for the estimates are reviewed, two factors emerge that account for at least part of the discrepancies. First, some large projects included resettlement and rehabilitation provisions for people who had already been evicted and impoverished by previous projects. One resettlement specialist pointed out that India's project numbers "grew" in part because these past oustees were "rolled over" into new projects. For example, as public pressure grew, people who had been evicted long before by Maharashtra I and II were included as "add-ons" to Maharashtra III.[41]

A second major reason for the discrepancies in the figures involves the basic "unit of analysis" for resettlement planning. Instead of surveying the total affected population, many India projects had used a hypothetical "household" as the basic unit on which to base resettlement estimates, an arbitrary assumption that the average household had five members. This assumption had two very serious problems. First, families are on average much larger in many regions, and second, many households include multiple extended families, such as those of landless "major sons." In response to the task force's insistence on actual numbers of individuals affected, in several cases the India Department simply adjusted their arbitrary assumption of family size from five to six people, which explains why some of their project estimates increased by 20 percent increments. As a result, the final published numbers may well still undercount the actual affected populations. These different "technical" issues are mere reflections, however, of the broader underlying reason why approximately half a million people were officially ignored by the original project plans: the lack of public accountability of both the Indian government agencies and the World Bank authorities responsible for the projects.[42]

The conflict over policy compliance in India is an extreme case and is not representative of the Bank's resettlement portfolio in terms of numbers of projects. But India does account for a very large fraction of the *people* displaced by Bank projects worldwide; therefore, the egregious violations by a relatively small number of old guard staff loomed disproportionately large in the overall social impact of Bank operations.

The Final Report

The final report was characterized by discreet but fierce bargaining over both form and content. Once acceptable data was secured, conflict focused on the presentation of the information. Numerous internal Bank documents contest the definition of the shared goal: to be "factual yet balanced." The implication of this formulation was that too many uncomfortable facts presented too directly could lead to the appearance of "imbalance"—with the glass embarrassingly half empty rather than half full.

Internal documents consistently show task force members engaged in bureaucratic trench warfare to defend specific points, sections, tables, and boxes. They believed that serious dilution of the "lessons learned" would only be detrimental to the Bank, exposing it once again to repeated cycles of promises to improve, followed by noncompliance, public protest, scrutiny, and internal damage control. Task force members saw the review's intellectual integrity and frankness as being in the Bank's broader institutional interest. Like reformists in a wide range of institutions, they were willing to challenge what they saw as the shortsightedness of those old guard staffers whose recalcitrance threatened the interests of the institution as a whole.

Among senior management, the editorial debate focused on the "Executive Summary," which set the tone for the report as a whole. Some significant sections were removed, but the task force felt that their key findings were reflected in the final version. *Resettlement and Development* was released to the public on April 8, *before* it went to the board for presentation and approval, thus setting an important precedent. Although the Bank's new information disclosure policy created momentum for public release, the task force still had to overcome significant internal resistance. Their internal credibility was reinforced, however, by the fact that the report was not leaked to the public, as other critical internal reports had been.[43]

NGO activists anxiously awaited the report and criticized the lack of formal consultation with affected populations about either the data gathering or the draft of the report itself.[44] German and French NGOs were especially effective at organizing public pressure for its public release. Ever since confidential World Bank evaluations of failed projects made German headlines in 1993, the issue of forced resettlement provoked strong concern throughout German society and across the political party spectrum. German citizens sent thousands of postcards to their Ministry of Economic Cooperation and to the World Bank president, each bearing the famous quote from leading resettlement expert Thayer Scudder: "Forced resettlement is about the worst thing you can do to a people next to killing them." These preprinted postcards called on the Bank to

hold World Bank staff accountable for not complying with the Bank's guidelines on resettlement, take retroactive measures to rehabilitate those already impoverished through Bank projects, put all projects that will entail forced resettlement on hold, until alternatives are examined, rehabilitation measures are developed with affected peoples and monitoring systems are installed which ensure compliance with Bank guidelines, and make public the draft bankwide resettlement review, so that NGOs and affected peoples can have input before the document comes before the Board.[45]

The impact of these postcards on the World Bank was reportedly significant because it was the first such broad-based citizen campaign from a large western European donor, and Germany's executive director was paying close attention.

Although the NGO campaign did not manage to broaden the process of public debate of the review, the pressure appears to have helped to prevent internal Bank critics from possibly vetoing its public release. Task force members urged the Bank to respond directly to the NGOs in order to avoid the appearance of having something to hide.

The final report gives significant credit to Bank critics for contributing to improved performance:

the Bank shares the views of those critics who deplore bad resettlement operations. Their concern for the welfare of the displaced populations is fully justified—and germane to the Bank's own mandate and policies. In practice, criticism of resettlement failures by NGOs and other interest groups frequently has helped improve the Bank's policies and operations. Through its very decision to adopt a policy based on equitable principles and sound approaches, the Bank has delivered the sharpest criticism of bad displacement practice that cause impoverishment of those displaced. (p. 4)

This explicit recognition of the positive contribution of external scrutiny remained in the report in spite of strong objections from several very high-level officials who explicitly feared giving their opposition too much ammunition.

"Spin control" in the final report attempts to buffer both the external criticism and the internal backlash that were sure to follow its release. This framing strategy follows three main tracks. The first insists that the "glass was half full" because of improvement over time. The report concludes that policy implementation had been below Bank standards, but stresses that project planning had improved after 1991–1992. The second track is the so-called "small tail on the big dog" approach, which stresses the relatively small role Bank projects had played in displacement worldwide, the latter accounting for an estimated 2 to 3 percent of the total displaced people during the period studied. The third line stressed that treatment of affected populations was better in Bank-funded projects than in non-Bank projects.[46]

NGO reaction was mixed—supportive of the report process itself, but focused on the more critical findings (the half-empty glass).[47] The Environmental Defense Fund, one of the U.S. NGOs most active on this issue, praised the report "for its thoroughness and candor, with a strong urging that future Bank exercises to improve project quality follow the standard the report has set," but then noted that "the major finding of the report [pervasive noncompliance with policy] is not even mentioned in the Bank's Press Release nor in the letter from the Bank's President submitting the Report to the Board."[48] As major NGO Oxfam (United Kingdom and Ireland) put it, "The latest Bank report, though it tries to put a brave face on it, is a dismal catalogue of failure.... On the evidence of the report, the policy is not working.... The report makes a number of practical and useful recommendations [but] Oxfam believes that the Bank should not fund any new projects involving resettlement until such provisions are in place."[49]

Once *Resettlement and Development* was presented to the Bank's board, it received strong backing from several executive directors, including those representing the United States, Germany, Britain, France, the Netherlands, and Japan. For example, as the May 3 statement of the U.S. executive director noted:

> The question of resettlement has been a continuing, chronic concern. . . . Until the Bank does a better job in implementing [its] policies, it will not be a credible agent for change in this important area. . . . The Bank-wide review is a good start. We found the report to be thorough and candid—although we would have expected greater attention to indigenous people—and felt that it provides a good basis for future action. . . . This is a model for the Bank overall.

Though some executive directors were concerned that the report was released before their approval, the executive directors of major donor countries expressed relief that the report responded to the concerns of "critical publics" in their countries. Pro-reform executive directors then focused on reinforcing the push for remedial actions and follow-up.

Bank management agreed to produce a regular annual report, known as the regional remedial action plan, to institutionalize regular reporting on problem projects. Follow-up reporting on resettlement implementation, however, was left in the hands of the project task managers rather than left to an independent body. As the remedial action plan itself candidly notes, "positive ratings by Task Managers may be somewhat optimistic in light of the Bank's earlier experiences."[50] This limitation makes it difficult to draw strong conclusions from the follow-up data on policy compliance.

The remedial action plan reported significant improvements in project design, especially for the larger projects. Improving performance of ongoing projects was more difficult in the short term, and the Environmental Defense Fund has argued that several of the specific projects cited as having few problems actually remain quite controversial.[51] The original remedial action plan itself was heavily revised on orders from high-level managers in between its original presentation to the board in May 1995 and its eventual public release in November 1995. The revision and delay in publication suggest that it might not have been made public if pressure had not been exerted by letters of concern cosigned by the Environmental Defense Fund, the National Wildlife Federation, and the Sierra Club. In other words, the internal influence of resettlement specialists remains contested.

Conclusions

The resettlement review task force was able to make its precedent-setting breakthrough because it combined high degrees of both autonomy and authority. It would have been easy to imagine an evaluation unit with *either* autonomy *or* authority, but this group's unique feature was that it was able to exercise *both* the autonomy and authority needed to (1) extract controversial information from sometimes extremely reluctant operational staff and (2) make critical findings public. Unlike any other Bankwide review to date, internal or external, the resettlement review is the only one to cross-check systematically the information produced by inherently interested parties—the Bank staff and government agencies responsible for the projects themselves.

In reflections on the contending external pressure and internal-learning approaches to change in public sector organizations, internal learning must be disaggregated in terms of who learns what within an institution. Some recalcitrant staff reportedly did "learn," though others merely adapted. Optimistic reformers stressed the role of education as well as debate and confrontation in their work: "You simply cannot under-estimate how uninformed the old guard was. I honestly don't think the majority didn't care; they had no idea how their poor management of resettlement was affecting people's lives. When confronted with the evidence, they naturally wanted to cover up, but they also wanted to clean up the mess." A more independent assessment of staff motivations would require extensive ethnographic research.

The insider reformists also learned, but only after many years dedi-cated to what one might call "learning without leverage." Their fore-warnings and documentation of social disasters had little effect on the operational apparatus and borrowing governments until quite recently. The resettlement review finding that policy compliance for new projects improved significantly only in 1991–1992 suggests that learning about how to avoid or mitigate mass suffering, by itself, did little to make re-settlement policy a prority for much of the operational apparatus. The NGOs' political threat to donor government contributions added a new and intangible set of disincentives in 1991–92 for operational staff, thus increasing the potential cost of ignoring reform policies. After the Morse

Commission published its findings and the Inspection Panel was instituted, the need to head off potential external criticism reinforced the reformers' internal education and lobbying efforts.

Learning was also political: the reformers learned to isolate the most anti-reform elements within the Bank and borrowing governments, and at the same time tried to avoid being perceived as disloyal to the institution. Unlike many NGO critics, they firmly believed that their institution was reformable, but they agreed with the critics that resettlement was a critical test case. Many reformers—accustomed to fine-tuning their own internal critique and struggling to gain credibility with skeptical Bank colleagues—rejected what they saw as rhetorical, misdirected, and sometimes exaggerated criticisms of the Bank by U.S. and European advocacy NGOs. The more radical external critics, whose discourse treats the Bank as a monolithic institution, tarred internal reformers with the same brush as those most directly responsible for "problem projects," thus causing further resentment. Although dealing with NGO critics can provoke ideological and professional dissonance, insider reformists are nevertheless well aware that advocacy groups create an enabling environment that bolsters their own leverage. For all their differences, external critics and insider reformists agree that public transparency can be a major force for institutional accountability.

In conclusion, approaches that focus on external pressure alone ignore the diversity of interests and ideas within large bureaucracies. External pressure can change the internal balance of power within an institution so that reformers are at least sometimes *heeded* by those who actually control the money. Those who advocate internal-learning approaches, on the other hand, need to recognize the power of insider-outsider synergy and the central role of conflict, both inside and outside the organization. The World Bank's resettlement policy experience of the early 1990s suggests that it is the *interaction* between external pressure and internal reform initiatives that encourages public accountability.

Acknowledgments

This study did not receive official cooperation from World Bank staff. Four Bank social and environmental specialists did generously provide incisively critical comments and corrections on earlier drafts. Thanks also to Sanjeev Khagram for very insightful comments. None bear responsibility for the authors' findings or intepretations. Jonathan Fox's research was carried out while he was an International Affairs Fellow of the Council on Foreign Relations, with partial support from grants from the C. S. Mott and Ford Foundations to the Institute for Development Research. Eva Thorne, Ph.D. candidate in the Massachusetts Institute of Technology Political Science Department, provided excellent research assistance, supported by grants from the Institute for the Study of World Politics and the Center for International Studies at the Massachusetts Institute of Technology.

Notes

1. World Bank, Environment Department, *Resettlement and Development: The Bankwide Review of Projects Involving Involuntary Resettlement, 1986–1993* (Washington, D.C.: World Bank, April 1994). It was later reformatted and published by the Social Policy and Resettlement Division as *Environment Department Paper,* no. 032 in March 1996. (Washington, D.C.: World Bank). Quotations in the text cite the page numbers in the more recent edition.

2. Key civil society actors on resettlement issues include project-specific local protest movements and their international allies: the transnational Narmada Action Committee; Northern environmental and development NGOs groups such as the Environmental Defense Fund, the Berne Declaration, Urgewald, and the Oxfam network; North-South NGO bridging coalitions such as the International Rivers Network, World Rainforest Movement, Friends of the Earth, and INFID (Indonesia); developing country networks such as the Third World Network; as well as indigenous and human rights groups such as Survival International, Cultural Survival, and Human Rights Watch.

3. Thanks to Steven Van Evera for discussions about these alternative approaches. For a classic assessment in a broad public policy context, see Aaron Wildavsky, "The Self-Evaluating Organization," *Public Administration Review* 32, no. 5 (September/October 1972).

4. For a comprehensive discussion of the institutional-learning approach, see Ernst B. Haas, *When Knowledge is Power: Three Models of Change in International Organization* (Berkeley: University of California Press, 1990). *Learning* refers to "situations in which an organization is induced to question the basic beliefs underlying the selection of ends" (p. 36). Haas's broad overview of diverse international organizations finds that "adaptive behavior is common, whereas true learning is rare" (p. 37). He argues that the World Bank is one of these exceptions (although his analysis seems ambivalent at times). For a more recent

formulation, see Peter M. Haas and Ernst B. Haas, "Learning to Learn: Improving International Governance," *Global Governance* 1 (1995).

5. It is not clear to what degree this discourse is accepted at face value or is simply used to mask conflict between factions (for example, between those who want to implement social and environmental reforms and those who do not).

6. For conceptual elaboration on an interactive approach to the dynamics of reform, see Jonathan Fox, *The Politics of Food in Mexico: State Power and Social Mobilization* (Ithaca: Cornell University Press, 1992).

7. On the networking between social analysts inside and outside the Bank during this period, see Nuket Kardam, "Development Approaches and the Role of Policy Advocacy: The Case of the World Bank," *World Development* 21, no. 11 (1993).

8. Among the vast literature on resettlement, see Michael Cernea's annotated bibliography of World Bank research publications, *Sociology, Anthropology and Development* (Washington, D.C.: World Bank, Environmentally Sustainable Development Studies and Monographs Series no. 3, 1994); as well his many articles, including: "Social Science Research and the Crafting of Policy on Population Resettlement," *Knowledge and Power* 6, nos. 3–4 (1993); and "Social Integration and Population Displacement: The Contribution of Social Science," *International Social Science Journal*, 143, no. 1 (1995). See also Scott Guggenheim, *Involuntary Resettlement: An Annotated Reference Bibliography for Development Research* (Washington, D.C.: World Bank, Environment Working Paper, no. 64, February 1994). On the role of World Bank–sponsored research on resettlement, see Michael Horowitz, "Victims Upstream and Down," *Journal of Refugee Studies* 4, no. 2 (1991). Most of the literature on displacement is empirical, but one of the contributions to the early U.S. literature on urban "renewal" offers a conceptual discussion of the determinants of local resistance that turns out to be directly relevant to the current debate on the nature of social capital. See, among others, Mark Granovetter, "The Strength of Weak Ties," *American Journal of Sociology* 78, no. 6 (1973). On anti-dam movements more generally, see among others, Patrick McCully, *Silenced Rivers: The Ecology and Politics of Large Dams* (London: Zed Books, 1996); Anthony Oliver-Smith, "Involuntary Resettlement, Resistance and Political Empowerment," *Journal of Refugee Studies* 4, no. 2 (1991); and Sanjeev Khagram, "Dams, Democracy and Development: Transnational Struggles for Power and Water," Ph.D. dissertation, Stanford University Political Science Department, 1998.

9. On the role of Brazil's Sobradinho and the Philippines Chico River dam conflicts in the Bank's policy process, see Cernea, "Social Science Research" and "Social Integration."

10. Cited in "Anthropological and Sociological Research for Policy Development on Population Resettlement," in Michael Cernea and Scott Guggenheim. eds., *Anthropological Approaches to Resettlement* (Boulder: Westview, 1993), p. 24. See World Bank *Involuntary Resettlement in Development Projects: Policy Guidelines for World Bank Financed Projects* (Washington, D.C.: World Bank Technical Paper no. 80, 1988). They were published by the Bank in English,

French, and Spanish, and translated and published independently in China, Indonesia, and Turkey.

11. Cited in Lori Udall, "The International Narmada Campaign: A Case Study of Sustained Advocacy," in William F. Fisher, ed., *Toward Sustainable Development? Struggling Over India's Narmada River* (Armonk, N.Y.: M.E. Sharpe, 1995), p. 206. See also Udall's chapter in this volume.

12. See Bradford Morse and Thomas Berger, *Sardar Sarovar: The Report of the Independent Review* (Ottawa: Resource Futures International, 1992), p. xii. Bruce Rich, *Mortgaging the Earth* (Boston: Beacon Press, 1994); Udall, this volume; the discussion in several chapters in Fisher, *Toward Sustainable Development?*; Amita Baviskar, "Development, Nature and Resistance: The Case of Bhilala Tribals in the Narmada Valley," Ph.D. dissertation, Rural Sociology Dept., Cornell University, Ithaca, N. Y., 1992, and *In the Belly of the River* (Delhi: Oxford University Press, 1995). Vasuhda Dhagamwar, "Reflections on the Narmada Movement," *Seminar* 413 (1994); Jean Drèze, Meera Samson, and Satyajit Singh (eds.) *The Dam and the Nation: Displacement and Resettlement in the Narmada Valley* (Delhi: Oxford University Press, 1997). Sanjeev Khagram, "Dams, Democracy and Development ...") and his "Transnational Coalitions, World Politics and Sustainable Development: The Case of India's Narmada River Valley Projects," in Kathryn Sikkink, Sanjeev Khagram, and Jim Riker (eds.) *Restructuring World Politics: The Power of Transnational Agency and Norms* (Minneapolis: University of Minnesota Press, forthcoming). Rahul Ram, "Muddy Waters: A Critical Assessment of the Benefits of the Sardar Sarovar Project," New Delhi: Kapavrikshi (1993); and Joseph Schechla, "The Price of Development: Housing, Environment and People in India's Narmada Valley," Mexico: Habitat International Coalition (1992), among others. On the internal Bank politics of the Narmada project, see Robert Wade, "Greening the Bank: The Struggle over the Environment, 1970–1995" in Devesh Kapur, John P. Lewis, and Richard Webb, eds., *The World Bank: Its First Half-Century* (Washington, D.C.: Brookings Institution, 1997). On follow-up, see "Lessons from Narmada," *OED Précis* 88 (May 1995). See also the responses from the Narmada Bachao Andolan (letter from Shripad Dharmadhikary to World Bank executive directors, 2 June 1995) and from Narmada Bachao Andolan, "The Narmada Struggle: International Campaign After the World Bank Pull-Out," unpublished mimeo (1995). For theoretical discussions involving Narmada as the key case, see Brett O'Bannon, "The Narmada River Project: Toward a Feminist Model of Women in Development," *Policy Sciences* 27 (1994), as well as Roger Payne and Brett O'Bannon, "Environmental TSMOs and the Narmada River Dam: Rethinking Complex Interdependence," paper presented at the Workshop on Transnational Social Movement Organization, Notre Dame University, April 1994.

13. Morse and Berger, *Sardar Sarovar*, pp. xv–xvi.

14. Note that the Bank's India Department continued to exclude the population affected by the canal from the official "count" even after the loan was canceled (see table 9.1). According to internal Bank records, by the early 1990s staff began

to pay attention to the canal issue. They started to design a project called "Sardar Sarovar Canal," scheduled to begin in 1997 and officially estimated to affect 120,000 people. This project was dropped from the pipeline along with the cancellation of Sardar Sarovar (along with design plans for another massive dam along the same river, called the Narmada Sagar, once scheduled for 1996).

15. Morse and Berger, *Sardar Sarovar,* pp. xxiv.

16. Michael Cernea, interview by the author, Washington, D.C., 6 September 1995.

17. *Ibid.*

18. Note that these figures refer to people *to be* displaced, rather than those actually evicted during that period.

19. Projects are considered "closed" when loan disbursements are complete or canceled. Construction and resettlement could be incomplete, as in the case of several "problem projects" that were officially "closed" during 1993 and therefore dropped from the scope of the review.

20. One World Bank study estimated that fifteen families benefited for each family displaced (*India Irrigation Sector Review* [Washington, D.C.: World Bank, 1991]), but this study was probably based on the systematic undercounting revealed by the resettlement review. A nuanced independent study of one such huge irrigation project—the Bank-funded Indira Gandhi Canal—suggests that their benefits can be both exaggerated and socially concentrated. See Michael Goldman, "'There's A Snake On Our Chests:' State and Development Crisis in India's Desert," Ph.D. dissertation, Department of Sociology, University of California, Santa Cruz, 1994.

21. The review focused primarily on the issue of compliance with policies regarding the treatment of the displaced, much more than on the broader issue of the development logic of the investments themselves versus possible alternatives. The review text moves from the general principle that some projects justify resettlement to the implicit assumption that most Bank-funded projects justified the resettlement. For comprehensive environmental critiques of hydroelectric dams, see, among others, Edward Goldsmith and Nicholas Hildyard, *The Social and Environmental Effects of Large Dams,* 3 vols. (Wadebridge: Cornwall: Camelford Ecological Centre, 1984–1991); McCully, *Silenced Rivers*; as well as regular coverage in the *World Rivers Review.* One Bank resettlement expert points out that smaller dams are not always lower impact: "The Gujarat Medium Irrigation II, a project with more than 20 medium-size dam projects throughout the state, for example, generated less power, irrigated less land, displaced more people and left them in worse condition than would the controversial Narmada Sardar Sarovar Project" (Scott Guggenheim, "Development and the Dynamics of Displacement," paper presented at Workshop on Rehabilitation of Displaced Persons, Institute for Social and Economic Change, Myrada, Bangalore, India, n.d.), p. 28. For analyses of the trade-offs by the World Bank's senior environmental analyst, see Robert Goodland, "Environmental Sustainability and the Power Sector,"

Impact Assessment 12, no. 4 (1994); "The Environmental Sustainability Challenge for the Hydro Industry," *Hydropower and Dams* 1, no. 1 (1996); and "How to Distinguish Better Hydros from Worse: The Environmental Sustainability Challenge for the Hydro Industry," *The International Journal on Hydropower and Dams* (summer/fall 1996).

22. For an extreme case, note the experience of Guatemala's Chixoy Dam. According to recent testimonies of survivors, in 1982 the Guatemalan Army massacred 369 of the people to be displaced, more than half of the village of Rio Negro. See Julie Stewart et al., *A People Dammed: The Impact of the World Bank Chixoy Hydroelectric Project in Guatemala* (Washington, D.C.: Witness for Peace, 1996). The first of two loans predated the Bank's resettlement policy (1978), but a follow-up loan was given in 1985.

23. At the time of the review, India's power company reforms were new and their implementation untested. The Philippine government is also mentioned as having reformed its urban displacement laws in 1992, but implementation remains questionable. For example, a subsequent recent Japanese citizen fact-finding mission found that the Philippine government routinely ignored the law in Manila and Cavite (Nelson Badilla, "Tokyo Asked to Stop Aid to RP for Violation of Rights of Poor," *Manila Times*, 7 April 1996).

24. Human rights activists have criticized China's record with resettlement, primarily involving dams not funded by the Bank. See "The Three Gorges Dam in China: Forced Resettlement, Suppression of Dissent and Labor Rights Concerns," *Human Rights Watch/Asia Report* 7, no. 2 (February 1995); Dai Qing, ed., *Yangtze! Yangtze!* (London and Toronto: Probe International and Earthscan, 1994)—the Chinese edition was banned in 1989; and Lawrence Sullivan, "The Three Gorges Project: Dammed if They Do?" *Current History*, 94, no. 593 (September 1995).

25. Possibly related to the Bank's review process, resettlement policies were adopted in 1994 in Bangladesh and the Sindh Province in Pakistan (both governments had very poor track records).

26. Further on, the report notes, "to the extent that 'incrementalism' is used as a substitute for the resettlement planning defined by Bank policy, the field record of its failure ... is clear" (p. 157). In the Upper Krishna II project, however, "prompt action by Bank management [in suspending disbursements] sent the Borrower a clear signal that resettlement performance counted as much as performance on other project components.... [I]nsisting on full compliance with the Bank's benchmarks, rather than hoping for incremental improvements, led to major improvements in the Borrowers' approach" (p. 158).

27. World Bank, *Annex V: Resettlement Supervision*, rev. version (Washington, D.C.: World Bank, July 1993), pp. 2–3.

28. *Ibid.*, p. 3. This review also confirmed that the absence of Bank funding for resettlement components "reinforces the 'externality' of resettlement in relation to the project" (p. 4).

29. Michael Cernea, interview by the author, Washington, D.C., 6 September 1995.

30. Cernea contrasted the resettlement review with the Wapenhans Report on portfolio management, which did not involve staff in the field and therefore, in his view, did not change staff behavior.

31. Anti-displacement grassroots movements in Brazil are perhaps the most notable. See, among others, Leinad Ayer de O. Santos and Lucia M. M. de Andrade, eds., *Hydroelectric Dams on Brazil's Xingu River and Indigenous Peoples* (Cambridge: Cultural Survival, 1990); Barbara Cummings, *Dam the Rivers, Damn the People: Development and Resistance in Amazonian Brazil* (London: Earthscan, 1990); Anthony Hall, "From Victims to Victors: NGOs and the Politics of Empowerment at Itaparica," in Michael Edwards and David Hulme, eds., *Making a Difference* (London: Earthscan, 1993); Mark D. McDonald, "Dams, Displacement, and Development: A Resistance Movement in Southern Brazil," in John Friedmann and Haripriya Rangan, eds., *In Defense of Livelihood: Comparative Studies on Environmental Action* (West Hartford: Kumarian Press, 1993); Maria Stela Moraes, "No rastro das aguas: organiçao, liderança e representatividade dos atingidos por barragens," and Franklin Daniel Rothman, "A emergencia do movimento dos atingidos pelas barragens da bacia do rio Uruguai, 1979–1983," both in Zander Navarro, ed., *Política, protesto e cidadanía no campo* (Porto Alegre: Universidade Federal do Rio Grande do Sul/MAB-RS/CE-TAP, 1996); Lygia Sigaud, *Efeitos sociais de grandes projetos hidreletricos: As barragens de Sobradinho e Machadinho* (Rio de Janeiro: Federal University of Rio de Janeiro, National Museum, 1986); and Aurelio Vianna, *Etnia e territorio: Os Poloneses de Carlos Gomes e a luta contra as barragens* (Rio de Janeiro: CEDI, 1992).

32. The most controversial ongoing Bank-funded project involving displacement in Latin America is the Yacyretá Dam between Argentina and Paraguay. NGO and Bank assessments of resettlement issues continue to conflict (though the Bank is also critical of project performance). See, among others, Gustavo Lins Ribeiro, *Transnational Capitalism and Hydropolitics in Argentina: The Yacyretá High Dam* (Gainesville: University Press of Florida, 1994), and Ramón Fogel, "La represa de Yacyretá: Beneficiados y perjudicados," paper presented at the Latin American Studies Association, Washington, D.C., September 1995.

33. One Bank resettlement expert cited the following cases: Upper Krishna Irrigation II (suspended in 1992, suspension lifted in 1994, and suspended again in 1995); Karnataka Power I and II (suspended and then canceled in 1993); Maharashtra Composite Irrigation III (postponed closing four times and restructured to "retrofit" past resettlement problems); Gujarat Medium Irrigation II (postponed closing five times to "encourage proper R&R"; the Bank pushed very hard and succeeded in getting the government of Gujarat to hire SEWA, a participatory and effective NGO, to assist the R&R, especially in microenterprise work with women); as well as the retrofitting of the Farakka Thermal, Dudichua Coal, and Jharia Coal projects. (Interview with author, Washington, D.C.: May 1996.)

34. The result was the confidential *Resettlement and Rehabilitation in India: A Status Update of Projects Involving Involuntary Resettlement,* 2 vols. (Washington: World Bank, 1994).

35. On displacement issues in India, see, among others, Walter Fernandes and Enakshi Ganguly-Thukral, *Development, Displacement, and Rehabilitation: Issues for a National Debate* (New Delhi: Indian Social Institute, 1989); Enakshi Ganguly-Thukral, ed., *Big Dams, Displaced People: Rivers of Sorrow, Rivers of Change* (New Delhi: Sage Publications, 1992); Walter Fernandes and Samyadip Chatterji, "A Critique of the Draft National Policy," *Lokayan* 11, no. 5 (March/April 1995); Smitu Kothari, "Developmental Displacement and Official Policies: A Critical Review," *Lokayan* 11, no. 5 (March/April 1995); McCully, *Silenced Rivers*; Hari Mohan Mathur, ed., with Michael Cernea, *Development, Displacement, and Resettlement: Focus on Asian Experiences* (New Delhi: Vikas, 1995); S. Parasuraman, "The Anti-Narmada Project Movement in India: Can the Resettlement and Rehabilitation Policy Gains be Translated into a National Policy," Institute of Social Studies, Working Paper Series, no. 161 (1993); Jai Sen, "National Rehabilitation Policy: A Critique," *Economic and Political Weekly,* 4 February 1995; and "Displacement and Rehabilitation," *Economic and Political Weekly,* 29 April 1995.

36. Morse and Berger, *Sardar Sarovar,* pp. 54, 56.

37. Cited in Janet Mancini Billson, "Complexities of Involuntary Resettlement in World Bank Projects: Task Manager Focus Group Report," (Washington, D.C.: World Bank, 1993), p. 29.

38. See World Bank, *Early Experience with Involuntary Resettlement: Overview* (Washington, D.C.: World Bank, 1993), p. v. As Walter Fernandes and Nita Nishra, notable experts on Indian resettlement, note: "Despite the displacement of nearly twenty million persons during the last four decades, till today the country as a whole does not have a rehabilitation policy. The human factor, i.e., the welfare of the Displaced Persons/Project Affected People itself has begun to be taken into consideration only in recent years, mainly because of pressure from human rights and environmental activists, as well as from external agencies like the World Bank" (p. 3). Compensation is still considered welfare rather than a right. In terms of responsiveness, they find that "till now the governments, project authorities as well as the Bank seem to have paid little management attention to R&R except in so far as it threatened to hinder project implementation or where people were in danger of being drowned or agitated" (p. 32). See "The Impact of World Bank Resettlement Guidelines on the Processing and Quality of Bank-Funded Projects in India," unpublished paper, September 1993. For contrasting overviews of Bank-India relations, see S. Guhan, *The World Bank's Lending in South Asia* (Washington, D.C.: Brookings Institution Occasional Papers, 1995), and Public Interest Research Group, *The World Bank and India* (New Delhi: PIRG, 1995).

39. For the most comprehensive external analyses of how these internal accountability problems limit reform policy implementation, see Bruce Rich,

"Memorandum: The World Bank After 50 Years: No More Money Without Total Institutional Reform," (Washington, D.C.: Environmental Defense Fund, 1993; and U.S. House, Bruce Rich, speaking on behalf of the Environmental Defense Fund, National Wildlife Federation, Sierra Club, and Greenpeace, *Hearing before the House Committee on Banking and Financial Services, Subcommittee on Domestic and International Monetary Policy, Concerning the World Bank Effectiveness and Needed Reforms* (27 March 1995).

40. Billson, *Complexities of Involuntary Resettlement*, pp. 5, 16, 18.

41. Note that the MCIP III estimate in table 9.1 for December 1993 is higher than the final figure. This discrepancy was the result of a typographical error in the Asia regional report, which simply added a zero to the original very low estimate by accident.

42. One of the most striking internal conflicts over information presentation unfolded around the publication of the India Department's separate resettlement portfolio review, scheduled for submission to the Board at the same time as the task force's Bankwide report. Two particular issues are noteworthy: the radically different tone of the two reports and the protectiveness with which the India Department appears to have treated its own portfolio review. The major point of contention was the issue of population displacement data provided (or not) by the India Department both to their own Asia technical staff and to the task force. The report was scheduled to go to the board in late April 1994. In early April, the Bank's South Asia Region submitted new data for several projects to the task force, the ramifications of which were quite serious. The task force had been making complex assessments based on Asia Technical Department figures that were then four months old. It was not until after the Bankwide review had been completed that the task force received the new numbers, involving ninety thousand "new" oustees in at least sixteen projects in India. Some projects increased by twenty to twenty-five thousand people (see table 9.1). Given the public scrutiny of Bank activities, as well as the special interest generated by a Bank-executed review, two simultaneous reports with different data would have cast doubt over the entire *process*, thereby calling into question the credibility and seriousness of the review effort.

43. Bank senior management did fear that advocacy NGOs had managed to get a copy of a late draft, causing stress levels among insiders to peak. Bank managers had been forwarded an indiscreet NGO e-mail message that hinted that a copy of the report was already in their possession. Top managers therefore suspected that if the report were not released, it would be made public without the "spin control" needed to strike the appropriate balance between good and bad news. This fear led management to overrule internal pressures to keep the report confidential (Bruce Rich, Environmental Defense Fund, interview by author, Washington, D.C., May 1996).

44. Some called for postponing its presentation to the board until a broader consultation process was held. The April 13, 1994, letter from the World Rainforest Movement, signed by NGO leaders from seventeen countries in both North

and South called on the Bank to make the review public "in the languages of affected groups, so that affected peoples and NGOs can have input before it comes before the Board."

45. The German Economic Cooperation Ministry was swamped by this new tactic because its procedures required them to send out individual responses (Heffa Schucking, Urgewald, interviews by author, Washington, D.C., October 1995 and June 1996). Schucking noted that Germans responded well to the campaign in part because of their direct historical experience with massive forced resettlement. They followed up with direct action and the delivery of twenty thousand solidarity thumbprints to symbolically represent evicted people. In France, the human rights organization FIAN, together with environmental groups, distributed more than forty thousand preprinted cards with the headline "Forced from Their Lands by the World Bank" and the subhead "A Policy with No Effect" (Susanne Hildebrand, personal communication, 9 November 1995). The campaign was timed to send the cards to the French executive director by April 7 to influence the upcoming board meeting. This effort was a new level of campaigning in France, where Bank reform efforts are less developed than elsewhere in Europe.

46. These two affirmations are key to the report's public presentation, and they may well be true, but because the study focused only on Bank-funded projects, its statements about other projects involving resettlement are based largely on the existing literature, specialists' inferences, and extrapolation of broad trends.

47. The effort to craft this combined message produced friction when a draft of the main joint/internal NGO letter included criticisms of the proposed creation of a "resettlement industry" within the Bank. The intent was to focus on the issue of reducing resettlement rather than simply mitigating its impact, but the choice of words left Bank reformers aghast at the questioning of their effort to increase Bank resources devoted to improving resettlement implementation. The NGOs left in the critique of a growing "resettlement industry" because they contended that the Bank should stop funding projects that involved forced resettlement. More Bank social science supervision of resettlement, in this view, at best merely mitigated an illegitimate process.

48. Environmental Defense Fund, "The Bankwide Resettlement Review: Issues and Questions," unpublished memo, Washington, D.C., 2 May 1994.

49. Patricia Feeney, "Oxfam's Response to the World Bank Report: *Resettlement and Development*," unpublished memo, Oxford, U.K., 27 April 1994.

50. World Bank Environment Department, *Regional Remedial Action Planning for Involuntary Resettlement in World Bank Supported Projects: A Report on One Year of Follow-Up to Resettlement and Development, the Report of the Bankwide Resettlement Review*, (Washington, D.C.: World Bank, 1995), p. 37.

51. Few improvements in public participation or information access were claimed. According to the Environmental Defense Fund, "Even more disturbing is the fate of some 632,000 people in 22 projects that were closed or cancelled in the year

period between the issuance of the Bankwide Resettlement Review and the completion of the [Remedial Action Plan] ... although most of the people affected have been, or will be, deprived of their livelihoods" (for example, Thailand's Pak Mun Dam, Argentina's Yacyretá I and II loans, Cote D'Ivoire Forestry, India's Upper Krishna II, Upper Indravati Hydro, and the NTPC [Singrauli] power loan). Cited in Environmental Defense Fund, "Memorandum: Final Report of the World Bank on Remedial Action Planning for Involuntary Resettlement," Washington, D.C., November 1995 draft.

10

Reforming the World Bank's Lending for Water: The Process and Outcome of Developing a Water Resources Management Policy

Deborah Moore and Leonard Sklar

Water scarcity and pollution problems are pervasive around the world. Since 1940, water use has quadrupled while the global population has doubled in size.[1] To meet growing demands, governments, development institutions, communities, and nongovernmental organizations have invested in various water development projects to increase supplies available for irrigation; urban, rural, and industrial uses; and hydropower generation. Other water projects have focused on flood control, water treatment, sanitation and sewage, navigation and ports, and fisheries development.

The World Bank is the single largest source of funds for water projects in the world and thus plays a significant role in determining the kinds and number of water projects identified and financed. The water sector comprises a significant portion of World Bank lending, accounting for about 15 percent of the Bank's cumulative lending.[2]

Despite this large investment in water development projects, water problems have continued to worsen. In 1990, more than 1.2 billion people lacked access to adequate drinking water supplies, and more than 1.7 billion people lacked access to adequate sewage and sanitation services.[3] Contaminated water causes more than 80 percent of diseases in the developing world and more than one-third of all deaths, including those of 4–5 million children per year from diarrhea.[4] About 125,000 hectares of irrigated land becomes uncultivable annually due to waterlogging and salinization, despite the world's need to expand agricultural production.[5] Wetlands and fisheries have been seriously degraded worldwide. These problems will be further exacerbated by the greater demands of larger populations and the prospects of reduced water availability due to global climate changes.

Given our inability to meet adequately human and environmental needs for water—in spite of significant financial investments in the sector—it is important to evaluate whether both water and financial resources are being allocated and used in the most effective ways possible. To meet existing and future needs for water and to sustain aquatic ecosystems and fisheries, changes in how water resources are managed are necessary—at local, national, regional, and international levels.

This chapter examines the World Bank's new water resources management policy and the efforts of nongovernmental organizations (NGOs) to influence the policy development process and outcome. The policy was developed partly in response to criticism of the Bank's water-lending priorities and practices by NGOs as well as by governments and internal Bank reviews. For NGOs, the water policy presented an opportunity to try to move beyond efforts to change Bank water lending on a project-by-project basis by addressing the sector more comprehensively, as a whole. And by pressing the Bank to conduct a meaningful consultation with them on the water policy, NGOs also hoped to set a precedent for establishing lasting guidelines for future Bank-NGO consultations. We describe internal and external elements of the consultative process, the initial guidelines for the process, and the ultimate outcomes. In order to place the water policy in context, we examine the types and distribution of water projects funded by the Bank and describe the factors that led the Bank to develop the new policy. Lastly, we evaluate the first few years of the Bank's efforts to implement the new water policy and examine its influence on water projects now in the Bank's portfolio.

The Water Sector

Freshwater is supplied in a variety of ways, depending on the sources of water available, the type of use, and the economic and technological resources accessible: tapping into groundwater aquifers via shallow wells and hydraulic pumps is most common; water can be diverted directly from rivers and lakes; dams and reservoirs provide storage capacity to capture water to supply over longer periods; and, less commonly, desalination plants and wastewater-reuse systems are used to supply freshwater.

Worldwide, agriculture is the largest user of water supplies, accounting for about 69 percent of water diversions and about 85 percent of consumptive use of water.[6] Domestic and municipal uses account for about 8 percent, and industrial and energy uses account for 23 percent. In many areas, current levels of water consumption are unsustainable, withdrawing more water from underground aquifers, lakes, and rivers than can be replenished. Moreover, water delivery systems and water uses are often wasteful and inefficient: In developing countries, irrigation efficiency is typically 25–30 percent, compared to 50 percent and higher under best practices.[7] These figures indicate that more water is diverted and delivered to the field than is needed to grow the crop, thereby reducing the available water supplies for other uses. Similarly, lost and unaccounted for water can exceed 50 percent of the available supply in some cities, such as Manila in the Philippines.[8] Although some of the lost water may reappear as return flow to streams and aquifers, and may be available for other users, the return flow is of lower quality and may be unavailable at needed times.

Many of the most easily accessible water resources have already been developed, thus making it more expensive and more difficult to locate, develop, and finance additional water supplies. In addition, more competition over the scarcer remaining supplies and previously ignored or not understood ecological consequences to water development further complicate water sector issues. Better management and conservation of the world's existing water supplies can, however, extend our ability to meet people's needs without developing new water sources. One water expert estimates that with existing technologies we could save 10–50 percent in irrigation uses, at least 30 percent in municipal and domestic uses, and 50–90 percent in the industrial sector without reducing our standard of living.[9] Saving "consumptive losses" can reduce existing uses of water, thereby freeing up existing water supplies to meet new demands.

Trends in World Bank Water Sector Lending

Water projects, particularly large-scale dams, have been part of the World Bank's portfolio of projects since its inception. In 1949, the

Bank's first full fiscal year of lending outside of Europe, 79 percent of Bank loans were for construction of large dams.[10] In 1993, the World Bank estimated that its cumulative lending for water projects since its inception amounted to $40 billion, in nominal terms: $19 billion for irrigation and drainage, $12 billion for water supply and sewerage, and $9 billion for hydropower projects.[11] Approximately $28 billion of these loans were to support 527 dam-related loans for 604 dams in 93 countries,[12] which amounts to $58 billion when adjusted for inflation (in 1993 dollars).

The 1980s were marked by the recognition that a shift in focus toward improving access to safe, clean water supplies and sanitation was needed. The United Nations declared the 1980s the International Decade on Water Supply and Sanitation; the goals of the decade, established at the HABITAT Conference in 1976, were to provide "service to all" and to extend safe, clean drinking water supplies and sanitation services to everyone by 1990.[13] The international agreement to focus on these critical problems was forged at the Mar del Plata International Conference in 1977, where an action plan was also adopted. Together, the International Decade and the Mar del Plata action plan were intended to increase investments in the drinking water and sanitation sectors, to coordinate better the activities of donor agencies, and to reform water management practices at local, national, and international levels.[14]

During the International Decade, 1981–1990, total global investment in the water sector was more than $100 billion, with national governments contributing about 65 percent and international agencies, such as the World Bank, contributing the remaining one-third.[15] World Bank loans and credits for all water projects for this period were about $27 billion (nominal), or $35 billion when adjusted for inflation (1993 constant dollars).[16] Despite rhetorical support for the new priorities, the Bank's lending practices throughout the International Decade reflected a business-as-usual approach. The breakdown of Bank loans and credits during the 1980s across the subsectors is shown in figure 10.1. Approximately $9.7 billion of the $35 billion was for support of combined rural and urban water supply and sanitation projects, with another $3 billion for multipurpose development projects of which water supply was a component. Despite the enormous need for water services in rural areas,

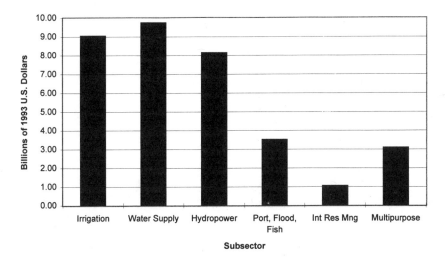

Figure 10.1
World Bank lending for water projects (1981–1990), allocation by subsector.
Source: See note 17. *Note:* "Int Res Mng" designates integrated resource management projects—e.g., soil/water management and water basin management projects. "Multipurpose" designates development projects in which one component is for water.

only about $1.2 billion was allocated directly for support of rural water supply projects—or about 3 percent of the Bank's total support for water projects. More than $17 billion of the loans and credits were for irrigation and hydropower projects, with large dam-related loans accounting for about $12 billion. Almost half of the World Bank's funds were allocated to irrigation and hydropower projects, in spite of the consensus to focus on the water supply and sanitation subsector.

It is also important to examine the 1981–1990 allocation of resources among the types of projects—including new infrastructure, expansion of existing infrastructure, rehabilitation of existing infrastructure, and alternative water management projects, such as small-scale irrigation, water conservation, and watershed development projects (as shown in figure 10.2). The vast majority of Bank loans and credits supported new infrastructure and expansion projects, amounting to about $23.7 billion or 67 percent of total water sector lending for 1981–1990. Rehabilitation and operations and maintenance projects amounted to about 18 percent,

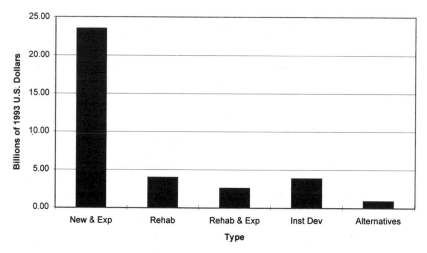

Figure 10.2
World Bank lending for water projects (1981–1990), allocation by type of activity.
Source: See note 17. *Note:* "Alternatives" include watershed development, water conservation, and small-scale projects. "Inst Dev" includes technical assistance, sector reform, and institutional development projects.

or $6.3 billion. Alternatives such as small-scale irrigation, watershed development, and water conservation projects amounted to only $815 million, or 2.3 percent of lending between 1981 and 1990. Technical assistance and institutional development efforts, many for new infrastructure, accounted for the rest.[17]

Performance of World Bank–Financed Water Projects

During the International Decade drinking water supplies were extended to 1.3 billion people and sanitation services were provided to about 750 million people.[18] Global investment in the water sector did result in tangible gains. However, many wasteful practices and inefficient water management patterns persisted in inhibiting efforts to extend water supplies further and in contributing to environmental degradation.

World Bank–financed projects in the water sector have contributed to both the gains and the inefficiencies of the sector. Water supply and irrigation projects have been performing poorly for years, according to the World Bank's own internal reviews. Its Operations Evaluation Department (OED) found that for twenty-one irrigation projects approved between 1961 and 1978, agricultural production actually declined after project completion in 43 percent of these projects, and 70 percent of projects had a lower economic rate of return than estimated at completion.[19] Cost overruns were, on average, 40 percent of the project's estimated cost at appraisal, which contributed to lower economic returns. Most of the projects were less durable than anticipated and would not meet their expected useful life without major rehabilitation, meaning that additional investments would be necessary to projects already performing poorly. Water efficiency was lower than anticipated in 60 percent of the projects.

Another OED report showed that for 129 water supply and sewerage projects reviewed, more than half suffered from cost overruns, almost all had economic rates of return below 10 percent, and 64 percent had excessive levels of lost and unaccounted-for water.[20] Operation and maintenance activities were neither adequately funded nor carried out by utilities, governments, nor the Bank.

In India, one of the Bank's largest borrowers during the 1980s, the irrigation sector was found to be grossly mismanaged. The Bank's *India Irrigation Sector Review* pointed to "poor sector planning and financial management, and inadequate water management and maintenance, as the main causes of poor performance." The review continued:

"Over the past decade the situation appears to have worsened: lack of financial discipline and accountability [e.g. corruption], neglected maintenance, and construction abuses have become endemic. Past investment in high-cost construction of new schemes has led to a substantial backlog of incomplete works. With rare exception, there is no justification in the medium term for new surface irrigation investments."[21]

In contradiction to the Bank's stated mission to alleviate poverty, Bank-funded water projects have often not targeted poor communities or small farmers. Only two of the 129 water supply and sewerage projects reviewed by the OED demonstrably succeeded in improving conditions for poor households; fifteen others failed; twenty claimed success but offered no means of measuring it; and fifty-two projects did not address poverty (twenty-one of which were started before the Bank's new commitment to poverty alleviation). In many projects, poverty issues were simply not incorporated into project design because of the lack of information about the poor and because of pressure for services from influential groups.[22] In 43 percent of the twenty-one irrigation projects reviewed, large farmers were found to have captured the bulk of the benefits.[23] Increases in incomes associated with irrigation in India were 84 percent for marginal farmers, compared with increases of 130% for large farmers.[24] In general, irrigation and water supply projects have tended to subsidize the middle and wealthier classes, who are better able to pay for public services than the poor.[25]

Water projects, especially large dams and irrigation projects, have entailed the resettlement of millions of people. Ongoing and completed World Bank projects between 1986 and 1993 are forcibly resettling 2.5 million of the world's poorest people (see Fox, chapter 9 in this volume),[26] two-thirds of whom are being displaced by water projects. The Upper Krishna Irrigation Project and the Subernarekha Dam in India, the Akosombo Dam in Ghana, the Bayano Dam in Panama, and hundreds of other Bank-financed dams have turned poor and indigenous people into refugees—"oustees" who rarely improve their standard of living and most often experience a diminished standard of living.

Large dams, agriculture, navigation, and other water development projects are primary causes of the loss and degradation of more than 50 percent of the world's wetlands.[27] Many commercial and subsistence fisheries are on the verge of collapse because of dams, river development projects, and pollution. The Bank has rarely done adequate environmental assessments on its dam projects. According to a 1989 review of the Bank's environmental performance in lending for large dams, its assessments often have been done after-the-fact, have not been thorough, do not include economic analysis of environmental factors affecting the performance of the dam, or show an incomplete understanding of the significance of some of the impacts from dams.[28] Lastly, the Bank has maintained a continued focus on "mitigation" of environmental impacts, rather than a focus on the avoidance of negative impacts in the first place via changes in project design, operation, or siting, among other possible changes.

According to the Wapenhans Report, a general review of the Bank's entire portfolio, project performance is of serious concern.[29] The share of projects with "major problems" increased from 11 percent in 1981 to 13 percent in 1989, and 20 percent in 1991. For 1991, 43 percent of water supply and sanitation projects and 42 percent of agriculture projects in their fourth or fifth year of implementation reported major problems. The number of projects judged unsatisfactory at completion increased from 15 percent of those reviewed in 1981 to 37.5 percent of those reviewed in 1991.

The poor performance record of Bank-financed water projects—in terms of water delivered, agricultural production, efficiency, and economic rates of return—should force both donor and borrower countries to examine the effectiveness of the Bank's approach in meeting its stated development and poverty alleviation goals. In general, too great a focus has been placed on construction, while the project's operations are neglected.[30] Distortions throughout the project process include biases toward capital-intensive projects, lack of local input, lack of accountability, and failure to budget for operation and maintenance activities.[31] For example, 75 percent of the financial covenants were not in compliance in Bank water supply projects financed between 1967 and 1989,[32] an indication of a lack of accountability that contributes to the Bank's poor project performance. In addition, the "pressure to lend" contributes to the bias toward large-scale projects so as to maximize the flow of aid.

Moreover, the failure to properly and promptly perform *ex post* evaluations of projects inhibits learning from past mistakes. This lack of evaluation "helps perpetuate the illusion of productivity of aid projects and avoids the embarrassment of letting the donor country home folks know the fate of the projects."[33]

These poor performance trends are of serious concern, providing a context that helps to explain the controversies associated with individual Bank-financed water projects discussed in other chapters in this volume. Indeed, as cited above, the Bank recognizes these failures in its own internal evaluations:

failure to address water resource issues in a comprehensive manner; improper attention to financial covenants and inadequate cost recovery; lack of accountability, autonomy, and flexibility in water management; inadequate investment in sewage treatment and in drainage systems; inadequate concern for poverty relief;

neglect of operations and maintenance; delayed and poor-quality construction; lack of consideration of environmental assessments and pollution control; inadequate concern for project sustainability; and lack of programs to address erosion problems in upstream watersheds.[34]

In 1990, due to the controversies surrounding specific Bank-financed dam projects, the internal evaluations documenting poor performance, and the general trend within the Bank to formulate sectoral policies, the Bank embarked on developing a "water resources management policy." The new policy was intended to guide the Bank's investments in the water sector and to institutionalize reforms to correct the sector's poor performance.

The Process of Developing the World Bank's Water Resources Management Policy

The Bank had known for some time about many of the problems with poor project performance, but it had not communicated these lessons effectively to relevant agencies and officials in borrowing and donor country governments. NGOs, among others, have tried to bring the Bank's activities "to light" and to increase public awareness and scrutiny of their activities and use of public—that is, taxpayers'—funds. Because NGOs are not directly involved with regulating the Bank, as are government officials, and because they are often not direct recipients of Bank funds (although many of them are directly involved with implementing Bank-financed projects), they can be freer to question the purposes and results of Bank efforts. In addition, NGOs can often give the Bank and government officials a "ground-level" perspective on the outcomes of development projects because they work directly with or represent project beneficiaries and people at the local level. For all these reasons, NGOs have important qualifications, information, and interests that are relevant to World Bank policies.

For NGOs concerned about the negative environmental and social impacts of Bank-supported projects, reform of Bank policies has offered a potentially more effective focus than opposing individual destructive projects because a single policy applies to all relevant Bank projects. In addition, Bank policies often become international standards to which borrowing governments, and other international institutions and corporations, can be held accountable, to some degree. However, so far the experience with Bank policies intended to reform the most egregious aspects of lending practices has been disappointing. For example, first established in 1980, the Bank's policy on involuntary resettlement has been poorly understood by Bank staff and widely unenforced, as documented in a 1994 internal Bank policy review.[35]

In practical terms, the primary value to NGOs of the Bank's environmental and social policies has been in using them to gain the attention of donor governments and the Bank's executive directors when challenging the Bank's financing of particular environmentally and socially destructive projects that violate key aspects of the policies. Although NGOs working on the water policy campaign hoped to see far-reaching changes in the Bank's approach to water resources lending, a more realistic objective was to win the inclusion of a few key points in the water policy specifically—requiring, for example, consideration of alternatives and valuation of environmental "uses" of water, which could strengthen NGOs in future battles over Bank-funded projects. Key passages in other policies—such as the stipulation in the Bank's policy on involuntary resettlement that people forced to move to make way for Bank-funded projects should not, as a result, suffer a decrease in income—have been very useful in pressuring donor governments to oppose funding for potentially destructive projects that involve large population displacements.

In recent years, the World Bank has increasingly talked about "participatory development": promoting consultation and cooperation between NGOs, the Bank, and borrower governments in the implementation of Bank-funded projects.[36] In much more general terms, the Bank has also publicly accepted the value of maintaining a "policy dialogue" with NGOs,[37] which can help the Bank to view the impacts of its investments from the perspective of the intended beneficiaries and from the point of view of the local communities most affected. Moreover, NGOs are often more able and willing to speak in frank and practical terms than the academics and government officials that the Bank otherwise relies upon for outside feedback.

Nevertheless, the Bank does not have any guidelines or established procedures for how or even whether to involve NGOs and civil society actors in the process of developing a new policy.[38] Not coincidentally, the Bank has a poor record of working with NGOs in the process of developing new policies. Its revised forestry policy, developed in 1990 immediately prior to work on the water policy, is a case in point.

The Bank held what it called a "consultative meeting" with NGOs to discuss the draft forestry policy in late April 1991, less than a month before the policy document was scheduled to be formally presented to its board of executive directors for approval. It was unwilling to fund the travel and other expenses required to convene the meeting, however, thus forcing NGOs to turn to a private foundation for financial support. No translations of the draft policy were made, limiting the ability of non-English-speaking NGO participants to critique the document. And during the two days of discussions, NGOs discovered that they were not even in possession of the most recent draft of the policy and that the Bank staff attending the meeting had completed their work on the document, with responsibility for final editing having been transferred to another department not represented at the meeting.

When it became clear that the Bank did not come to the meeting with the intention of incorporating NGO views into the policy document, Bank staff claimed that the meeting was merely an "airing of views" and was always intended to be only advisory. In contrast, the Bank's final published forestry policy paper claims that consultations with representatives of nongovernmental organizations "have led to substantial changes in the Bank's policy and approach."[39]

NGO efforts to encourage genuine consultation on the World Bank's water resources management policy were led by two U.S.-based organizations, the International Rivers Network (IRN) and the California office of the Environmental Defense Fund (EDF), two members of whose staff are also coauthors of this chapter. IRN and EDF's efforts resulted in participation by more than fifty other NGOs from thirty-two countries. The experience of both organizations legitimated their interest in both the process and the content of the new World Bank policy. The Berkeley, California–based International Rivers Network founded in 1986, serves a constituency of eight hundred environment and development NGOs in ninety countries by providing information and advocacy for socially and environmentally sound management and restoration of rivers and watersheds.[40] Since its inception, IRN had closely monitored and attempted to influence World Bank practices, particularly those related to large-scale river development schemes. The 300,000-member Environmental Defense Fund, a U.S.-based national environmental organization founded in 1967, researches and promotes solutions to an array of environmental problems, both domestic and international.[41] Since the early 1980s, EDF's international program had initiated campaigns to reform specific multilateral development bank-financed projects, including the Sardar Sarovar Dam in India, and monitored the environmental and social effects of U.S. foreign aid via bilateral and multilateral institutions. Both organizations collaborated extensively with other NGOs around the world.

EDF and IRN's combined efforts focused not just on the substance of the Bank's policy, but also on the process by which the Bank interacted with NGOs as it developed the policy document and the degree to which the concerns and views of NGOs involved in water issues would be reflected in the policy itself.

The Ford Foundation supported both EDF and IRN to encourage diverse international input on the water policy, to promote a participatory and constructive policy-making process with the Bank, and to influence the substance of the Bank's water policy. IRN invited participation of its wide network of members specifically concerned with river and watershed development; it also informed its members of progress on the policy through its two publications, *World Rivers Review* and *BankCheck Quarterly*. EDF brought to bear the results of its research of alternatives to large dams and new approaches to water management, as well as the involvement of its NGO partners, who had collaborated with EDF on more general reforms of the multilateral development banks. Together, IRN and EDF were able to bring together a network of NGO contacts, substantive knowledge about Bank-financed water projects, and ideas for more environmentally and socially sustainable approaches to water management.

Four major goals of the water policy campaign were identified: (1) to influence the content of the policy document itself and the subsequent policy implementation; (2) to raise the profile of water resources as an issue in the overall NGO efforts to improve the policies and practices of the World Bank; (3) to consolidate and broaden the informal but worldwide network of NGOs who were working on international water policy issues; and, equally importantly, (4) to compel the Bank to conduct a genuine and meaningful consultative process. This fourth goal, if accomplished, would set an important precedent for the establishment of formal consultative processes, a necessary step toward legitimate participation by communities directly affected by Bank projects.

Our organizations' approach differed significantly from previous attempts by the NGO community to participate in the policy-making process: they challenged the Bank to take a series of concrete steps that could be viewed as the absolute minimum requirements for what could legitimately be called a "consultative process" with NGOs. The steps included:

• Bank involvement of NGOs early in the policy development process so that NGOs could contribute to the goals of the policy and influence its overall scope and outline.

• Bank provision of the funds necessary to convene consultative meetings with NGOs, covering travel and logistical expenses, as well as financial support for NGO staff time required to critique and respond to the policy documents provided by the Bank.

• The holding of *regional* consultative meetings with NGOs—for example, in Sao Paulo, Harare, Bombay, and Jakarta, rather than exclusively in Washington, D.C.—to facilitate the increase in the number and diversity of NGO participants.

• The provision of relevant policy documents to NGOs at least six weeks prior to any consultative meeting to allow them adequate time to prepare comments and alternative positions. The six-week period should constitute a blackout period for changes in the draft by the Bank.

• Provision by the Bank of translations of all relevant documents provided to NGOs in order to maximize the range of NGO participation.

• Clear definition by the Bank, in advance of consultative meetings, of the manner in which NGO contributions will be considered and/or incorporated into the policy.

• Provision by the Bank of written documentation of its understanding of NGO input, as well as written confirmation of verbal commitments made during NGO consultations.

IRN and EDF sought not only to bring greater openness and accountability to Bank-NGO policy consultations, but also to apply those principles to relations among NGOs, particularly in the context of NGOs in the North reaching out to NGOs in the South. Specifically, we established a number of ground rules intended to strengthen participation of borrowing country NGOs and to clarify the role of NGOs in donor countries:

• To move away from what had become the standard practice of involving only a small and select circle of NGOs in policy dialogue, and instead, to contact, inform, and be informed by as broad a range of NGOs as possible. (This ground rule meant that the policy dialogue would not be restricted to those NGOs already experienced in policy dialogue at the World Bank—that is, those with significant resources who could be counted on to respond quickly and, often, predictably.)

• To refuse to act as an intermediary between other NGOs and the Bank. IRN and EDF would neither represent the Bank in reaching out to NGOs, nor speak on behalf of other NGOs when addressing the Bank. (This step was intended to stress the importance of direct dialogue between the Bank and NGOs in borrowing countries, thus recognizing the capability and competence of those NGOs.) IRN and EDF's role, rather, would be simply to inform its network of NGOs that the policy was being developed and to suggest the opportunities for direct communication between NGOs and the Bank.

• To encourage other NGOs to articulate their own views, instead of seeking their endorsement of the positions of IRN and EDF or of other NGOs in the North. The goal was to spread information as widely as possible, but to leave it to each individual organization to decide whether and how to respond.

• To document in writing all interaction with the Bank and to provide copies of that documentation and any correspondence to NGOs around the world, in an effort to provide adequate background information so that borrowing country NGOs could evaluate not just the substance, but the process of the attempted consultation.

The World Bank first publicly announced its intention to develop a new water resources management policy at the April 1991 meeting with NGOs on its forestry policy. Responsibility for developing the Bank's water resources policy paper was given to the Agriculture and Rural Development Department, under the direction of Michel Petit, with the senior advisor for water resources, Guy Le Moigne, as senior author.

In late June 1991, the Bank organized a five-day workshop for representatives of more than a dozen member governments to present their perspectives on water resources issues in their countries. Those discussions—which focused on intersectoral water allocation and pricing, privatization of water services, environmental and health issues, and international river basins—were intended to outline the scope of the water policy paper. Government participants were commissioned to prepare "issues papers" on various topics.

The first NGO-Bank discussion regarding the general process to be used to develop the water policy and the design of a consultation process between the Bank and NGOs took place in July 1991 and included EDF, the Bank Information Center (BIC), and World Wildlife Fund International (WWF). At that time, the Bank was generally quite open to input from NGOs and to improving the consultation process—an openness based, in part, on the lessons learned from the forestry policy. For example, Bank staff agreed that early input would be better in order to avoid protracted and controversial debates later and that they should consider holding a number of regional meetings with NGOs.

Given that a "consultation" of sorts had already been held with governments at the June workshop, EDF, BIC, WWF, and IRN were encouraged that broader consultation with more NGOs would soon follow, so we provided Bank staff with a more comprehensive list of NGOs in different regions who could assist in organizing consultations outside of Washington, D.C., and encouraged the Bank to develop additional means of soliciting NGO input. However, within a month, we were informed that the Bank had decided not to consult with NGOs because, in the words of one senior Bank staffer involved in the water policy, "NGOs don't have anything useful to offer, water resources management is a highly technical issue."[42]

In September 1991, EDF and IRN wrote to the Bank vice president for Sector Policy and Research, protesting the Bank's decision not to consult with NGOs on the water policy. Responding for the vice president, Mr. Le Moigne wrote back that although the Bank had no plans to meet with NGOs, the staff preparing the draft policy paper would accept written statements from NGOs. Even as it offered this limited opening, however, the Bank was still unwilling to make any draft materials available to NGOs on which to base written comments.

During this period, IRN and EDF began planning outreach that would serve the overall goals defined above. In the first phase, a mailing to NGOs included information on the Bank's planned water policy and on how to contact the responsible Bank staff, as well as copies of the correspondence between the Bank and U.S. NGOs. The cover letter outlined the goals and strategy in initiating this work and encouraged recipient NGOs to respond in whatever manner they saw fit. Despite limited resources for translation, outreach materials in English, Spanish, and Portuguese were produced.

IRN and EDF informed Bank staff of plans to solicit input from NGOs, in effect letting the Bank's policy-writing team know that they would soon be hearing from a wide variety of NGOs, and again requested that the Bank release copies of the draft policy paper so NGOs could more effectively target their comments. In response, Bank staff agreed to release an outline that summarized the working draft policy paper then circulating within the Bank.

In March 1992, the first mailing went out to more than eight hundred NGOs in ninety countries, with the Bank's policy outline enclosed. Spanish and Portuguese materials were sent to nearly 250 organizations in Latin America. The list of NGOs to receive the mailing was compiled from IRN's database of NGOs that work on river and water issues, supplemented by suggestions from EDF, the Bank Information Center, and NGO networks such as the African Water Network and the Friends of the Earth–International network. In addition, the documents included in the mailing were posted on international computer networks linked by the Association for Progressive Communications (Econet).

Over the next three months (March–May 1992) the World Bank received more than fifty letters regarding the water policy—written in at least three different languages[43]—from NGOs in thirty-two countries, with eight from NGOs in India alone. Altogether, twenty-four borrowing country NGOs responded, accounting for forty of the fifty-four letters that the Bank received. NGOs responding ranged from indigenous rights groups and trade unions in Brazil to environmental research institutes in Malaysia and Bangladesh. Some groups sent only a short statement requesting that the Bank convene a meeting with NGOs to discuss the water policy, whereas others drafted long and detailed position papers.

The letters reflected a wide spectrum of attitudes toward the Bank: from a Prague-based ecology institute that politely invited Bank staff to a conference on watershed restoration, to an Indian action-research organization that called for a "people's tribunal—à la Nuremberg"—to put Bank officials on trial for crimes against the peasants of Rajasthan committed in the implementation of Bank-funded irrigation schemes. In general, the NGO submissions were strongly critical of the Bank's past lending for water projects and called for a complete reorientation away from construction of new large scale infrastructure and toward support for locally-based and smaller-scale alternatives, efficiency improvements, and environmental restoration.

Following receipt of the NGO comments, the Bank announced in May 1992 that it would convene a consultative workshop with NGOs at the Bank's expense, to be held in Washington, D.C., later that same month. It is likely that the international input the Bank received was a factor in resolving positively an internal conflict over whether to hold such a meeting. By showing to Bank management that many NGOs held views similar to those of some Bank staff, the letters may have helped strengthen the positions of those Bank staff who argued in favor of holding a consultation.

While IRN, EDF, and other NGOs active in the water policy campaign were pleased with the Bank's offer of a meeting, the short notice posed a considerable problem. It would be difficult to arrange to bring a representative group of NGOs to Washington with less than three weeks notice. And there was another conflict: the planned consultation was scheduled for May 28–29, 1992—one week prior to the opening of the greatly anticipated Earth Summit (the United Nations Conference on Environment and Development) in Rio de Janeiro, for which most NGOs interested in the water policy were busily preparing. Moreover, preparation for the water policy consultation was already suffering because the Bank still had not released a draft of the policy paper, contradicting one of the concrete steps NGOs had identified as a prerequisite for meaningful consultation. Without it, NGOs would arrive at the meeting with no opportunity to have viewed the very document under discussion.

The Bank paid to bring nine representatives of borrowing country NGOs to Washington, who were joined at the meeting by eight representatives of donor country NGOs. Most of the groups invited had already sent written statements to the Bank on the water policy as a result of IRN's outreach effort, and a few others had been identified by EDF as having especially relevant experience to share. Several, it appeared, were chosen by the Bank based on its previous experience with them from past NGO-Bank meetings. NGOs from ten countries participated, including Bangladesh (two), Brazil, Czechoslovakia, Ecuador, Hungary, India, Indonesia, the United Kingdom, the United States (seven) and Zimbabwe. The Bank also invited representatives of four water industry lobby groups, three of whom, although participating in the meeting on behalf of private industry associations, were actually full-time employees of government water development agencies in Pakistan, Egypt, and the United States. In addition to the seventeen NGO and four industry representatives, about thirty-five Bank staff participated in the two day meeting.

Unlike in previous Bank-NGO consultations on Bank policies, the Bank hired professional facilitators to run the meetings. The facilitators developed an agenda similar to internal Bank workshops, which featured long segments devoted to small-group discussions focused on specific issues suggested by NGOs. NGO participants looked at the presence of the facilitation team as a positive development because it provided a relatively neutral authority. Though the agenda did promote spontaneous debate among the participants, the format did not particularly encourage attentive listening by the Bank staff to NGO views. Divided between many small discussion groups, NGO representatives were outnumbered by Bank staff and industry representatives. Borrowing country NGO representatives for whom English was a second language were even more in the minority in the smaller groups.

By the conclusion of the first day, NGO participants were frustrated with the lack of focus on NGO concerns and called for NGO participants to be given a block of time to present their views to the Bank staff responsible for drafting the water policy. Furthermore, they asked for additional time to meet among themselves first to discuss the issues and organize a presentation. Bank staff agreed to suspend the planned agenda and to focus the second and final day of the meeting on the NGO presentation and a response by the Bank.

For NGOs reviewing the Bank's draft water policy paper—both NGOs participating in the consultation and others not present—their overarching question was: How would this policy prevent the Bank's financing of environmentally and socially disastrous water projects, such as the Sardar Sarovar Dam on India's Narmada River? Although the draft policy paper did clearly recognize the failure of the Bank's water-lending program to address adequately the most pressing problems in the water sector (such as the widespread lack of access to safe water supply and environmental degradation of freshwater resources), it contained very few specific prescriptions for changes in the operational practice of the Bank itself. Nowhere was there even a statement of the goals and objectives of the new policy.

The NGO presentations suggested that for the policy to have any meaningful effect, the document must state clear goals and identify how Bank practice would have to change to accomplish these goals. NGOs further suggested that the policy include binding guidelines that would apply to Bank lending operations staff in four broad areas of NGO concern:[44]

1. *Prioritization of alternatives.* The Bank's new water policy should shift the Bank's investment focus away from costly new large-scale water development projects toward programs to manage existing water systems more efficiently and toward smaller-scale, more environmentally appropriate projects that could be implemented and managed by local users. Furthermore, the Bank should directly make the needs of poor people a priority (i.e., emphasize domestic water supply and sanitation over industrial and agricultural supply projects). To accomplish this shift in priorities, it would be necessary for the Bank to address its own institutional bias toward large-scale and supply-side projects (staff are rewarded for the quantity of money loaned and not for the quality of project outcomes). One way to promote alternatives on a broad scale would be for the Bank to provide sector loans for disbursement to many small-scale, community-controlled projects.

2. *Public participation.* The policy paper should make a clear commitment to a participatory planning approach that would seek to promote community control and management of water resources. The Bank

should require open public access to information and participation in planning on both a sector-wide level, and a project-specific level. The term *public* needed to be broadly defined to include water users, community associations, NGOs, and affected communities. Legal rights, including customary and traditional water rights, needed to be recognized and protected.

3. *Sustainability*. The Bank's water policy should explicitly recognize the right of natural ecosystems to adequate supplies of clean water and require the water planners to value ecological water "uses" such as fisheries and wetlands. The Bank should prioritize environmental restoration and utilize an "ecosystems approach" that would seek to maintain the ecological integrity of entire river basins. Pollution prevention should be prioritized over treatment. Cost recovery, to cover at least system operation and maintenance, should be required, and "lifeline" rates, particularly for the poor, should be used to guarantee equal access to water supply.

4. *Policy implementation*. Effective implementation of the Bank's water policy would require a commitment to transparency at all levels of decision making and would bring a much greater measure of accountability to Bank operations. For example, all Bank loans for water projects should have clearly stated goals, specific performance criteria and targets, and a detailed monitoring program with enforceable sanctions for noncompliance with design standards.

Both surprised by what they termed the NGOs' "constructive tone" and defensive about charges of bias, Bank staff in return accused NGOs of small-scale bias. They believed that some of the conflicting views arose from differing readings of the same passages in the draft policy paper. Bank staff argued that most of the NGOs' concerns were actually dealt with in the paper, just not in the sort of direct language NGOs prefer. Bank staff participating in the consultation were not all united in their reaction to the NGO positions. Some, mostly from the Environment Department, were quite supportive of NGO criticisms and suggestions. Others, including many engineers from both the Policy Department and the country departments, strongly defended the Bank's record, arguing for example that the Narmada Valley project, which NGOs said exemplified the worst of the Bank's many water blunders, was in fact a model of exactly the type of project that the new policy should promote. Another group, primarily economists familiar with urban water supply issues, agreed with NGOs that the Bank needed fundamental changes, but offered a very different direction for change, toward privatization of government services and allowing market forces to solve the problems of the water sector.

However, Bank staff were nearly all in agreement on the issue of conditionalities. Although arguing that the general policy had the implied force of a conditionality, they maintained that it would be impossible to get approval from Bank management and executive directors for any specific conditionalities that would tie the hands of the operations division, to paraphrase their argument. They argued that conditionalities were strongly disliked by borrowing governments, who viewed them as infringements on sovereignty, and that the country departments would know best how to tailor the policy to meet the needs of individual countries. When NGOs claimed that the Bank violated the "spirit and letter" of existing policy conditionalities, such as those in the resettlement policy, Bank staff objected, arguing that the policy directives were not truly binding rules, but rather goals to be strived for but not always attained.

At the conclusion of the second day, Bank staff in charge of the meeting were eager to find areas of agreement between NGOs and the Bank. Eight areas of agreement were eventually defined, including reference to many of the key concerns identified in the NGO presentation: stakeholder participation; clearly stated policy goals and objectives; consideration of *all* alternatives for water supply, including demand management; improved enforcement of existing policy directives; an ecosystem approach to planning; clearer statements of how the water policy would affect Bank practice; greater emphasis on community-based water management; and training for Bank and government staff in participatory techniques.

Before agreeing to adjourn the meeting, NGOs asked the Bank for a commitment to continue the consultative process with them, maintaining that NGOs should be given copies of later drafts and be invited to provide further comments before a final draft was presented to the Bank's board of executive directors. Bank staff assured the NGOs that the next draft would reflect the items of agreement noted above, that the Bank would continue to accept written statements from NGOs, and that a memo summarizing the Bank's understanding of NGOs' concerns would be circulated along with the next draft so that Bank management and others could also learn from the consultation with NGOs. In addition, following many demands for clarification, the Bank staff chairing the meeting agreed to distribute copies of later drafts to NGOs prior to submission to the board of executive directors for approval.

After the consultative meeting, the NGO campaign on the World Bank's water policy entered a new phase. Bank staff who were drafting the policy paper had announced their intention to present the paper to the board of executive directors in October 1992, five months after the consultative meeting with NGOs. During the Washington, D.C., meeting, NGO representatives had agreed to draft a consensus document, based on the NGO presentations and written submissions, to use in informing and urging the executive directors to request changes in the draft policy that would further address NGO concerns and make the policy more operationally specific for Bank activities. EDF took responsibility for drafting the consolidated NGO position paper,[45] IRN produced another mass mailing to NGOs with a report on the consultative meeting and overall progress, again urging interested NGOs to become involved by communicating their views to the Bank and to their country's executive directors. The NGO position paper was posted on Econet and distributed to Bank staff involved in rewriting the policy paper.

Bank staff made good on their commitment to produce written documentation of the consultation, releasing both an edited transcript of the two-day meeting and a three-page memo entitled, "Lessons from the Bank-NGO Consultation on Water Resources Policy."[46] As promised, the memo listed the key NGO positions as Bank staff understood them and the changes the Bank had agreed to make in the draft policy document. Having such documentation of the process and specific agreements Bank staff and NGOs had made was useful in ensuring that positions, agreements, and commitments could be honestly portrayed.

Then the process came to an abrupt halt. Contrary to what appeared to be an agreement at the consultative meeting, the senior Bank official responsible for the water policy informed NGOs that he could not release any subsequent drafts of the document until after the final draft was approved by the executive directors because the policy paper would be considered a "public" document only once the board had approved it. The policy paper had become hostage to a larger battle within the Bank over public access to information, an issue for which yet another new Bank policy was being developed. According to the Bank official, executive directors of borrowing countries (in particular the Bank's two biggest borrowers, India and China) were increasingly frustrated by the ability of NGOs to obtain project environmental assessments and other documents—nominally the property of the borrowing governments—through the Bank and donor governments and were demanding that the Bank restrict even further the public's access to Bank documents. At the same time, in response to pressure from donor governments, Bank management was increasingly adopting the rhetoric of greater accountability and transparency in Bank operations. Until new policy guidelines for information disclosure were in place, the later drafts of the water policy would remain internal to the Bank and confidential.

IRN quickly distributed a third mass mailing with news of the apparent end to the consultative process, urging NGOs to write to their executive directors to request that Bank staff be instructed to honor their commitment. There were many fewer NGO responses to this appeal than to the previous outreach effort, with fewer than a dozen (and mostly donor country) NGOs contacting their executive directors. By September 1992, Bank staff sympathetic to many of the NGO positions had anonymously leaked copies of the revised draft policy to NGOs, who were pleased to find that many of their ideas had, in fact, been incorporated in the document.

Meanwhile, EDF had worked with the Foreign Operations Subcommittee of the U.S. Senate Appropriations Committee—which has responsibility for appropriating funds for foreign aid (including the funds for the multilateral development banks)—to include a passage in the Foreign Appropriations Act of 1993 regarding sustainable freshwater development.[47] The passage requires the U.S. government to press the World Bank, which the act recognizes as the largest single source of financing for water projects worldwide, to adopt a comprehensive water policy with the goal of protecting aquatic ecosystems and watersheds; to implement a least-cost approach to planning and investing in water projects; and to train staff in and promote the use of the full range of water supply activities, including water demand management and small-scale systems. This passage provided another source of pressure on the Bank from a donor country government to move forward with adopting a water policy.

Soon after receiving copies of the revised water policy, NGOs learned that the revised draft policy had been rejected by senior Bank management and was being rewritten—at the urging of the Bank's chief economist, Lawrence Summers—by staff in the Operations Division of the Bank. This new development represented an internal ideological debate between agriculture/irrigation staff, many of whom are engineers, and the urban water supply/policy staff, many of whom are economists.

Removed from the newest draft were not only the passages on ecosystems approach and participatory development most important to NGOs, but also much of the integrated intersectoral approach that formed the core of the policy-writing team's original draft. The "comprehensive approach" was replaced instead by reliance on privatization and market forces to solve the problems of the water sector. As NGOs were relegated to the sidelines while the battle over the water policy raged within the Bank, one Bank staff engineer close to the policy-writing team commented that "at this point the engineers have more in common with the environmentalists than we do with the economists, at least we both have hands-on experience with the actual resource."[48] All staff involved seemed to agree that market failures and externalities—such as environmental impacts of water development—were part of the problem, but the disagreements were about the appropriate instruments or mechanisms to address the externalities. For the engineers, the move toward comprehensive basinwide planning was a recognition of the state of the art in water resources management, but the economists felt this approach was too reminiscent of centralized state control of the economy and contrary to the neoliberal economic restructuring that the Bank was promoting around the world.

Some Bank staff noted that it was more appropriate to resolve the debate within the Bank because the policy was, after all, a Bank policy that Bank staff would have to feel comfortable in implementing. In this light, they said, the debate is not relevant to NGOs—even though the internal debate directly affected elements of the policy that NGOs had succeeded in convincing Bank staff to adopt.

The revised water policy paper that emerged from the internal Bank debate was not presented to the board of executive directors until December 1992. As that event approached, NGOs, particularly those in the United States, were forced to resort to the traditional lobbying tactics that they had set out to avoid. They sent letters to selected donor country executive directors, requesting that the revised draft be released to NGOs as agreed to at the consultation and that the directors ensure that the eight points outlined in the "Lessons Learned" memo (which they attached to their letter) had been incorporated into the revised draft.

According to the secondhand reports that NGOs later received, the executive directors did in fact find the proposed policy paper too weak on environmental and operational issues and requested that revisions be made, some of which echoed the concerns of NGOs. Through the efforts of a sympathetic advisor to the U.S. executive director, EDF and IRN, along with a representative of Washington D.C.–based World Watch Institute, were invited to make a presentation to a group of advisors to donor country executive directors in March 1993, before the next board meeting slated to discuss the water policy.

In their presentation the NGOs offered specific language on key points, which they hoped the executive directors would insist be inserted in the policy paper. Based on the longer, consensus-based NGO position paper but endorsed by only seven U.S. NGOs because of time constraints, a five-page position paper was further distributed to selected executive directors.[49] Another draft of the water policy was presented to the board in April 1993 and essentially became the final draft approved by the board in May. It was not until September 1993 that the final, board-approved water resources management policy paper was publicly released by the World Bank.

The Outcome: The World Bank's Water Resources Management Policy

The core of the Bank's new water resources management policy consists of the adoption of a comprehensive policy framework and the treatment of water as an economic good, combined with decentralized management

and delivery structures, greater reliance on pricing, and fuller participation by stakeholders.[50] The objectives of the policy are to continue to
extend coverage for drinking water supplies, increase efficiency in the
water sector, manage water resources sustainably, increase cost recovery,
and protect ecosystems.

The Bank's new water policy includes the following general element.

• A comprehensive approach to planning and management will be used,
with the river basin as the unit of analysis, to evaluate intersectoral
components of water supply and management.

• National policy reforms and country-specific national strategies for
water resources management will be developed and supported.

• Appropriate pricing policies will be promoted to improve cost recovery
and efficiency, taking into account lifeline policies for the poor.

• Decentralized management—via capacity building of water users'
associations and institutional reform of water utilities and irrigation
agencies—will be used to improve performance of water systems, increase cost recovery from users, and stimulate local participation.

• Rehabilitation and improved operations and maintenance of existing
water systems will be a priority.

• Water conservation, demand management, and other alternative
options will be evaluated and considered as potential new sources of
water supply.

• Environmental protection and mitigation will be an integral part of the
comprehensive approach.

These general themes are analyzed and discussed throughout the policy
paper. More explicit objectives are identified in the policy.

• For industry, extensive water conservation and protection of groundwater sources; controlling pollution to help reduce the quantity of water
used per unit of output;

• For water supply and sanitation, more efficient and accessible delivery
of water services and sewage collection, treatment, and disposal—with
the ultimate goal of providing universal coverage—to be achieved by
extending existing supplies via water conservation and reuse and by
other sustainable methods, as well as great involvement of the private
sector, NGOs, and user groups;

• For irrigation and hydropower, modernized irrigation practices,
greater cost recovery, measures to reduce pollution from agricultural
practices, operation and maintenance improvements in existing systems,

and investments in small-scale irrigation and various water-harvesting methods. Particular attention will be given to the needs of small-scale farmers. Greater priority should be given to managing the demand for energy, identifying small-scale and renewable energy alternatives, promoting watershed conservation practices, and retrofitting and enhancing dam facilities;

• For the environment and poverty alleviation, more rigorous attention to minimizing resettlement, maintaining biodiversity, and protecting ecosystems in the design and implementation of water projects; water and energy supplies gained through conservation and improved efficiency can be used instead of developing new supplies to extend service to the poor and maintain water-dependent ecosystems; the water supply needs of rivers, wetlands, and fisheries will be considered in decisions concerning the operation of reservoirs and the allocation of water. (Goals extracted and paraphrased from water resources management policy paper, page 12.)

All of these changes are intended to correct the poor performance of water projects and achieve the Bank's goals of poverty alleviation and sustainable development. Clearly, these objectives have important implications for the Bank's own role in the water sector. Chapter 4 of the water policy, entitled "The Role of the World Bank," contains a section called "Implications for Bank Operations." Although the stated objectives and the chapter on the Bank's role present an optimistic picture of how the water sector will be positively transformed into a "sustainable system," the policy does not state in clear language how the Bank's operations will change, how the proposed activities will be implemented and enforced, or how to measure what progress, if any, is made. Out of the 140 pages of the policy paper, the three and a half pages of this section including the two pages that follow on "Procedures, Staffing, and Training" are the only ones remotely specific about what the Bank will actually do. These sections state that the Bank will

• assist governments to identify and formulate priority policy and institutional reforms and investments and determine their appropriate sequencing;
• highlight these priorities in the country assistance strategy and use the priorities to guide the sectoral lending program;
• "deal with issues such as (a) the appropriate incentive framework and pricing, (b) service delivery to the poor, (c) public investment priorities,

(d) environmental restoration and protection, (e) water resource assessment and data requirements, (f) a comprehensive analytical framework; and (g) legislation, institutional structures, and capacities;" (page 76)

• monitor progress in implementing the identified priorities through normal Bank interactions with the country; and "if the absence of adequate progress on priority actions is judged to produce serious misuse of resources and to hamper the viability of water-related investments, Bank lending in this area will be limited to the provision of potable water to poor households and to operations designed to conserve water and protect its quality without additionally drawing on a country's water resources;" (page 77) (such operations include sanitation, waste treatment, water reuse and recycling, abatement of pollution, drainage, and rehabilitation of the distribution system);

• require "assessment of environmental impacts of projects, environmental assessments of the entire river basin for significant water-related projects, and full consultation with affected people and local organizations" (page 77) (as already required by other policies and operational directives);

• make arrangements within Bank operations to ensure that the Bank's water-related activities for a country are treated comprehensively, through the development of country teams, cross-sectoral water teams, or a central water unit, for example; (page 78)

• implement the new water policy via preparation of guidelines and best-practices papers, development of coordination mechanisms, preparation of regional water strategies, upgrading of staff skills through workshops, seminars, and training programs, use of pilot projects to experiment with newer aspects of the policy; and (pages 78–79)

• review implementation of the policy in two years. (page 78)

Most of these operational guidelines fall short of what NGOs hoped to see in the policy. These guidelines will be very difficult to enforce or implement effectively because they are so broadly defined and because they are not "conditionalities" that will be required and enforced for every water-related loan. The item that comes closest to a loan condition is the guideline that limits the Bank's lending to providing potable water for the poor and various water-conserving efforts in cases where countries are not making adequate progress in meeting their priorities for reform (page 77 of the policy paper, as quoted above). NGOs could use this point to question many of the Bank's new investments for water-related projects because most countries are not "making progress" and because most large-scale infrastructure projects do "produce serious misuse of resources." (page 77) However, the criteria for evaluating whether a country is "making progress" are not well defined, and even this policy point will be difficult to enforce.

Originally, NGOs had expected the Bank to develop an operational directive (OD) based on the water policy paper, as it had for past policies, which would delineate more specifically what would actually be required of Bank staff in operations. Many NGOs' interpretations of operational directives are that they are analogous to regulations, which define how broadly defined legislation is to be implemented. However, in line with the general trend within the Bank to stop writing ODs, the Bank produced a one-page "policy note" in July 1993 that broadly summarizes the water policy. Unlike an operational directive, this "operational policy" has no mandatory elements and merely serves as a best-practices guide.

Evaluating the Bank's Implementation of the New Water Policy as of March 1996

The new water policy was under development for almost three years and has been in effect for more than two years (as of this writing). It is still too early to evaluate the exact effect of the water policy on the performance of projects funded since it was passed (May 1993), as most of these projects are yet to be fully implemented. We can, however, assess the policy's effect on the design and selection of projects to be funded by the Bank. In addition, we can examine the internal mechanisms and training programs the Bank has used to ensure that the policy's new approaches are adopted by staff in operations.

As of March 1996, we identified 191 water projects in the World Bank's pipeline of loans and credits.[51] Figure 10.3 shows the distribution of lending across the water subsectors. Compared to the distribution in the 1980s, as shown in table 10.1, there is greater investment in the water supply subsector (urban and rural) and a smaller share for the irrigation and hydropower subsectors. There are also more multipurpose urban and rural development projects that include water as a component of larger projects.

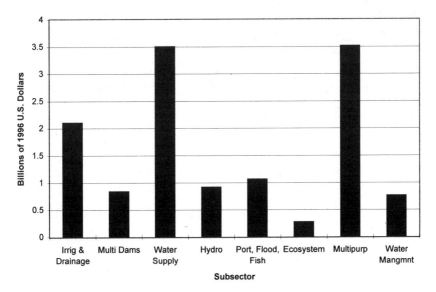

Figure 10.3
World Bank water projects in the pipeline (as of March 1996), allocation by subsector. *Source:* See note 17. *Note:* "Multipurpose" designates development projects in which one component is for water.

The types of projects being considered in the water sector are shown in figure 10.4, similar to the types shown previously in figure 10.2 for the 1980s. The largest investments continue to go to new infrastructure projects, about 44 percent of the $13 billion represented in the pipeline.[52] One of the significant differences in these investments, when compared with the lending in the 1980s (as shown in table 10.1), is the growth in investment for projects to rehabilitate and expand existing water supply projects. During the 1980s, rehabilitation projects represented about 11 percent of the resources, and rehabilitation and expansion projects represented about 7 percent. In 1996, the shares of resources allocated in the pipeline were about 13 percent for rehabilitation projects and 19 percent for rehabilitation and expansion projects. A small increase was made in the amount allocated to institutional reform and capacity-building types of projects. These changes would seem to be consistent with several elements of the Bank's new water policy, including the priority for efficiency and the focus on decentralization and water users' associations to improve management.

Table 10.1
Comparison of allocation of World Bank resources for water projects 1981–1990 versus March 1996 pipeline

By subsector	1981–1990	1996
Irrigation	26.2%	16.2%
Water supply	28.2%	27.1%
Hydropower	23.6%	7.1%
Multipurpose dams	—	6.4%
Port, flood control, and fisheries	10.2%	8.2%
Integrated resource management	3.0%	5.8%
Multipurpose with some water	8.9%	27.1%
Ecosystem	—	2.1%
By type	1981–1990	1996
New construction and expansion	68.0%	44.3%
Rehabilitation	11.3%	13.1%
Rehabilitation and expansion	7.3%	18.9%
Institutional development/sector reform	11.0%	12.7%
Environment	—	5.3%
Alternatives	2.4%	5.7%

Note: See note 17.

Although more watershed management and small-scale projects are in the pipeline, at 5.7 percent of the total resources allocated to the sector, these "alternative" types of projects remain a small share. There is an increase in loans and credits for projects directed specifically at protecting water-dependent environments, representing about 5 percent of the resources. These projects include ecosystem management and restoration projects, such as the Lake Victoria Basin and Ecosystem Management Project and the Aral Sea Restoration Project.

Another change in the types of projects being supported is the significant increase in projects that promote private sector involvement in water—which was negligible during the 1980s, but now represents about 13 percent of the $13 billion of loans and credits in the pipeline. This change is consistent with elements of the water policy; however, it is un-

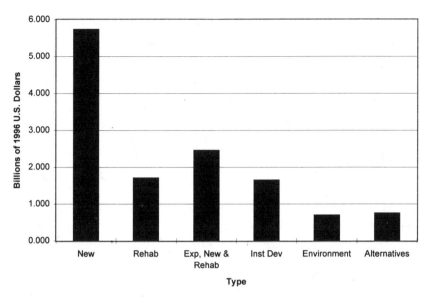

Figure 10.4
World Bank water projects in the pipeline (as of March 1996), allocation by type of activity. *Source:* See note 17. *Note:* "Alternatives" include watershed development, water conservation, and small-scale projects. "Inst" includes technical assistance, sector reform, and institutional development projects.

clear whether privatization is pursued as a means to achieve a specific goal of the policy—such as efficiency and conservation—or privatization itself is the goal, separate from particular performance criteria. Apparently, though, if the Bank chooses to make changes in its allocation of resources toward certain objectives, it can and does do so. In contrast to the small increase of about 2 percent in resources devoted to small-scale alternatives, the Bank increased the resources devoted to privatization types of projects by 13 percent.

Examining individual water projects in the pipeline and projects on which the board has recently voted, we continue to see several large-scale infrastructure projects that appear not to follow the intent of the new policy, some of which are being opposed by NGOs or other organizations. Such projects include the Xiaolangdi Dam in China, the Arun III Dam in Nepal, the Lesotho Highlands Water Project in southern Africa, the Inland Waterways Project in China, and the Pripyat River Flood Control Project in Belarus, among others.[53] For NGOs, the overarching question about the water policy has been whether the new policy would prevent the Bank's financing of environmentally and socially destructive projects that harken back to the Sardar Sarovar Dam in India. Although there have been shifts in the allocation of Bank resources in the water sector away from new infrastructure toward rehabilitation projects and institutional capacity building, the large-scale projects in the pipeline indicate that the policy has not completely prevented the continued financing of projects that have high potentials for negative environmental and social impacts, and that thus face significant opposition.

For example, the Xiaolangdi Dam in China is estimated to displace almost two hundred thousand people in a country where more than three million people previously displaced from reservoir development have not been successfully resettled.[54] The World Bank provided loans for both the dam and an associated project aimed at mitigating the dam's social impacts on displaced communities. The project to build another large-scale dam, the $1 billion Arun III Dam, was the first case brought before the World Bank's Inspection Panel on the basis of the Bank's failure to comply with its own economic evaluation and environmental assessment policy guidelines.[55] The Inspection Panel recommended a more in-depth investigation of the Bank's role in the project and its failure to explore adequately alternative means of supplying electricity.[56] In a move that may signal other changes in the Bank, James Wolfensohn, the Bank's new president appointed in 1995, canceled the Arun III Dam in the summer of 1995, stating that the scale and cost of the project would overwhelm the government of Nepal's capacity to implement the project successfully.

In the case of the Lesotho Highlands Water Project—a massive hydropower and water pipeline project estimated to cost $2.4 billion for phase 1A and up to $8 billion for the entire project—an environmental impact assessment for phase 1A has never been done and an environmental action plan was produced in 1990 after construction had already begun. Although physical construction and engineering components proceed at a rapid pace, the rural development plans, resettlement efforts, and environmental aspects lag far behind.[57]

Some projects, such as the Pripyat River Flood Control Project in Belarus, were removed from the portfolio before they ever reached an in-depth evaluation stage. Prescreening and early evaluations can and do affect project identification and selection, and the water policy has the potential to influence this process positively. However, as of March 1996 the portfolio continues to include new, megascale water development projects that are of concern: the Wanjiahzhai Water Transfer Project and the Inland Waterways Project in China, the Dai Ninh Hydropower and Inland Waterways Projects in Vietnam, and several flood control projects in Indonesia and Argentina all show significant environmental and social impacts. Several of these projects have been in the pipeline for nearly two years and have most likely passed through the early screening process. The World Bank is also providing unspecified "preappraisal" support for a series of feasibility and environmental studies of the Nam Theun II hydroelectric dam in Laos, which would be one in a series of dams planned in the Mekong River Basin that are being questioned by many communities and organizations because of the costs and impacts.[58] In this case, the Bank is taking a slower and more cautious approach even before the project is considered for appraisal, but it does not seem to be investigating alternatives to building the dam and in fact has recently decided to go forward with preparation of a loan package.

Although there may be fewer large-scale infrastructure projects in the pipeline, many of the ones that remain continue to display serious problems such as bias against alternatives, inadequate attention to resettlement plans, and lack of adequate environmental assessment. Such projects show that the Bank continues to have significant difficulties in enforcing its policies.

In addition to examining individual large-scale projects in the pipeline, we also interviewed fifteen World Bank task managers (TMs) who were in charge of different kinds of water projects (new infrastructure, rehabilitation, etc.) in different regions (Asia, Africa, etc.) to ascertain the degree to which the water policy is being followed by staff responsible for project operations.[59] Interviews were held in Washington, D.C., in September 1994. When asked what the impact of the new water policy was, about half of the TMs said that "it was nothing new" or "it was not new, but was still an important document." The other half of TMs thought the policy was a significant step in the right direction and was needed to improve the Bank's performance in the water sector, which was "extremely problematic." When asked whether the water policy would have an impact on project lending, the TMs who thought the policy was "nothing new" thought that it would not affect project lending. TM responses to other questions about comprehensive planning, national water strategies, and stakeholder participation also showed a lack of clear understanding about key principles of the policy.

The water policy highlighted the need for water management to occur within the context of a national water strategy. Most task managers we interviewed stated that the borrower countries for their water projects did not have a national water strategy (or did not have one *yet*) to guide investment decisions, or that such national strategies were not really relevant to rehabilitation or institutional reform projects because they did not require new construction. These situations indicate that the "comprehensive approach," which is at the core of the new water policy, is not being practiced in evaluating or selecting water projects for Bank funding. In particular, the water policy states that the Bank would assist governments in identifying and formulating priorities and strategies for water resources management that would guide the Bank's sectoral lending program.

Generally, these responses show the inadequacy of internal Bank mechanisms to change the practices of staff in the operations departments. It is unlikely that the water policy will be effective if Bank staff themselves do not believe that the policy will affect project lending.

Regarding other elements of the new water policy's recommendations for changes within the Bank, the Bank did produce a report, *A Guide to the Formulation of Water Resources Strategy*, with the United Nations Development Program (UNDP), as it stated it would in the water policy paper.[60] The report is aimed at policymakers in developing countries and is intended to help build a country's capacity for developing institutional, abilities to plan and manage resources and for designing national water resources strategies. Although the foreword states that the report was reviewed and discussed by participants at the World Bank's Tenth Water Resources Seminar in December 1993, which IRN and a few other NGOs attended, it was never reviewed by these NGO reviewers. It does contain chapters on "stakeholder participation" and "environmental and health considerations," which include many of the points raised by NGOs during the consultation with Bank staff on the water policy.

There do not appear to be many "water units" or "cross-sectoral water teams" established in country departments to coordinate the Bank's development, appraisal, evaluation, or approval of water-related loans and credits, although a "water" staff person is assigned to most water projects. Country and regional departments have tried to improve coordination in different ways. For India and China, for example, water projects all fall within the same division, so the division chief is responsible for ensuring the coordination of water projects from different sectors—such as irrigation or urban water supply—being planned for the same river basin.

The Bank's annual Irrigation and Drainage Seminar has been expanded and is now the Water Resources Seminar, which includes staff from both irrigation and urban water supply areas. This seminar provides an opportunity for staff in different subsectors of the water sector to discuss issues such as "integrated approaches to water." However, the result thus far has been that the seminar is divided into separate sectoral "tracks," with a few plenary sessions where more comprehensive issues are presented.

Although internal training mechanisms appear to be weak, the Bank has incorporated training workshops related to water management into the Economic Development Institute's (EDI) training workshops for government officials and other decision makers. EDI's program now includes a component called "Reforming Water Policies: National Water Strategy Formulation, River Basin Management, Privatization, Participatory Irrigation Management."[61] Workshops, seminars, and study tours have been conducted in Asia, Africa, and Latin America. Some workshops have included "role-playing" exercises aimed at incorporating water conservation efforts as alternatives to development of new water supplies in the context of designing national water strategies.

Lastly, the water policy states that the Bank would conduct a review of the policy's implementation in two years. As the two-year mark approached, Bank staff requested and the board approved an extension of the deadline for the review until fiscal year 1998 (starting in July 1997), stating that two years was not a long enough period for evaluation. The period for evaluation will thus be about four years.

Overall, changes in the Bank's lending for water projects lean toward rehabilitation, privatization, and, to a lesser degree, smaller-scale projects. It is difficult to determine, however, what the driving factors are in these changes. The final policy is more descriptive than prescriptive, and the actual "words on the page" may therefore have less impact on how decisions are being made about which water projects to finance. For example, the smaller share for the irrigation subsector is the continuation of a declining trend that started in the early 1980s, a trend that was, perhaps, further institutionalized by the water policy.

The process of developing the policy *has* had an influence on individual Bank staff whose minds were changed during the period of debate over the appropriate approaches and strategies for the Bank to use in the water sector. In addition, some Bank staff are convinced that NGOs had a significant and positive contribution to the substance of the policy and helped to strengthen the hands of staff working for such changes from within. The ultimate evaluation of the Bank's progress will be determined by whether project performance actually improves, which it is still too early to assess.

Conclusions

If we look back on the more than two-year-long campaign to influence both the process and substance of the Bank's water policy, NGOs can claim to have made meaningful progress toward all four of the goals we originally established. First, the letter-writing campaign demonstrated to the Bank that NGOs had developed a substantive critique of Bank lending for water and an alternative vision that should not be ignored. Many of the NGOs' comments and positions were incorporated into the policy, although often not verbatim, and some of their specific language suggestions were used. Clearly, the policy contains a better analysis of the Bank's poor past performance in the sector, more explicit objectives, greater attention to environmental and poverty issues, and slightly more specific sections on the Bank's role in the water sector than did the draft policy prior to the NGO consultation. However, because NGOs were shut out of the drafting process after the consultation, they were not able to press in an informed manner for additional changes to the policy. In addition, the internal debate within the Bank that followed the consultation with NGOs was not anticipated and was difficult for NGOs to influence.

Water issues are becoming of increasing interest to NGOs, governments, and international institutions for a variety of reasons—including greater scarcity and pollution, a recognition of growing health and environmental crises, and a growing understanding that safe, clean, and reliable water supplies are a precondition for sustainable economic development. The process of developing the Bank's water policy enabled NGOs to raise the full array of water issues, beyond the debates surrounding large dams, with Bank staff, U.S. government agencies and congressional staff, and the media, thus achieving our second goal. The policy itself was not "attention getting" to the wider public, but the fact that a policy was being developed provided an opportunity for NGOs to raise the profile of water issues and to discuss both serious problems and potential solutions. Third, NGOs now better recognize water issues as fundamental to sustainable development, as indicated by the "50 Years Is Enough" coalition in the multilateral development bank campaign incorporating water issues into its campaign.

Our fourth goal was to pressure the Bank to conduct a meaningful consultation process with NGOs and establish new precedents for Bank-NGO working relations. The trend in developing Bank policies was toward more consultation, not less, due to pressure from NGOs, donor country governments, and unflattering media accounts of the Bank's poor performance and secrecy. Therefore, in the end, after several flip-flops, the Bank did agree to hold a consultative meeting with NGOs. The consultation did not occur as early in the process as NGOs had requested, and the Bank eventually balked at the concept of having regional consultations, but it did provide funding for Southern NGOs to participate and hired professional facilitators to lead the consultation. The draft policy paper was not distributed in a timely fashion, however, nor was it translated, and subsequent drafts were not released. The documentation of the Bank's understanding of NGOs' positions and the agreements made during the consultation (represented in the "Lessons

Learned" memo and the facilitators' report) was a positive step forward in increasing the Bank's accountability to agreements made during the consultation.

Yet before the final policy was adopted, both Bank staff and NGOs resorted to more conventional methods of approaching the differences in their positions. From the NGO perspective, the lobbying tactics with executive directors were necessary because NGOs had been shut out of any formal discussions and were denied access to information about both the direction of the process and any changes in the policy paper. The Bank returned to its claim that discussions over Bank policy were most appropriately resolved internally and that NGOs did not have rights to information or revised drafts. Despite the precedents for an improved standard of Bank-NGO policy consultation, the Bank still has no clear guidelines for conducting NGO consultations, and the level of NGO access to Bank policy development remains arbitrary.

Lastly, the NGO coalition did not become as tightly knit as hoped for and did not stay in close contact as a group following the consultation. Although it succeeded in broadening the diversity and types of NGOs participating in the dialogue, the development of a closely allied working group to sustain mutually beneficial activities and campaigns was hampered by the diversity of interests and organizational programs. For example, the coalition succeeded in reaching out to NGOs primarily focused on providing water services, not on policy and advocacy, which helped to provide on-the-ground experience to the policy debate. However, these organizations have limited resources and can not necessarily sustain international advocacy campaigns that compete with their service-providing responsibilities to local communities. Accountability among the NGO coalition was generally high in terms of keeping NGOs informed during the policy development process. Yet, when the policy was quickly headed for final adoption by the Bank's executive directors, Northern NGOs could not mobilize the coalition as a whole quickly enough and resorted to lobbying on other NGOs' behalf, to some degree. Southern NGOs did not, in some cases, follow through on commitments to respond to later actions, in part because of distance, time, and competing demands. Despite the loose-knit nature of this coalition, individual working relationships between subsets of the group have emerged in response to particular Bank-funded projects. In addition, the outreach effort helped to make more NGOs aware of the policy aspects of water issues, the role of the Bank in the water sector, and the array of NGOs working on water issues around the world.

The new water policy's impact on and its ultimate utility regarding Bank lending in the water sector are still unclear and have not yet really been tested by NGOs. Although some positive changes in the allocation of the Bank's resources in the water sector have worked more toward rehabilitation, institutional reform, and, to a small degree, alternatives, the largest share of resources continues to go to fund new water development infrastructure projects. Several large-scale, environmentally and socially damaging water projects are in the pipeline, indicating that the Bank's new policy is not (or not yet) preventing the approval of investments in projects of questionable merit. In addition, there are few indicators that the Bank is actively implementing the policy, as shown by the lack of internal mechanisms to assess water projects in a comprehensive manner, to train staff about the water policy, or to use nationally identified priorities to guide the Bank's investments. The policy's strongest language—which, in cases where countries are not making adequate progress at meeting their priorities for reform, would actually limit the Bank's lending solely to the provision of potable water for the poor and to various water-conserving efforts—has not yet been used by Bank staff or by NGOs to stop or limit Bank lending on a particular project or to a particular country.

Water problems around the world are continuing to worsen; governments, NGOs, and international institutions will be challenged to find better means of providing water to meet human needs while protecting water-dependent ecosystems. Some elements in the Bank's new water policy—combined with other Bank policies on information disclosure, independent inspections, forestry, energy, environmental assessments, and resettlement—could be used to strengthen challenges to Bank financing of water projects if more NGOs can be made aware of the water policy and of how to use it effectively in their own campaigns or coalitions.

Although the World Bank's new water policy does provide a conceptual framework for a new Bank strategy that could address many of the current failings of the Bank's lending for water projects, it remains to be seen whether the policy will improve project performance significantly. The policy provides justification for investments in rehabilitation and water conservation improvements, in water supply projects for poor communities, and in restoring and preserving aquatic ecosystems; however, the Bank's decision not to establish a binding operational directive, which would require specific changes in the design and selection of Bank projects by the operations staff, is not grounds for optimism. Furthermore, the high degree of internal disagreement over fundamental aspects of the policy, such as whether to adopt a comprehensive planning approach, coupled with the low level of interest in the policy shown by water project task managers, indicates that without concerted effort by concerned governments and NGOs, the policy and its new approach are fragile and may never be fully implemented. In addition, other factors that influence the design, selection, and approval of water projects remain unchanged. For example, the "pressure to lend"—which creates incentives for Bank staff to favor large loans and credits for big projects—has not been removed and continues to work against aspects of the water policy that recommend greater attention to smaller and cheaper alternatives. Lastly, there is no clearly organized constituency—either inside or outside the Bank—strong enough to ensure that the water policy will be implemented and enforced.

For NGOs concerned about how to reform Bank policies and allocate limited resources, the water policy experience has a mixed message. Although the campaign demonstrated that a broad-based coalition could be organized and could influence the outcome of the policy to some extent, the Bank's lack of progress in delineating clear guidelines for future Bank-NGO policy consultations, the often arbitrary nature of World Bank internal decision-making processes, and the nonbinding nature of the water policy may lead many NGOs to conclude that policy dialogue with the Bank is not the best investment of their time and resources. The case study indicates that for Bank-NGO partnerships to work effectively, groundrules and the consequences for breaking them must be clearly defined and enforced, particularly regarding access to information, the binding nature of the final outcome or policy document, and the consultative process.

Notes

1. Robert Engelman and Pamela LeRoy, *Sustaining Water: Population and the Future of Renewable Water Supplies* (Washington, D.C.: Population Action International, 1993), p. 11.

2. The World Bank, *Water Resources Management: A World Bank Policy Paper* (Washington, D.C.: World Bank, 1993).

3. Engelman and Leroy, *Sustaining Water.*

4. "Water, Sanitation: Earth Summit Priorities," *Water Quality International* No. 4 (1991) (London: International Association on Water Pollution Research and Control), p. 24.

5. "More Food from Finite Resources," *Water Quality International* No. 4 (1991), p. 28.

6. Engelman and Leroy, *Sustaining Water,* p. 11.

7. World Resources Institute, *World Resources 1992–93: A Guide to the Global Environment* (New York: Oxford University Press, 1992), p. 161.

8. The World Bank, *World Development Report 1994: Infrastructure for Development* (New York: Oxford University Press, 1994), p. 27.

9. John Briscoe, "Poverty and Water Supply: How to Move Forward," *Finance and Development* 29, no. 4 (December 1992), p. 17.

10. Sandra Postel, *Last Oasis: Facing Water Scarcity* (New York: W.W. Norton, 1992), p. 23.

11. Leonard Sklar and Patrick McCully, *Damming the River: The World Bank's Lending for Large Dams*, International Rivers Network, Working Paper no. 5 (Berkeley, Calif.: IRN, 1994).

12. World Bank, *Water Resources Management.*

13. Sklar and McCully, *Damming the River.*

14. Joseph Christmas and Carel de Rooy, "The Decade and Beyond: At a Glance," *Water International* 16, no. 3 (September 1991), pp. 127–34.

15. John M. Kalbermatten, "Water and Sanitation for All, Will It Become Reality or Remain a Dream?" *Water International* 16, no. 3 (September 1991), pp. 121–26.

16. Christmas and de Rooy, "The Decade and Beyond," p. 131.

17. Mafruza Khan, "The World Bank and Investments for Developing Water Resources: Past Trends, Current Policies and Practices, and Future Directions for Reform," unpublished professional report done for the Environmental Defense Fund as partial requirement for Master of City Planning, Department of City and Regional Planning, University of California, Berkeley, 1995. Using the methods developed by Khan to monitor and categorize the Bank's water-lending portfolio, Environmental. Defense Fund scientist Deborah Moore monitors and analyzes the Bank's portfolio on an ongoing basis. Additional unpublished analysis of

trends in Bank lending was done by Deborah Moore, with Julene Freitas, in the form of an Excel database "Trends in World Bank Lending for Water," (updated as of March 1996, Berkeley, California). The database and analysis is based on World Bank data obtained from the Monthly Operational Summary, the World Bank, Washington, D.C., published monthly (the most recent one used in analysis was March 1996). Additional information is from Project Information Documents and Environmental Data Sheets, World Bank Public Information Center, Washington, D.C.

18. Christmas and de Rooy, "The Decade and Beyond," p. 128.

19. Operations Evaluation Department, the World Bank, *Annual Review of Evaluation Results: 1989,* report No. 8970 (Washington, D.C.: World Bank, 1990), p. 4-2.

20. Operations Evaluation Department, the World Bank, "Urban Water Supply and Sanitation," *OED Précis* no. 29, June 1992.

21. Agriculture Operations Division, India Department, the World Bank, *India Irrigation Sector Review* (Washington, D.C.: World Bank, 1991).

22. OED, "Urban Water Supply."

23. World Bank, *Annual Review* (1990).

24. World Bank, *India Irrigation.*

25. Briscoe, "Poverty and Water Supply," and World Bank, *World Development Report 1992: Development and the Environment* (New York: Oxford University Press, 1992), p. 100.

26. Environment Department, the World Bank, *Resettlement and Development: The Bankwide Review of Projects Involving Involuntary Resettlement 1986–1993* (Washington, D.C.: World Bank, April 1994).

27. World Resources Institute, *World Resources 1992–93.*

28. Sklar and McCully, *Damming the River.*

29. Willi A. Wapenhans et al., *Report of the Portfolio Management Task Force,* a report to the president of the World Bank (Washington, D.C.: World Bank, July 1992).

30. Charles W. Howe and John A. Dixon, "Inefficiencies in Water Project Design and Operation in the Third World: An Economic Perspective," *Water Resources Research* 7 (July 1993), pp. 1889–94.

31. *Ibid.* and World Bank, *Annual Review* (1990).

32. Wapenhans et al., *Report,* p. 8.

33. Howe and Dixon, "Inefficiencies," p. 1893.

34. World Bank, *Water Resources Management,* p. 66-7.

35. See World Bank "Involuntary Resettlement," operational directive 4.30, *Operational Manual* (Washington, D.C.: World Bank, 1990), for a description of the policy itself; and World Bank, *Resettlement and Development,* for documentation of the Bank's failure to implement the policy.

36. Bhuvan Bhatnagar and Aubrey Williams, eds., *Participatory Development and the World Bank: Potential Directions for Change*, World Bank discussion paper no. 183 (Washington, D.C.: World Bank, 1992).

37. Lawrence Salmon and Eaves, "World Bank Work with Non-Governmental Organizations," World Bank Policy, Planning, and Research Working Paper no. 305 (Washington, D.C.: World Bank, 1989).

38. World Bank, *How the World Bank Works with Nongovernmental Organizations* (Washington, D.C.: World Bank, 1990).

39. World Bank, *The Forest Sector: A World Bank Policy Paper* (Washington, D.C.: World Bank, 1991).

40. International Rivers Network, membership brochure, n.d. Berkeley, California.

41. Environmental Defense Fund, *Annual Report 1994–1995* (New York: EDF, 1995).

42. This was an off-hand statement made to an NGO representative in a conversation with a Bank staffer regarding the status of the Bank's plans for organizing a consultation with NGOs.

43. IRN had requested that NGOs writing to the Bank also send a copy of their letter to IRN. Altogether, IRN received fifty-four letters; others, however, may have been sent to the Bank, of which IRN did not receive copies.

44. Environmental Defense Fund, International Rivers Network, World Rainforest Movement et. al., "Reforming the World Bank's Lending for Water: An NGO Critique of the World Bank's Draft Water Policy Paper" unpublished (Oakland, Calif., September 1992). Also see *World Bank: NGO Consultation Meeting on Draft Water Policy Paper,* unpublished facilitator report (Washington, D.C.: World Bank, 1992). (Facilitators: Lorenz Aggens, Jerome Delli Priscoli, Judy Nicholson, and Bill Werick.)

45. EDF, IRN, WRM, et al. Reforming the World Bank's Lending For Water.

46. The transcript is the facilitator report, *World Bank*. The memo was written by Bank staff: "Lessons from the World Bank/NGO Consultation on Water Resources Policy," 8 July 1992, Washington, D.C.

47. U.S. Senate, *Committee on Appropriations, Report Regarding Foreign Operations, Export Financing, and Related Programs Appropriation,* Bill 1993, report no. 102–419, 102d Congress, 23 September 1992.

48. Anonymous, World Bank irrigation staff engineer, phone interview by Leonard Sklar, October 1992.

49. Deborah Moore and Leonard Sklar, "Reforming the World Bank's Lending for Water: NGO Comments on the World Bank's Water Resources Management Policy Paper," position paper endorsed by Bank Information Center, EDF, Friends of the Earth-US, IRN, National Audubon Society, National Wildlife Federation, and the Sierra Club, 3 March 1993. Unpublished paper (Oakland, Calif.: Environmental Defense Fund).

50. World Bank, *Water Resources Management*.

51. Moore, "Trends in World Bank Lending for Water," see note 17 for additional information.

52. *Ibid.*

53. Information about projects in the pipeline is from the *Monthly Operational Summary* (Washington, D.C.: World Bank, March 1996) issued by the World Bank, which describes the project, the environmental classification, and the potential amount of the loan, among other elements. Specific information about three projects is from Elliott Mainzer, "Moving Lesotho's Water," *World Rivers Review* 8, no. 4 (1993); Joe Karten, "Xiaolangdi," *World Rivers Review* 9, no. 1 (1994); and Janet Bell, "Seeking Alternatives to Arun III," *World Rivers Review* 8, no. 2 (1993).

54. Khan, "World Bank and Investments for Water."

55. Lori Udall, "Complaint Filed at World Bank, NGO's Charge Policy Violations at Arun III," *World Rivers Review* 9, nos. 2/3 (1994).

56. "Panel Calls for Arun III Probe, Bank Management Tries for Delay," *Bank-Check Quarterly* 10 (December 1994).

57. "Herd of White Elephants: The Bank Has Plenty of Boondoggles to Answer For," *Newsweek*, 9 October 1995, p. 43. Ratsin Mahao, acting coordinator, Highlands Church Action Group, letter to Mr. James D. Wolfensohn, president, the World Bank, 16 August 1995.

58. Andrew Nette, "Nam Theun II Stranded with No Buyer for Its Power," *World Rivers Review* 12, no. 2 (1997).

59. Khan, "World Bank and Investments for Water."

60. Guy Le Moigne, Ashok Subramanian, and Sandra Giltner, *A Guide to the Formulation of Water Resources Strategy* (Washington, D.C.: the World Bank and United Nations Development Program, July 1994).

61. François-Marie Patorni, *Training Strategy in the Water Sector, FY 1994–96* (Washington, D.C.: the World Bank Economic Development Institute, June 1993).

11

The World Bank and Public Accountability: Has Anything Changed?

Lori Udall

At the heart of the campaign to reform the World Bank is an attempt by citizens' groups, nongovernmental organizations (NGOs), and local communities to make the Bank more publicly accountable and transparent. This effort is part of a larger, worldwide NGO effort to reform the projects, policies, and programs of all multilateral development banks in order to promote democracy and development alternatives that are socially just and environmentally sound. The World Bank has been specifically targeted because it is the most influential of the development banks, lending over $22 billion a year and setting trends for the regional banks.

The multilateral development bank (MDB) campaign has used a two-track approach of reform or "damage control" of specific MDB-financed projects along with the promotion of broader policy reform within the institutions. NGO coalitions have worked on the basis that policy reforms are critical to broadening public access to the decision making process and to promoting development alternatives. They believe that if development bank project planning and design were open and transparent, and involved democratic processes such as public hearings and mechanisms to challenge projects, fewer disastrous projects would be approved and a greater opportunity to promote development alternatives would exist.

For years, World Bank–financed projects and programs have been planned and designed with little opportunity for public input or scrutiny. In the Bank's borrowing countries, local communities directly affected by Bank-financed projects have often been denied access to the most basic information about the project, particularly at the time that it is most crucial—at the project-planning and design stage. At the same time, tax-payers in the World Bank's major donor countries, whose money supports and underwrites the Bank, also have little knowledge or influence over how their money is being used. Moreover, executive directors (EDs) who represent the Bank's shareholder governments are also often denied access to critical information even though they are responsible for approving virtually all Bank policies, projects, and programs.

The failure to disclose information on a timely basis has resulted in a lack of local consultation and public participation at the project level, ill-conceived and poorly planned projects, and deteriorating project quality. Intense local opposition to projects such as the Sardar Sarovar (Narmada) Dam in India, the Kedung Ombo Dam in Indonesia, the Pak Mun Dam in Thailand, and more recently the (since shelved) Arun III hydro-electric dam project in Nepal is a stark reminder of this problem.

In August and September of 1993, the World Bank board of executive directors took the unprecedented step of approving the creation of an independent appeals panel (the Inspection Panel) that will allow citizens adversely affected by Bank projects to file claims regarding violations of the Bank's policies, procedures, and loan agreements.[1] In addition, the board took steps to allow the release of more information about Bank projects and programs by revising its information policy.[2] Both of these reforms took place under conditions of intense public international pressure and under the threat of reduced funding by the U.S. Congress. From Narmada to Arun III, this chapter examines the origins and implementation of these two crucial public accountability reforms.

A major catalyst in the development of both of these policies was the struggle and public debate that emerged over the Sardar Sarovar (Narmada) Dam over a seven-year period from 1985–1992. Prompted by Indian grassroots activists, an international campaign emerged around Narmada whose public pressure ultimately led to the formation of an unprecedented independent commission (the Morse Commission) that reviewed the project with regard to the Bank's adherence to its own environment and resettlement policies. The Morse Commission report underscored the problem of project planning and design without informed public participation and access to information; also, the existence of the commission itself reinforced the NGO proposal that an independent mechanism was needed where adversely affected people could air their grievances about Bank projects. The international Narmada campaign and subsequent International Development Association (IDA) 10 campaign brought pressure to bear on the Bank via the executive directors, the U.S. Congress, donor governments, and the media to institute the two accountability reforms.

Expanded public access to information and the creation of a mechanism of recourse for citizens affected by Bank projects are two important steps forward in grassroots groups' and NGOs' struggle to reform the World Bank and to ensure that Bank practices are more responsive and accountable to communities in borrowing countries. By increasing transparency through greater access to information, problem projects may be caught, halted, or restructured relatively early in the pipeline through citizen input. The Inspection Panel gives teeth to a variety of important Bank environmental and social policies (e.g., those concerning environmental assessment, resettlement, and indigenous peoples) in that it provides affected citizens the opportunity to file complaints if the policies are not properly implemented and enforced by Bank management. The reforms complement each other by creating incentives for Bank management to increase openness, public participation, and public debate in the development process.

This chapter also looks at the Arun III project in Nepal as the first test case of the implementation of these two reforms. Access to information was a key element of the international campaign that emerged around Arun III, and the case illustrates how the Bank's information policy affects the relationship between democracy and development policy in borrowing countries, as well as between NGOs and the Bank. Arun III was also the first claim filed with the Inspection Panel and the only one, to date, that has undergone a full investigation by the panel.

The Origins of the Twin Accountability Reforms: From Narmada to Washington

The Sardar Sarovar (Narmada) Dam and Power Project in western India is a classic example of the Bank's failure to provide information to and consult local populations, which resulted in growing opposition to the project. The Narmada campaign became a major catalyst for the two accountability reforms. The grassroots opposition to Narmada is rooted in support of the basic human right of indigenous and rural poor people to have control over their own future and any development underpinned by democratic processes. Worldwide, Narmada came to symbolize an outdated, top-down, centralized development paradigm that continues to be promoted by the World Bank.[3] The sustained international campaign that developed around Narmada brought sufficient pressure on the Bank to appoint the independent Morse Commission to investigate the project. The Morse Commission's report was highly critical of the Bank's performance on environment and resettlement; it highlighted the Bank's failure to involve local communities in the development process and to take seriously the complaints and problems of affected people. After the release of the report, NGOs active in the Narmada campaign lobbied the U.S. Congress to place "conditions" on funding for the International Development Association (IDA), the Bank's soft-loan window, until the Bank developed mechanisms that would increase its public accountability and transparency. Later on, pressure from the public and the U.S. Congress, as well as the support of several donor country executive directors, ultimately forced the Bank to make the reforms.

Given the centrality of the Narmada project as a catalyst in the establishment of the reforms, a brief history of the Narmada campaign and related developments follows.

The Narmada Campaign

Approved by the World Bank in 1985 with a loan of $450 million,[4] the Sardar Sarovar (Narmada) project was fraught with environmental and resettlement problems from the outset. Although the project was slated to forcibly displace more than 150,000 people[5] ("oustees") from their homes and villages, most of those people did not have access to the most basic information about their impending resettlement—such as basic entitlements, timetables, or locations where they would be resettled. Many were not even informed that they had to move; instead, local government officials installed markers in their villages that signified the submergence level of the reservoir. Over time, NGOs and individuals such as Medha Patkar (a social worker originally from Bombay) who started working with the oustees began demanding access to information about resettlement plans and timetables. They also demanded environmental studies, project appraisals, and cost/benefit analysis from both the Bank and the Indian government. Although such information was rarely officially released, documents obtained by NGOs through leaks and informal sources gradually revealed a project that not only had environmental and social problems, but was also ill-conceived, poorly designed and appraised, and economically questionable. The Indian activists also discovered that a comprehensive resettlement and rehabilitation plan, which according to Bank policy was supposed to be completed before project appraisal, had not been finished even five years after project approval. Project authorities did not even know the total number of people to be directly affected by the project. Environmental impact studies and an environmental work plan stipulated by project legal agreements to be completed by December 1985 were still incomplete. Throughout the late 1980s, violations of World Bank policy and loan agreements worsened.

The failure of the Bank and the Indian government to respond to charges regarding these omissions spurred the broadening of the resistance to the project. Medha Patkar and other grassroots activists working with the affected people in India began expanding their campaign to challenge the underlying assumptions and validity of the project, and by November 1988 they had declared their total opposition to the project.[6]

The grassroots movement against the Sardar Sarovar Dam in India—a movement now known as Narmada Bachao Andolan (NBA, Save the Narmada Movement)—is one of extraordinary vision, perseverance and vitality. It is also one of the largest, most sustained grassroots movements against a World Bank–funded project anywhere in the world. The movement grew from a small number of villages in 1986 to a majority of the "oustees" by 1992. In the Gandhian tradition of nonviolent resistance, villagers refused to cooperate with project authorities who came to conduct surveys or offered to show them resettlement land. By 1992, several thousand individuals, leaders, and village heads had vowed to drown rather than move from their villages.

Medha Patkar and other Indian activists lived in the valley and organized the villagers to defend their rights to their land, their lives, and cultures.[7] They reached out to international NGOs for support in targeting both the World Bank and the Japanese government for their involvement in the project. Medha Patkar and key NBA activist Shripad Dharmadikary traveled to important international meetings in the United States, Europe, and Japan, and sent out a steady stream of information, updates, and action alerts to international activists. Although their primary commitment was to the local people and oustees, they also understood the importance of an international campaign to pressure the World Bank, Japan, and the Indian government.

The activists also knew that Narmada was symbolic of what was happening on a larger scale to indigenous people affected by destructive development in India and around the world. In 1989, in a letter to a World Bank executive director, Medha wrote: "We will not leave the valley even if it is actually flooded … as our alternatives are [a fate] worse than death, a lesson we have learned from the experience of millions of others who have been devastated by so-called development projects in the past."[8]

By 1988, an informal network of individuals and NGOs in the World Bank major donor countries was working together to influence their legislators, finance ministries, and executive directors on Narmada.[9] Most of the people in this network had spent time in the Narmada Valley with the oustees and activists, and as a result had developed a deep long-term personal commitment to the movement. Over a period of years, members of the network convinced legislators to raise questions in parliaments, held press conferences, launched letter-writing campaigns, and hosted seminars and international meetings on Narmada.

A turning point in the international campaign was when the World Bank's involvement in Sardar Sarovar became the subject of a special U.S. congressional oversight hearing led by Congressman Jim Scheuer (D-NY) in 1989.[10] Medha Patkar, human rights lawyer Girish Patel, and economist Vijay Paranjpye all came from India to testify.[11] Jim Scheuer was also chair of Global Legislators for a Balanced Environment (GLOBE), which includes a number of prominent Japanese Diet members.[12] After the hearing, Scheuer met with GLOBE members and shared his concerns about Narmada. The fact that Narmada had been the subject of a U.S. congressional hearing was a catalyst for growing government concern in other major shareholder countries, particularly Japan.

Aside from being one of five World Bank major shareholder countries, Japan was also funding the project bilaterally. In 1990, the next spring, Japanese NGOs held the first International Narmada Symposium, in which several Indian activists associated with Narmada Bachao Andolan participated. The symposium had unprecedented exposure in the mainstream Japanese media, and within a month the Japanese government announced that, based on inadequate resettlement and local opposition, it would not consider further funding of the project.[13]

As a result of international pressure from their own countries, many donor country executive directors began sending their questions to Bank management, and in 1989 Paul Arlman of the Netherlands began holding informal executive director meetings on Narmada. By the time Japan pulled out, many of the EDs were questioning Bank operations staff's version of developments in the Narmada Valley because information they received from the affected people was so widely divergent from Bank operations staff reports. EDs did not have access to "back-to-office" supervision reports or other Bank files on Narmada, which in many cases were quite critical of the project. They had to rely on oral briefings from operations staff or briefing documents specially prepared for their consumption.[14]

By September 1990, executive directors began discussing a possible independent review of the project, but at the time it was unclear what the character of the review would be. There was some discussion about the review being done by the Operations Evaluation Department (OED), which normally only conducts evaluations of completed projects, and about a trip for executive directors to the valley itself. NGOs warned the EDs that, in their view, neither option would be sufficiently independent.

In December 1990–January 1991, an extraordinary event took place in the valley: tribal people, farmers, and the rural poor from all three states affected by the project began marching toward the dam site to stop the construction of the dam through a nonviolent sit-in. The marchers were stopped at the Gujarat border by police, military, and state government officials, and a thirty day stand-off ensued. Stuck on the border, Medha Patkar and a few other activists started a twenty-six-day fast, demanding a comprehensive review of the project. By mid-January, several executive directors led by Eveline Hefkins (successor to Paul Arlman) and then-Senior Vice President Moeen Quereshi convinced then-President Barber Conable of the need for an independent review of the project. Although Bank officials did not publicly announce it during the fast, they indicated to Washington-based NGOs that a review panel would be announced as soon as a chairperson was chosen. This message was then conveyed to Medha Patkar and other members of Narmada Bachao Andolan in India, who ultimately called off their fast.

The Independent Review of Sardar Sarovar

The decision to hire an independent review team, although announced and sanctioned by President Conable, would never have occurred without the pressure and support of some executive directors. It clearly signaled an institutional crisis in which many board members felt they could no longer trust the Bank's India operations staff or regional management for accurate information about the situation on the ground in the Narmada Valley. Because most of the issues raised over the years were on environment and resettlement grounds, the review team was charged with conducting an investigation into the Bank's adherence to its own environmental and resettlement policies, procedures, and loan agreements. The team, now known as the Morse Commission, was headed by a former director of the United Nations Development Program (UNDP), Bradford Morse.[15] The deputy chair was Canadian human rights lawyer Thomas Berger.

Appointed in September 1991, the Morse Commission conducted an intensive ten-month investigation into the project and issued a report that concluded that the project was "flawed, that resettlement and rehabilitation of all those displaced by the project [was] not possible under prevailing circumstances and that environmental impacts ... [had] not been properly considered or adequately addressed."[16] The commission questioned the viability of underlying assumptions regarding the project and stated "there is good reason to believe that the project will not perform as planned."[17] The report also condemned the Bank's "incremental strategy" of accepting piecemeal improvements in resettlement, saying that it was a failure and counterproductive: "Nor is it a question of applying more intense pressure to [the state governments] Maharashtra or Madhya Pradesh in order to improve resettlement policies.... A further application of the same strategy, albeit in a more determined or aggressive fashion, would fail."[18] The commission recommended that the Bank "step back"[19] from the project; citing hostility in the valley and local opposition to the project, it further stated that "progress will be impossible except as a result of unacceptable means"[20] and indicated that the project could only be implemented with police force.[21]

The Bank's Response to the Independent Review

After the release of the independent review, Bank management attempted to justify its support of the project by issuing a document to the board of executive directors in which it detailed the Bank's and Indian government's plan to improve the project's record.[22] This document, "Next Steps," called for the Bank to continue funding the project during a six-month conditional period—at which time, it claimed, construction would be tied to improvements in resettlement. It downplayed, however, the possibility that villages would be submerged in the 1993 monsoon season due to the water rising behind the partially constructed dam. For NGOs that had monitored the Bank's involvement for years, this document was just another attempt to continue the "incremental strategy" that was so clearly condemned by the Morse Commission.

During the World Bank/IMF Annual Meeting in September 1992, Shripad Dharmadikary of the NBA came to Washington to meet with World Bank executive directors, their staff, and U.S. Treasury Department officials. Shripad severely criticized the "Next Steps" document, discussing in detail how the document failed to acknowledge the on-the-ground reality in the villages, the level of local opposition to the project, and the seriousness of the rising waters during the monsoon. He also exposed the Bank's false distinction between "temporary" and "permanent" submergence, explaining to the executive directors that for a tribal person living in a wood and mud hut, even a one-day temporary submergence would completely destroy the house, personal property, and annual harvest.[23]

The international NGO community was equally outraged at "Next Steps" and with the help of a few wealthy individuals ran full-page advertisements in the *Washington Post, New York Times,* and *Financial Times* that called for the World Bank to withdraw immediately from

Sardar Sarovar. On September 21, 1992, as finance ministers from more than 150 World Bank member countries of the World Bank were arriving in Washington, the ads ran a warning that if the Bank continued funding Sardar Sarovar after the Morse Commission's overwhelming evidence then "NGOs and activists would put their weight behind a campaign to cut off funding to the Bank."[24]

Morse Commission members were also disturbed by "Next Steps." On October 13, 1992, about ten days before the World Bank board meeting on Narmada, Bradford Morse and Thomas Berger sent a letter to President Lewis Preston and the executive directors, charging that "Next Steps" was misleading: "it ignores or misrepresents the main findings of our review."[25] They further stated that the Bank had "ignored our conclusion that the Bank's incremental strategy greatly undermines prospects for achieving successful resettlement and rehabilitation" and that as a result "the well being of tens of thousands of people will continue to be at risk."[26] Subsequently, Morse Commission members flew to Washington to meet with the board and set the record straight.

In an attempt to address the attacks on "Next Steps," salvage credibility, and pacify executive directors, Bank management proposed a series of "benchmarks" in order to justify continued funding—benchmarks that actually represented a series of conditions the Bank had already failed to enforce over a six-year period.[27]

On October 23, 1992, the World Bank board voted to continue the project for six more months, despite the objections of six executive directors (42 percent of the vote) who called for a suspension of the project—the largest number of executive directors ever to vote against a project that was also being opposed by NGOs.[28] Three of the executive directors who were calling for a suspension represented the Bank's largest shareholders—the United States, Germany, and Japan. The U.S. executive director, Patrick Coady, who was thoroughly convinced of the unsoundness of the project, made an impassioned plea for the Bank to suspend the project and charged Bank management with covering up the project's problems. In the board meeting he stated that if the Bank continued with the project, "it would signal that no matter how egregious the situation, no matter how flawed the project, no matter how many policies have been violated, and no matter how clear the remedies prescribed, the Bank will go forward on its own terms."[29]

The IDA 10 Campaign

Coady's sentiment was shared by NGOs; they were furious that the Bank contravened the recommendations of the Morse Commission. To them, the Bank's misrepresentation of the Morse report and its continued funding of the project deeply undermined its credibility and caused them to question whether it was capable of real reform.

The main point of leverage that would enable NGOs to force Bank reforms was to attack the Bank's funding in the major donor countries. A group of Indian, Asian, and international NGOs who were part of the Narmada campaign moved immediately to launch a campaign opposing $18 billion dollars for the tenth replenishment of the International Development Association (IDA 10), "the soft-loan window" of the World Bank.[30] More than any other event, the campaign of this small but worldwide network of NGOs around the tenth replenishment of IDA led to the rapid introduction of new accountability reforms.

The campaign called for direction of IDA money to other multilateral and bilateral institutions that are more accountable, democratic, and participatory. Because Narmada was symbolic of many of the institutional problems and policy violations widespread inside the Bank, NGOs felt that if the Bank could not address critical issues in such a high-profile problem project, then reform in other projects and programs was unlikely. They also felt that problems in the review of Narmada underscored the need for a permanent appeals mechanism to independently investigate problem projects.

Contributing to this evidence was a leaked internal Bank report now known as the Wapenhans Report, which revealed that up to 37 percent of Bank projects were "unsatisfactory." This high-level study of deteriorating project quality was focused on a pervasive "approval culture" inside the Bank: staff perceived project appraisals as internal "marketing devices" for securing loan approval and achieving personal recognition.[31] Although the Morse Commission had found the appraisal process in one particular project to be inadequate, the Wapenhans Report revealed that the credibility of the appraisal process *Bankwide* was in question.[32] It underscored what NGOs had criticized about Bank operations all along: that the pressure to lend money overwhelmed all other considerations—including project quality. Wapenhans states, "The portfolio is under pressure. The pressure is not temporary, it is attributable to deep-rooted problems which must be diagnosed and resolved. The cost of tolerating continued poor performance is highest not for the Bank, but for its borrowers."[33]

Because IDA money is mostly comprised of donor country taxpayers' money, U.S. and some European NGOs have increasingly used the IDA negotiations, as well as congressional and parliamentary approval and appropriations of IDA funds, as vehicles for promoting reforms and raising questions about the quality of World Bank projects and programs. IDA 10 negotiations were being finalized in November 1992, only a month after the board had voted to continue Sardar Sarovar.

By November 1992, a document entitled *International NGO Statement Regarding the Tenth Replenishment of IDA* and signed by 138 NGOs from twenty-two countries had been circulated to members governments, policymakers, and legislators.[34] Because IDA is the window of the World Bank that lends to the poorest countries, and because a large percentage of the funds go to debt-ridden sub-Saharan African countries, some international NGOs were opposed to this IDA 10 campaign; they believed that calling for a reprogramming of IDA funds would ultimately hurt the poor in Africa and do little to change the Bank. However, as the campaign mounted, the dramatic response from member governments and Bank management made it clear that the IDA 10 campaign was working.

In the spring of 1993, Washington-based NGOs testified in Congress that they would oppose funds to the Bank unless it could meet a series of tough new benchmarks. The two key benchmarks were (1) a complete revision of the Bank's information policy and (2) the creation of a citizen's appeals panel to give directly affected people access to an independent body to file complaints.[35] NGOs also proposed that the U.S. Congress begin redirecting funds for IDA to other more accountable and democratic mechanisms if these benchmarks were not met by June of 1994.

Congressman Barney Frank (D-Mass.)[36] led the initiative to pressure the World Bank to increase public access to information and to establish a panel to investigate citizen's complaints. Although he did not publicly state it, Frank basically refused to authorize the $3.7 billion replenishment to the Bank until the Bank had concrete, adequate written proposals on both reforms. The length to which the Bank went to pressure Frank and subcommittee staff was remarkable. As never before, it sent representatives directly to Capitol Hill to lobby, often even bypassing the U.S. Treasury, the agency that normally lobbies Congress for World Bank funds.[37] Drafts of the revised information policy and a resolution that created an independent "inspection panel" were also sent secretly to Congress for comments before they were presented formally to the Bank's board.

In the summer, Washington-based NGOs commented on several drafts of both the Inspection Panel resolution and the revised information policy and indicated to Congress those draft provisions that were unacceptable from an NGO standpoint.[38] Comments were sent to the U.S. Congress, the Treasury Department, and the U.S. executive director. European NGOs also sent comments to the executive directors representing their countries.

During the fall of 1993, after months of congressional hearings and debates on IDA's performance, and after the two reform proposals were approved by the executive directors, Congress took the unusual step of authorizing IDA for only two years, although it had previously authorized IDA in three-year increments.[39] It also cut the U.S. Treasury's pledge to IDA by $200 million and the pledge to the International Bank for Reconstruction and Development (IBRD) by $15 million. Although the amounts cut were small, a clear message was sent to the Bank that U.S. legislators would no longer tolerate the lack of public accountability and transparency that pervaded Bank operations.[40] In 1994, Congress continued to withhold the third-year authorization of IDA, and further congressional support was conditioned on implementation of the new reforms.

Together, NGO lobbying, grassroots protest, and Bank management's rejection of its own unprecedented review turned the Narmada project into a powerful symbol and catalyst. Once its flow of concessional funds was at risk, the Bank conceded on the two accountability reforms that had grown out of the public debate regarding the Narmada project and the Morse Commission's review.

The World Bank's Revised Information Policy

Before the 1993 information policy revision, the general principle of the Bank's information policy was *presumption in favor of disclosure of information in the absence of a compelling reason not to disclose.*[41] Despite this presumption, in practice the Bank consistently restricted almost every type of document regarding Bank projects, policy-based lending, economic programs, as well as much environmental and social information. Attempts to gain access to information—by the public and particularly by people directly affected by Bank projects and programs—were consistently met with refusals and red tape.

Then and now, the Bank's main justification for withholding project documents is that its deliberative and decision-making process must be protected from public scrutiny.[42] NGOs have maintained that it is precisely when critical decisions regarding the design of a project are to be made or discussion of alternatives arises that the public should have access to project documents and the deliberative process. They believe that the public's legitimate interests in project quality, sustainable development, and viable project alternatives or options must be balanced with the need for confidentiality.

Another justification for restricting information often used by the Bank is that the document is the property of the borrowing country; the Bank receives information from its borrower countries with the understanding that it will remain confidential. However, NGOs have maintained that this restriction could be easily and legally overcome if, when receiving a document, the Bank would simply indicate that it intended to release the document publicly.[43] Past experience has revealed that the Bank has often exaggerated the need for confidentiality.[44] Although there may be reasonable grounds to restrict interoffice memoranda or a small body of documents that contain confidential or sensitive information about the Bank's borrowers, most NGOs believe that a major portion of project information has not been legitimately withheld.

Following the revision of its information policy, the Bank created categories of documents and information that were supposed to be made available to the public. It also created Public Information Centers where the documents were to be made available. The story of the information policy reveals how both reform advocates and opponents viewed access to and control over information as critical. Pressured by the Narmada campaign, the international public, and the U.S. Congress to start releasing more information about its projects and programs, in November of 1992, the World Bank convened an internal working group to decide what steps it should take in revising its information policy. In March 1993, the working group issued recommendations for changing the policy that would ultimately have little practical effect at the project level in increasing transparency or openness.[45] However, inside the working group, a dissenting group of Bank staff issued an "alternative view" that recommended far-reaching changes in the Bank's policy.[46] If implemented, the alternative view would have resulted in the timely release of almost all project documents and all environmental and social information.[47] This view recommended the public release of early project documents—such as draft staff appraisal reports, which contain critical information regarding the basic justification for the project, key components of the project, the financing plan, and elements of the cost/benefit analysis. This approach was also promoted by the U.S. Congress and in particular by Representative Barney Frank.

In spite of the dissent, the mainstream working group recommendations formed the basis for most of the policy revisions which were sent to the World Bank's board of executive directors and eventually approved on August 1993. In an attempt to incorporate some of the alternative information policy proposals, especially public access to early project documents, the U.S. Congress, later supported by the U.S. Treasury, promoted the inclusion of an important provision for the release of "factual technical information."[48] The provision provides that upon request, the country department would release "factual technical documents or portions of documents" about projects in preparation. NGOs believed that "factual technical information" would include early project documents, though later, during implementation, Bank management's interpretation of its meaning was very restrictive (as discussed below). Despite that, the provision could still make the major difference between the old and new policies. Documents that are now publicly available under the new policy include project information documents, factual technical information, final staff appraisal reports, and environmental impact assessments.[49]

The Bank's New Information Policy: An Assessment

Project Information Documents

The Bank's Information Policy creates a new category of document called a project information document (PID), which is written specifically for public consumption. At a very early stage in the cycle of a project, right after the project is identified, a PID is supposed to be written by the task manager—the Bank staff member responsible for the project. This document is the main source of information available to the public at a critical stage even before the environmental assessment is available. NGOs monitoring the process were initially opposed to the PIDs because they believed the document would be "sanitized" and might obscure important or controversial aspects of a project. For this reason, they felt that the first project document used by Bank staff (called an "initial executive project summary") should be released instead. When a PID is first created, it should contain the same information as the initial executive project summary on the main elements of the project: the project objectives, expected or probable components, costs and financing, environmental issues, status of procurement and consulting services, studies to be undertaken, implementing agencies, and relevant contact points.

A review of selected project information documents produced shortly after the reforms revealed that many contained so little information as to be useless to an NGO or community that wanted to change or challenge a project.[50] PIDs are also supposed to be expanded and updated as the project progresses. However, those PIDs reviewed in 1994–1995 contained only the most basic information about a project and most were not updated or expanded even as the project progressed toward appraisal. Some project information documents did not even include the location of the project or the contact person for the project, which means that local people interested in having input in or challenging project planning and design have very little information to work from.[51] The quality of PIDs, however, has improved over time, and PIDs reviewed in 1996 are far better then in the previous two years.

"Factual Technical Information"

The new information policy allows for the release—before loan approval is given—of documents or parts of documents that are considered "factual and technical" in nature. However, documents that contain staff judgments or expose the Bank's decision-making process are considered confidential by the Bank. Factual technical documents are the main source of information available to the public between the time the project information document is released and the time the project is approved, when the final staff appraisal report is released. If properly implemented, this will allow for the release of information while the project is still in preparation—a critical time for NGOs and affected parties to have input in the design or to promote alternatives to the project. Factual technical information includes documents such as feasibility studies, detailed project design plans, technical studies underlying environmental impact analysis, and poverty analyses.

Without the information contained in early project documents and their annexes, public participation in project planning and design can only be superficial. Public scrutiny and debate about a project are also less likely. Yet the way many requests for information have been handled under the provision since the policy came into effect in January 1994 has not set a good precedent for information release. NGOs have been denied technical information contained in World Bank early project documents, such as draft staff appraisal reports and their annexes, feasibility studies, and baseline data. Nevertheless, NGOs see this provision as a critical avenue for information about projects in preparation, in addition to the environmental assessment process.

After much confusion about the precise meaning of "factual and technical," the Bank finally issued guidelines on June 20, 1994—exactly one day before a special oversight hearing of the Subcommittee on International Development, Trade, Finance, and Monetary Policy. Since the third-year authorization of the U.S. contribution to IDA ($1.25 billion) was contingent on the proper implementation of the new information policy and on the establishment of the Inspection Panel, the Bank was posturing to ensure that the subcommittee would authorize the contribution, which it finally did in late 1994.

Retroactivity

Another major shortcoming of the new information policy is that it does not apply retroactively. As a result, a large portfolio of projects that NGOs are interested in—all projects that were approved or negotiated before October 1, 1993—are not covered under the new information policy. Requests for these documents are handled on a case-by-case basis by the individual country directors and in the context of the information policy that was in effect at the time the project was negotiated, resulting in the inconsistent treatment of NGOs or individuals requesting information. The retroactivity problem, combined with the failure of the Bank to operationalize the "presumption in favor of disclosure" under the new policy, does not leave a very promising record for increased public accountability.

Access to Information, Democracy, and the Arun III Hydroelectric Project

The first crucial test case of the information policy was the Arun III Hydroelectric Dam Project. Arun III came to public attention because of highly controversial aspects of the project, such as its $1 billion price tag and the resignation of a high-level Bank staffer in protest over the project. The Arun III case is explored in some detail here because it has been, to date, the most important test case for both the new information policy and the Inspection Panel. This section first discusses Arun III and its relationship to the information policy. The issue of information highlights the failure of the Bank to explore alternatives and illustrates how the Bank's lack of transparency can undermine democratic institutions in borrowing countries.[52] A discussion of the creation of the Inspection Panel and an examination of the Arun experience with the panel are also included in this section. The emphasis here is on using the Arun III case to illustrate the impacts of the information and Inspection Panel reforms, and on highlighting the nexus between the two.

The Arun III Hydroelectric Dam Project in Nepal involved both a proposed massive dam project and a 122-kilometer road leading up to a remote pristine valley with unique biological diversity and some of the last remaining intact forest in the Himalayas. The valley is also home to

450,000 indigenous people.[53] The Arun III Dam was the first in a series of three dams to be built in this pristine area. Instead of considering the cumulative impacts of the three projects together, however, environmental assessments were being conducted in a piecemeal manner.

The World Bank was proposing to loan $170 million for the Arun III project, whose total cost was estimated at over $1 billion.[54] Rapid construction of the road into this remote region was a major environmental and social concern,[55] even more so than the dam itself (which was to be a run-of-the-river dam that would not create a reservoir).

Failure to Consider Alternatives

For a two-year period, local Nepalese and international NGOs opposed the Arun III project as designed. NGOs such as Alliance for Energy (Nepal) and Intermediate Technology (IT)[56] proposed alternatives to the project that they believed were much more appropriate for Nepal's long-term energy needs and focused on plans that would develop the country's existing capacity, both in the public and private sectors, move toward decentralization, promote local management and control of projects, as well as provide electricity to local and rural people.[57] These groups believed that Arun III was an unnecessary financial commitment that exceeded Nepal's capacity. The total project cost was approximately $1 billion—more than one year's national budget for Nepal. Local NGOs maintained, moreover, that approval of Arun III would have "crowded out" investments in critical social sectors such as health and education.

Although Alliance for Energy and IT put out their own studies and documentation,[58] they were consistently at a disadvantage because they had no access to World Bank documentation on either finances, alternatives, or cost/benefit analysis. Accordingly, it was difficult to counter the Bank's arguments that it had considered alternatives to Arun III when all Bank documents on alternatives were withheld. In early 1994, the Bank did hold some discussions with NGOs who were promoting alternatives, yet even in those cases it continued to withhold studies on alternatives and financial analyses of the project. Project approval at the Bank was slated for July 26, 1994.

The backdrop for the battle on Arun III was a newly established democratic government in turmoil; Prime Minister Koirala had resigned and disbanded the parliament in July 1994. Because the high cost of the project had implications for the whole country, Nepalese NGOs were trying to ensure that there was public debate about and scrutiny of the project from the outset. By refusing to release information, the Bank was, in effect, obstructing the democratic processes of public participation, debate, and consultation inside Nepal. Despite the fact that the Arun III project was becoming an issue in the upcoming Nepali elections on November 15, 1994, Bank management was still pressing ahead to send the project to the board before then.

As the loan negotiations and the board date grew closer, and as basic information was still being denied, Washington-based NGOs in support of Nepalese groups sent several letters requesting early project documents and alternatives studies to the Bank's Nepal Department and the vice president of its South Asia Division. Previously, Nepalese groups had made requests for information, but had received no response. The NGO letters requested factual technical information on Arun III, such as basic information contained in draft appraisal reports, analysis of alternatives, and feasibility studies. After receiving repeated requests for information over a two-month period, which involved an exchange of six letters and denials from Bank staff, on June 9, one day before a Washington "briefing" on alternatives to Arun III, the Bank publicly released a study on alternatives prepared by Argonne National Laboratory with technical annexes.[59] Other technical information related to the feasibility of the project, finances, and cost/benefit analysis has still not been released.

The report the Bank used to justify its choice of Arun III versus smaller-scale alternatives admits that the comparison is superficial. The Argonne National Laboratory report revealed that, compared to the proposed design, alternatives were not adequately studied: "Several of the Projects exclusive to Plan B [the alternative analyses] have not been studied sufficiently to place them side by side in credibility to the projects that are further along in investigation."[60]

When questioned by NGOs about this document during the briefing held by the Bank, Bank staff admitted that some alternative projects had been studied, but not past the feasibility stage. Several times when questions were asked during the public briefing, the task manager would look for his answer through the very document NGOs had previously re-quested be publicly released—the draft staff appraisal report.

Growing questions about the project forced the Bank to hold a "consultation" on alternatives and other issues with Nepalese and international NGOs on June 28 in Washington, D.C. However, going into the consultations, NGOs wondered whether the meetings would be futile because they had discovered that loan negotiations between the Bank and the Nepalese government for the project would already be finished or nearly completed by the time of the consultations.[61] Consequently, they feared that the consultation would become a public relations exercise rather then a serious effort by the Bank to consider any proposed alternatives. NGO representatives from Alliance for Energy, Intermediate Technology, and Arun Concerned Group flew to Washington for the consultation, which was followed by extensive meetings with World Bank executive directors, staff from the U.S. State and Treasury departments, and representatives of key U.S. Congressional subcommittees with jurisdiction over the World Bank. The Nepalese also had preliminary meetings with Richard Bissell, a member of the World Bank's new Inspection Panel, as well as the Japanese embassy and the Japanese Overseas Economic Cooperation Fund, one of the bilateral funders of the project.

These meetings raised serious questions in the minds of many executive directors not only about the viability of the project as currently designed, but also about whether there had been adequate public consultation about the project. In the meantime, the Japanese government told the Bank it did not want to be listed as one of funders of the project because it intended to conduct its own studies and field trips to the region. The combination of these developments, the resignation of Nepal's prime minister, and the disbanding of the Nepali Parliament caused executive directors to indicate to Bank management that the project should not be brought before the board on July 26. Within a few weeks, the board date had been postponed to November 3, 1994, ten days before the national Nepalese elections on November 15.

Despite Bank management's public defense of the project, behind the scenes a debate raged among Bank staff about the project, reportedly polarizing the entire South Asia regional staff, which included Nepal, Bhutan, and Bangladesh. In early June 1994, division chief for Population and Human Resources Martin Karcher resigned over Arun III because he felt the project was economically unsound and unnecessarily put Nepal at financial risk. In September, he released a public interview detailing his concern that the project would prevent critical social investments that would help alleviate poverty in Nepal, ostensibly the Bank's the primary objective there.[62]

The Bank's reluctance to release information also undermined the new democratic institutions within Nepal. In October, Bank management began stating publicly that they did not believe Arun III was an election issue because all political parties inside Nepal supported the project. Consequently, they did not feel obligated to wait until after elections to put the project before the board, still set for November 3.

"Even the Communist Party [the major opposition party in the elections] is not opposed to the scheme," stated Joe Wood, vice president for the South Asia Division,[63] four days *after* a letter from the general secretary to the Communist Party was sent to World Bank President Lewis Preston, specifically asking the Bank to wait until after the elections to allow for public debate and expressing his concern over the perceived attempt to approve Arun before a permanent government was in place. "Formal and meaningful discussion about the proposed project with the availability of basic project documents and information in advance has not yet taken place in parliament," he wrote, adding that Arun III "must be reviewed by the new government in light of the ongoing controversies before Nepal makes any commitment to such projects."[64] NGOs circulated the letter to all executive directors and donor governments, which created enough initial reaction to delay the board date once again, until after the elections. The board date was also delayed as a result of the claim filed by the Arun Concerned Group with the Inspection Panel on October 23 (see the section called "The Arun III Claim").

As the election approached, several public debates were held about the project. Although neither the Congress Party nor the UML Communist Party opposed the project outright, the Communist Party did express reservations about many aspects of it, especially its high cost. The Communist Party won the elections, and the new parliament debated Arun III on December 30, but did not resolve major questions raised about project costs and risks. The Arun Concerned Group met with the new Water Resources minister, Hari Prasad Pande, to attempt to convince him to consider alternatives.[65]

Enough doubts were raised about the project inside Nepal to prompt officials of the newly elected government to seek changes in Arun III in order to bring down the cost. In January 1995, a Nepalese government delegation led by Minister Pande came, to Washington, D.C., to propose changes in the conditions attached to the loan.[66] World Bank management was in no mood to change any of the conditions, and the Nepali government delegation left Washington without any concessions.

The World Bank Inspection Panel

Proposals for an independent appeals mechanism or body to investigate World Bank projects stem back to 1990. The first proposal was issued by the U.S. NGO Natural Resources Defense Council (NRDC).[67] Issued before the creation of the Morse Commission and the momentum of the Narmada campaign, the NRDC proposal had little practical impact, but it helped to lay down principles for later proposals issued during a period of greater momentum and influence. In June 1992, on the same day the Morse Commission issued its report, Washington-based NGOs called for an "independent appeals commission" to be set up permanently. Later that year, the Center for International Environmental Law (CIEL) and the Environmental Defense Fund (EDF) drafted and circulated a comprehensive proposal issued as a "resolution."[68] The CIEL/EDF resolution contained basic principles to promote the independent functioning of an appeals mechanism—such as an independent budget, access to Bank files, public release of reports at the time they are issued, and autonomy in deciding what claims to investigate.[69] In February 1993, four executive directors also circulated a proposal for an "independent evaluation unit," but this proposal was restrictive and contained provisions for Bank staff to be the reviewers of problem projects. As a result of intensive lobbying on the part of Washington-based NGOs during the congressional debate on IDA, the U.S. Congress and Treasury Department proposal adopted many of the principles in the CIEL/EDF proposal.[70] Washington NGOs also commented extensively on Bank draft proposals, which were sent to Congress during the summer of 1993.

On September 22, 1993, the World Bank board of executive directors passed a resolution that created an independent Inspection Panel.[71] On April 22, 1994, the World Bank announced the three panel members approved by the board: Ernst-Gunther Broder (Germany), Alvaro Umaña Quesada (Costa Rica), and Richard Etter Bissell (United States). The choice of these three people was partly dictated by the resolution that created the panel, which required that the members be of three different nationalities from three different member countries and which stated that they be chosen on "the basis of their ability to deal thoroughly and fairly with the requests brought to them, their integrity and their independence from the Bank's management, and their exposure to developmental issues and to living conditions in developing countries."[72]

The Inspection Panel officially opened on August 1, 1994, and issued its operating procedures on August 19,[73] thus setting an important precedent in international law by making an international financial institution directly accountable to citizens for the first time. The panel is empowered to investigate complaints submitted by people directly affected by Bank projects regarding violations of World Bank policy, procedures, and loan agreements. The panel will receive requests for inspection ("claims") from

any affected party in the territory of the borrower which is not a single individual.... The affected party must demonstrate that its rights or interests have been or are likely to be directly affected by an action or omission of the Bank as a result of the failure of the Bank to follow its own operational policies and procedures with respect to the design, appraisal and\or implementation of a project financed by the Bank ... provided in all cases that such failure has had or threatens to have, a material adverse effect.[74]

The affected party must also show that they have presented their problems to that Bank management and that Bank management did not respond adequately. The complaints can be filed by a local representative, but in an apparent attempt to limit the role of Washington-based NGOs and lawyers, other representatives are allowed only in "exceptional circumstances."

The Inspection Panel exercised early independence by issuing a strong set of procedures that clarified some of the ambiguity in the resolution, particularly in regard to the role and rights of the claimant and the public in the Inspection Panel process. For example, the procedures allow for members of the public to submit memoranda or opinions in support of or against a specific claim once the claim has been accepted for investigation. The procedures also allow for a very simple two-page form to be used by claimants when filing a claim and a simplified guide to the basic information required in a claim.

Although the procedures issued by the new Inspection Panel members have helped to further define and clarify some aspects of the panel, the resolution nevertheless has many serious underlying problems.[75] NGOs are concerned that the panel is too heavily controlled by the board of executive directors, that it lacks independence, and that panel reports and management's response to reports are not being released to the public in a timely manner.

Much of NGOs' early evaluation of the resolution, procedures, and developments involved in setting up the panel was based on the experience with the Morse Commission, in which the key lesson was the importance of the commission's independence from Bank management. This independence was demonstrated most clearly when the commission openly and publicly confronted Bank managers when they attempted to misrepresent the findings of the commission in "Next Steps."[76] With the exception of having access to all Bank information, the Inspection Panel, as currently codified in the resolution, has few of the features of the Morse Commission, however. Also, whereas Morse Commission members negotiated their terms of reference, budget, and work plan, Inspection Panel members were essentially faced with a fait accompli because the resolution creating the panel had already been approved when they were hired.

Nominations of Panel Members
The secrecy of the nomination process was also an element of concern. The names of approximately one hundred potential candidates were submitted by NGOs, donor and borrower governments, executive directors, Bank staff, and individuals to the Bank's managing directors in

then-President Lewis Preston's office. Even when the nominee list had been narrowed down to less than fifteen names, high-level Bank management did not release the names to the board of executive directors or to the public. Finally, under pressure from NGOs, it released the names to executive directors, and NGOs received them thereafter.

This process was problematic for several reasons. Aside from the resolution's vague criteria for choosing members, no one knew what selection process or criteria were being used by Bank management to sift through the names. Several highly qualified nominees promoted by NGOs and the U.S. Treasury were dropped from the list for no apparent reason.[77]

The Arun III Claim: The Panel's First Test

Just as Arun III became a major testing ground of the Bank's commitment to implementing its information policy, it also became the first test of the Inspection Panel and its independence from executive directors and management. On October 24, 1994, the Arun Concerned Group filed the first claim, charging the World Bank with violating its own policies and procedures on environmental impact assessment, resettlement, indigenous peoples, energy, economic evaluation of projects, and information.[78] The claim was filed in an attempt to block an impending November 3 board date and to force the Bank to look at alternatives to the project. The crux of the claim was that the Bank had failed to adequately consider viable alternatives to the project. It questioned the viability of the project as designed and used policy violations to illustrate the inadequate appraisal and planning process. A number of developments—such as attempts by Bank management to derail the Arun claim (discussed below) and the Bank general counsel's indirect comments on the eligibility of Arun III claimants in a January 3, 1995, memorandum to the board—raised concerns among NGOs about the ability of the panel to conduct its work independently and without interference. An investigation of the claim was ultimately approved, but was limited to investigating violations of environmental assessment, indigenous peoples' rights, and resettlement policies.

The problem of board control over the investigation surfaced quickly in the Arun III claim. After receiving the claim and reviewing management's response, on December 6, 1994, the Inspection Panel recommended to the board that there be a full investigation of several policy violations cited in the claim.[79] Four days later, D. Joseph Wood, vice president for the South Asia Division, launched into a defense of Arun III at the board meeting and attempted to derail the investigation by misrepresenting the panel's initial findings.[80] Wood stated that although "the panel had drawn attention in its initial report to the broad issue of alternative means of meeting Nepal's energy needs, it does not recommend that further work be done on exploring these alternatives." In contrast, the panel's report and recommendations made a clear prima facie finding that there was an "absence of a close investigation of alternatives."[81] The panel stated that if a less restrictive hydro resource assessment could be undertaken, "it would result in expanding the number of economically and environmentally acceptable options."[82] The panel indicated to executive directors that it was not necessarily in their mandate to investigate alternatives, but they believed that both Bank management and the Nepali government should do so.[83]

Wood also misrepresented the panel's findings by indicating to the board that the panel was seeking more "information and affirmation" only in the areas of the road alignment and compensation for land acquisition and a report on the plan regarding indigenous peoples. Contrary to Wood's assertions, however, the panel's initial report expressed a wide range of concerns, including risk analysis, poverty alleviation, and cumulative impacts of all projects in the Arun Valley.[84] Wood also implied that the investigation would merely require Bank management to supply more information to the Inspection Panel. However, because the investigation process allows for a comprehensive and participatory approach, the panel also later conducted site visits and interviewed local affected people, NGOs, experts, and government officials.

After receiving news of Wood's presentations to the Board, NGOs in the Arun Concerned Group were furious and immediately sent a letter to President Preston, accusing Bank management of interfering with the Inspection Panel process and threatening its credibility.[85] Several donor country executive directors were also concerned about Bank management's actions and reportedly made their disapproval known to upper-level Bank management.

By December, NGOs and some board members were also concerned that management was going outside the procedures in trying to influence the board decision on recommending a full-scale investigation. Management's unlimited access to the board gave it an advantage over the claimants, who did not have the opportunity to meet with the board after the claim was filed.[86]

The Bank's failure to disclose information also surfaced as a problem in the Arun III case: at a critical time for the claimants, neither the management's initial response to the claim nor the Inspection Panel's final report was made available to the claimants or the public when they were sent to the board.

The Arun claim, made public by the Arun Concerned Group, was the only document available to claimants until after the board approved the full investigation on February 2.[87] They did not even have a copy of Bank management's response to the claim until it was leaked to Washington-based NGOs. Nothing in the resolution or the procedures, however, prohibits the release of Bank management's initial response to the claim or the Inspection Panel's initial request for inspection. The Arun Concerned Group and U.S.-based NGOs such as the International Rivers Network and the Center for International Environmental Law sent letters to the panel and to Lewis Preston that complained of the secretive process.[88] However, nothing was released until after February 2, 1995. Even then, Bank management kept their response that the claim *was* "confidential."

Bank Management and Board Members Question Claimant's Standing in Arun III Claim

In another development related to board control over Inspection Panel investigations, management's behind-the-scenes response to the claim influenced board members to question the "standing" of the Arun III claimants. In formally responding to the claim, Bank management addressed only the substance of the claim, but did not question the claimants' standing. However, in back-door communications with the board, they clearly questioned the claimants' right to file a claim.[89] According to the resolution that created the panel, to have "standing" a party has to be (or is threatened to be) "directly, materially affected" as a result of the

Bank's failure to adhere to its own policies and procedures.[90] During meetings that led up to the approval of the claim and on the February 2 meeting that approved the claim, several board members indicated that they felt the panel had gone beyond the resolution by accepting the Arun claim[91] because the claimants' standing was not solid (based on a narrow definition of those "directly affected").

Although the Arun III claim presented an unusual case with regard to standing, the panel felt the case was sufficiently strong to recommend a full investigation. Two of the claimants, Gopal Siwakoti and Ganesh Ghimire, were members of the Arun Concerned Group and were based in Kathmandu. Ganesh Ghimire was from the Arun Valley and had family still living there; Gopal Siwakoti had no ties to the region. They claimed that their standing was derived from the fact that the high cost of the project would have a direct impact on the economy of the whole country and therefore all Nepali citizens, including themselves.[92] The claim also noted that crowding out of investments in social sectors would directly affect the claimants. The other two claimants represented by the Arun Concerned Group were anonymous. They were from the original hill route of the access road; they felt that they had received inadequate compensation for their land. Bank management argued that this area was not part of the project because the route of the road had been changed. The claimants, however, submitted that regardless of the road route, they were directly and materially affected by the project and by the Bank's failure to enforce its policies.

After a preliminary study of the claim and meetings in Kathmandu with the claimants and the Nepali government, the panel accepted the claimants' standing. Its December 16 report to executive directors recommends further investigation of the Arun III claim and states, "The Panel judged that the serious nature of the substance of the Request as a whole and its timing in relation to the project process outweighed outright rejection of the request on the grounds of doubts on the standing of the requesters.... Management apparently came to the same conclusion since, as noted before, it addressed the substance of the request without questioning its eligibility under the applicable terms of the resolution."[93]

During the February 2 board meeting, when the board gave its approval to the panel to continue to investigate the claim, several board members criticized the panel for going "beyond its mandate" by accepting the Arun III case and again questioned the standing of the claimants. However, when it came down to a vote, the board finally approved the investigation by consensus. The panel's desk work could progress, but its field work could commence only after the new government of Nepal made a formal request for Arun III project financing (a request made on March 16, 1995).

In the meantime, before the panel was even able to leave on its own investigation, Bank management seized the opportunity to take its own mission to Nepal and the Arun Valley in April 1995 and subsequently to issue pronouncements regarding a series of preemptive "remedial measures."[94] In fact, once panel members learned that management was going to Nepal, they postponed their own trip to "avoid duplicating efforts," which, as a result, put them in the position of having to respond to management's mission report and remedial measures, in addition to conducting their own field investigation and coming up with their own independent findings. After its May 27–June 1 field trip, the panel issued its final report on Arun III on June 22, 1995. Although the report reveals a number of violations of policies regarding environmental impact assessment and indigenous peoples, overall it is very deferential toward Bank management's proposed remedial measures. However, it also clearly reveals the panel's doubt as to whether these measures could be implemented within the proposed period: "There is a need to assess whether the measures can be implemented within the time\cost frame proposed for project construction."[95]

Despite its problems, the Arun III claim process contributed to the final withdrawal of the World Bank from the project. The panel process and investigation allowed time for NGOs and executive directors to raise further questions about the project. NGOs in Nepal used the time during the Inspection Panel investigation to educate the public, organize, and work with international NGOs to influence executive directors.

In the spring of 1995, even Bank management's previously vociferous defense of the project waned, and there was reportedly weakening support for the project inside the Bank. One reason for this waning support appeared to be that Heinz Vergin, the new director for South Asia, was reportedly not impressed with the economics or appraisal of the project. Another influencing factor was the June 1, 1995, appointment of James D. Wolfensohn as president of the World Bank. Wolfensohn was known as a straight shooter from Wall Street who was not happy about inheriting a public relations disaster like Arun III. According to Bank sources, he asked his environmental advisor, Maurice Strong, to look over the economics and other aspects of the project.

By August 1995, Wolfensohn announced to the board his decision not to proceed with financing for the Arun III project. The press release issued by the Bank revealed that the decision was based on the lack of institutional capacity in Nepal, the potential for crowding out of social sectors, and the decreasing support of key bilateral donors (mainly Japan and Germany). At the same time, Wolfensohn pledged to work with the Nepalese government to research and promote alternatives.

The Inspection Panel Two Years Later: An Assessment

During the years that NGOs have monitored the claims process since the Inspection Panel was established in August 1994, several issues have surfaced that are related both to weaknesses in the resolution and to the panel's actual practice, including its relationship to the Bank's board of executive directors and management: (1) problems of public access to the panel (e.g., Bank management's adversarial responses to claims, the panel's limited jurisdiction, retroactivity and lack of information disclosure), (2) the panel's lack of independence from Bank management and the board, and (3) a weakening of Bank policies.

NGOs aired these concerns with executive directors before the first board review of the Inspection Panel, in 1996. On February 11, 1996, in response to Bank management's behavior toward the panel process and in anticipation of the impending review of the Inspection Panel resolution, Washington-based NGOs sent a letter to President Wolfensohn, recommending a series of reforms in the resolution in order to strengthen the independence, long-term viability, and credibility of the panel. The recommendations in the letter focused on the following major areas of concern.[96]

Public Access to the Panel

Adversarial Responses of Bank Management   Since August 1994, five claims have been filed with the Inspection Panel; of these, only one, Arun III, has been fully investigated. Two of the claims, Arun III and Planafloro (see Keck's chapter in this volume for a discussion of Planafloro), were claims NGOs believe fulfilled all requirements of eligibility. Two other claims (from Ethiopia and Tanzania) involved expropriation and procurement issues apparently beyond the scope of the Panel's jurisdiction. The fifth claim, regarding the Bio-Bio Hydroelectric Project in Chile, involved the International Finance Corporation (IFC), a private sector branch of the World Bank that is not currently covered by the Inspection Panel resolution.

It is notable that in all five claims, Bank management's initial response was to challenge the claimants' eligibility.[97] According to NGOs monitoring the process, rather then seeking to find solutions to problems identified in claims, Bank management began treating the panel as though it were a court of law and setting out increasingly technical and legal arguments against presumptively valid claims.[98] For example, management opposition to the Planafloro claim on technical and legal grounds reflected an overly confrontational approach to the claim, as opposed to a more problem solving approach that would facilitate the panel's work in identifying and dealing with the issues raised by the affected parties. By responding as though the panel were a court of law, Bank management has embroiled the panel and the board in continuing debates over legal and technical interpretations of the panel resolution, resulting in a growing skepticism among affected groups and NGOs that view the panel process as too difficult and complicated.[99]

Limited Jurisdiction of Inspection Panel   Currently, the Inspection Panel has jurisdiction only to review claims on projects financed by the International Bank for Reconstruction and Development (IBRD) or the IDA, but not by IFC or the Multilateral Investment Guarantee Agency (MIGA)—the private sector branches of the Bank. Experience has shown that there are enough problem projects in the private sector branches to necessitate an avenue for review of those projects. The Bio-Bio Project in Chile, filed (and rejected) with the current Inspection Panel, is the best example of this. There is no compelling reason why the current Inspection Panel, with a separate set of procedures for private sector projects, should not have the mandate to investigate IFC and MIGA projects as well.

Retroactivity   Currently, the resolution states that the panel will not accept claims filed "after the closing date of the loan financing the project with respect to which the request is filed or after the loan financing the project has been substantially disbursed."[100] This provision could result in blocking valid and relevant claims, particularly those for which the Bank's legal responsibility is still in effect.[101]

Information Disclosure and the Inspection Panel   According to NGOs that have monitored the panel process, the panel resolution contains provisions that restrict information release to the public and the claimant at critical times during the inspection process. For example, both management's response to an initial claim and the panel's recommendation to the board (whether or not to investigate) are withheld from the public when public input could influence the board's decision about whether to approve the investigation. In fact, there is no current guarantee that the public or the claimant will ever see management's entire response to the initial claim, which has affected the public accountability of the claims process. Additionally, under the resolution, a copy of the panel's final report to the board is not released to the claimant or the public until two weeks after the board considers the report, which leaves little room for public input or scrutiny[102] and lots of room for Bank management manipulation or misrepresentation of the report. In this two-week period, Bank management are able to influence the board with their version of events without public knowledge of what they are saying or without the public or the claimants having any opportunity to influence the particular board members who represent their countries.

The experience the Morse Commission faced when Bank management tried, in "Next Steps," to misrepresent what the commission said illustrates the problem of this aspect of the resolution. Moreover, this misrepresentation occurred *even though* the independent review was publicly released, so the likelihood of misrepresentation or manipulation when panel reports and management's response are withheld from the public is clearly much greater.

Additionally, one of the key avenues for NGOs to influence decisions at the Bank is to contact the executive directors representing their countries. Without knowing the outcome of the report, the public cannot interact on the issue. This lack of due process keeps the claimants in the dark, about both the report and management's response to the report, until after their chance to influence the outcome or refute the report and response has passed.

Independence of the Inspection Panel

NGOs have maintained that the Inspection Panel's independence from the board and from Bank management is of critical importance to the panel's long-term credibility and viability. However, under the current resolution, the board determines whether or not the panel can investigate a claim. This means that the panel ultimately does not have independence to determine what cases it will investigate.

Because board members from the Bank's borrowing countries often feel that scrutiny of projects is an infringement of their countries' sovereignty, they may object to cases that originate in their country being brought to the panel. As a result, decisions about whether to take up a complaint may not be based on the merits of the case itself, but on the biases of individual board members. During the board discussions of the Arun III and Planafloro claims, there was evidence of a split between the donor country and borrowing country board members regarding approval of the claims, which polarized the board meetings and held up the investigation of valid claims. In the case of Planafloro, the Brazilian executive director was reportedly combative in his opposition to an investigation. The board ultimately chose not to authorize a full investigation of Planafloro, but instead promised another board review in six to nine months time.

Other independent offices inside the Bank do not have this board constraint. For example, although the Operation Evaluations Department (OED)—which conducts independent postmortem evaluations of Bank projects—reports to the board, it is free to make decisions about which Bank projects it will evaluate.

A problem related to independence concerns the role of general counsel and the Bank's legal department in the Inspection Panel process. The establishment of the Inspection Panel has created a structural problem for the Bank and its Office of General Counsel, which traditionally has simultaneously provided advice to both Bank management and the board in the day-to-day operations of the Bank. Although the resolution does not specifically require the board to seek the advice of general counsel on Inspection Panel matters, in practice the board receives advice from the counsel on how to interpret the resolution and the eligibility of claimants to file a claim. Panel claims inherently question the performance of Bank staff and indirectly of the general counsel as well because the counsel is involved in project preparation and loan negotiations, reviewing every loan prior to approval, and in project supervision. The general counsel also reviews or helps to write management's response to claims while providing advice to the board on how to address a claim. As a result, the public perception is that the general counsel is an interested party representing management's "side" of the issue and fostering an adversarial approach to panel issues, thus raising the question of conflict of interest and undermining public confidence in board decisions concerning the panel. Because the board will inevitably be asked to make decisions that are potentially adverse to Bank management's position, NGOs have proposed that the board seek independent advice on all matters relating to the Inspection Panel resolution and to individual claims.[103]

Weakening of Bank Policies

A key principle behind the Inspection Panel is to hold the World Bank accountable for implementing and enforcing its own policies, procedures, and loan agreements. In its first two years, there has been serious concern among NGOs and even some Bank staff that Bank policies are being weakened through a "streamlining process." While the Inspection Panel was being created, a parallel process inside the Bank began to revise and shorten Bank Operational Directives—statements of Bank policy and procedures that Bank staff are required to follow. Currently, the World

Bank's Operations Policy Department is revising the system of operational policies and procedures. As part of the revision, existing Operational Directives are being replaced by three categories of documents: (1) *Operational Policies*—short, focused statements of policy; (2) *Bank Procedures*—a common set of mandatory procedures for Bank staff to observe; and (3) *Good Practices*—advisory material for Bank staff. There is concern that in the revision process, Bank policies and procedures are being weakened and that stronger parts of the policies are being relegated to the category of good practices, which are advisory rather than mandatory. As a result, claimants may have far fewer standards to hold the Bank to, and ultimately, the scope of Inspection Panel's mandate will be greatly narrowed.

Conclusion

The importance of public accountability and transparency at the World Bank continue to be at the center of discussions both in NGO and donor government circles; certainly, public and donor government support for IDA will depend on progress in these areas. Although the Bank's new information policy and the establishment of the Inspection Panel are important steps in moving toward public accountability and transparency— and ultimately toward improved project quality at the World Bank—it is too early to tell whether these reforms will have a long-term or profound impact on project quality and on the Bank's overall project portfolio. Competing interests at the Bank—such as the pressure to lend money (identified as the "approval culture" in the Wapenhans Report) and the lack of career incentives to promote environmental and social quality in projects—may undermine or prevent full implementation of the reforms. As in the Arun III claim, the success of the Inspection Panel will also depend to a certain extent on NGOs' continued vigilance in watchdogging and commenting on the day-to-day functioning of the panel and its interaction with claimants and the board. Its success will also depend on its ability to maintain independence from the Bank and on the proposed NGO reforms in the Inspection Panel resolution. However, despite the fact that shortcomings in the resolution have been highlighted by NGOs and the Inspection Panel, it is not evident that strengthening the resolution is a priority for the board. Despite the Bank's public commitment to increased openness, implementation of the information policy has been far from smooth, and as the Arun III case illustrates, information disclosure is not systematic or uniform, even within country departments. The Arun III experience underscores the close links between information disclosure, policy violations, and the Inspection Panel process. Ultimately, the success or failure of either the information policy or the Inspection Panel will influence the other.

Notes

1. World Bank, "The World Bank Inspection Panel," resolution no. 93-10, resolution no. IDA 93-6 (Washington, D.C.: World Bank, September 1993).

2. World Bank, "The World Bank Policy on Disclosure of Information" (March 1994). See also "Disclosure of Operational Information," *Bank Procedures*, BP 17.50 (September 1993).

3. For an overview see William Fisher, ed., *Toward Sustainable Development? Struggling over India's Narmada River* (Armonk, N.Y.: M.E. Sharpe, 1995). See also Medha Patkar, Girish Patel, Vijay Paranjpye, Lori Udall, and Peter Miller, testimony before the House Subcommittee on Natural Resources, Agriculture Research, and Environment, 24 October 1989; Tata Institute of Social Sciences, "The Sardar Sarovar Project: Experiences with Resettlement and Rehabilitation," 1987–93; Paper presented at the Narmada Forum, Delhi School of Economics and Institute of Economic Growth, Bombay, December, 1993. Rahul N. Ram, *Muddy Waters: A Critical Assessment of the Benefits of the Sardar Sarovar Project* (New Delhi: Kalpavriksh, 1993); Lori Udall statement, published in *Authorizing Contributions to IDA, GEF and ADF*, Hearing before the House Subcommittee on International Development, Finance, Trade, and Monetary Policy of the Committee on Banking, Finance, and Urban Affairs, House Of Representatives, 103rd Congress, First Session, 5 May 1993, Serial No. 103-36 (Washington: U.S. Government Printing Office, 1994).

4. In 1985, the World Bank approved an IBRD loan for $200 million and two IDA credits totaling $250 million for Sardar Sarovar and related works.

5. Estimates on the exact number of oustees vary. Some estimates for the reservoir alone are as high as 200,000. Approximately 140,000 families would be adversely affected by the canal works; approximately 20,000 of those families would lose all or most of their land.

6. As early as 1983, Vahini Arch, an NGO based in Gujarat, was working on resettlement issues with their international partner Oxfam. However, Vahini Arch worked with only a small portion of the total Narmada oustees and did not launch an international campaign.

7. Key NBA activists based in the valley include Shripad Dharmadikary, Nandini Oza, Himmanchu Thakker, and Arundati Dhuru.

8. Medha Patkar letter to Paul Arlman, 1989.

9. This loose coalition later became known as the Narmada Action Committee. Key actors included Carol Sherman (AIDWATCH-Australia), Bruni Weisen (Action for World Solidarity–Germany), Yukio Tanaka (Friends of the Earth–Japan), Frank Brassel (FIAN-Germany), Aditi Sharma (Survival International–England), Paul Wolvekamp (Both Ends–the Netherlands), Juliet Majot (International Rivers Network–U.S.), Pat Adams and Peggy Hallward (Probe International-Canada), Christian Ferrie and Collectif Narmada (France), Maud Joahanssen and Joran Eklaf (the Swallows–Sweden), Ville Komsi (Committee on Environment and Development–Finland), and Lori Udall (formerly with the Environmental Defense Fund and International Rivers Network). The five major donors to the World Bank are the United States, Germany, France, England, and Japan. Other important donor countries include Canada, Sweden, Finland, Australia, Norway, Holland, and Italy.

10. Scheuer, chairman of the House Subcommittee on Natural Resources, Agriculture Research, and Environment, had held special hearings on the World Bank–financed Polonoreste project in Brazil in the mid-1980s and was convinced by Washington-based NGOs to examine the Bank's role in Sardar Sarovar.

11. Also testifying at the October 24, 1989, hearings were Peter Miller and Lori Udall, both of the Environmental Defense Fund. See also Lori Udall, "The International Narmada Campaign: A Case Study of Sustained Advocacy" in Fisher, *Toward Sustainable Development?*

12. GLOBE is an international coalition of legislators who collaborate on environmental issues.

13. The Japanese Overseas Economic Cooperation Fund had already lent $20 million and was apparently considering another proposal from the Indian government.

14. Executive directors have no access to Bank files on specific documents after project approval. The main document they receive before approval is the appraisal in its final form. If an executive director wants information about a specific project, he or she may request it from Bank management, which usually results in an "oral briefing" or a specially prepared update from management.

15. Other teams members were Deputy Chairman Thomas Berger and senior advisors Hugh Brody and Don Gamble.

16. Bradford Morse and Thomas Berger, *Sardar Sarovar: The Report of the Independent Review, Ottawa,* (Resource Futures International, 1992, p. xii).

17. Morse and Berger, *Sardar Sarovar*, p. xxii.

18. *Ibid.*, p. 355.

19. *Ibid.*, p. xxv.

20. *Ibid.*, p. 356.

21. The day the independent review was released—June 18, 1992—International Rivers Network and Environmental Defense Fund, Washington, D.C., issued a press release calling for a "permanent appeals commission" to be established to investigate claims of directly affected people regarding violations of Bank policy procedures and loan agreements.

22. World Bank, "Sardar Sarovar Projects: Review of Current Status and Next Steps" (Washington, D.C.: World Bank, September 1992).

23. Shripad Dharmadikary, Narmada Bachao Andolan, letter to World Bank executive directors, 21 September 1992.

24. *Financial Times*, 21 September 1992, p. 6.

25. Bradford Morse and Thomas Berger to Lewis Preston, 13 October, 1992.

26. *Ibid.*

27. World Bank, "Benchmarks," Washington, D.C., World Bank, October 1992.

28. The fact that the project went back to the board for a vote was also unprecedented. Normally, once a project is approved by executive directors, the board never "revisits" it because suspensions and cancellations are handled by Bank management. The fallout from the Morse Commission, Bank management's misrepresentation of the Morse report, and the resulting pressure on executive directors from international NGOs in effect forced the executive directors to make the decision about whether the project should move forward.

29. E. Patrick Coady, statement made at the meeting of the executive board, World Bank, Washington, D.C., 23 October 1994.

30. The International Development Association (IDA) was created in 1960 to provide no-interest loans on concessional terms to the poorest and least creditworthy Bank member countries. The stated objectives of IDA are poverty reduction, increased productivity, and economic development. During the IDA 9 replenishment, environmental protection was added to its objectives. IDA is financed primarily through aid from World Bank donor countries, and a smaller amount of additional money flows into it from borrower country repayments and from IBRD profits. Money from the donor countries is pledged to IDA in three-year increments. Since 1960, there have been ten replenishments of IDA. In 1992, donors pledged $18 billion for IDA 10, which runs from 1993 to 1996.

31. Willi Wapenhans, *Report of the Portfolio Management Task Force* (Washington, D.C.: World Bank, 1992, p. 14).

32. *Ibid.*, p. 12.

33. *Ibid.*

34. NGOs participating in the IDA campaign included Narmada Bachao Andolan, Public Interest Research Group, the Ecologist, Indian Farmers Union, National Campaign for Housing Rights–India, Development Group for Alternatives Policies, Environmental Defense Fund, FIAN-France, North-South Campaign–Italy, Campaign Committee for Human Rights–India, Urgewald-Germany,

Action for World Solidarity, Sierra Club, Integrated Environmental Organization–Sri Lanka, All Peoples Science Network–India, Appropriate Technology Association–Thailand.

35. See, for example, Barbara Bramble, testimony on behalf of National Wildlife Federation, and Lori Udall, testimony on behalf of Environmental Defense Fund at *Hearing before the House Subcommittee on International Development, Trade, Finance, and Monetary Policy,* of the Committee on Banking, Finance, and Urban Affairs, House of Representatives, 103rd Congress, First Session, 5 May 1993, Serial No. 103–36 (Washington, D.C.: U.S. Government Printing Office, 1994).

36. Frank was chairman of the House Subcommittee on International Development, Finance, Trade, and Monetary Policy (the subcommittee that authorizes U.S. funding for the multilateral development banks).

37. Frank was a major target, but so were Senator Leahy (D-Vermont) and Congressman Obey (D-Wisconsin), chairmen of key subcommittees that appropriate funds to the World Bank.

38. Environmental Defense Fund, Friends of the Earth, Sierra Club, National Wildlife Federation, Bank Information Center, and Center for International Environmental Law comments on World Bank draft paper, "Functions and Operations of an Inspection Function," 23 August 1993, submitted to the Clinton administration.

39. U.S. NGOs had requested Congress to authorize IDA year by year in order to create maximum leverage for reform. The U.S. Treasury had requested the regular three-year authorization.

40. IDA 10 authorization requirements by Congress in 1993 stated, "While the conferees recognize that the World Bank has adopted procedures in these areas and recently issued a resolution on an independent inspection panel, the reforms as written do not adequately address the conferees concerns. Therefore, the conferees have authorized the equivalent of only two thirds of the United States three year contribution to IDA-10. It is the conferees strong belief that the World Bank needs to progress further in these areas and that additional funding in support of IDA-10 will depend on the manner in which these new procedures are implemented and where necessary, broadened" (U.S. House of Representatives, *Congressional Record* 103rd Cong., 1st sess. [September 1993], H 7166).

41. World Bank and International Finance Corporation, *Directive on Disclosure of Information* (Washington, D.C.: World Bank and IFC, July 1989). This directive updated and improved Bank guidelines on information at the time.

42. See Ibrahim Shihata, "Some Legal Aspects of the Bank's Policy on Disclosure of Information," 24 July 1994; see also World Bank, *Policy on Disclosure,* p. 13.

43. This process is already occurring, for example, with Bank generated documents such as staff appraisal reports (SARs), in which Bank staff are required to inform the borrower during loan negotiations that the SAR will be released after approval by the board of executive directors. The borrower can then identify

sections it believes are confidential, and those sections can be excised before the SAR is released.

44. The most recent example of this type of exaggeration is in the Arun III case. Local and international NGOS had requested a copy of "Plan B," a study on alternatives to Arun III, but the Bank refused the request on several occasions on the grounds that the document was the property of the Nepal. However, when a high-ranking Nepalese official came to Washington, he told NGOs they were welcome to have the study and that it was not confidential.

45. World Bank, "Report of the Working Group on the Review of the Directive on Disclosure of Information," (Washington, D.C.: World Bank, 29 March 1993).

46. *Ibid.*, annex B.

47. The alternative view was written by high-level Bank officials Alexander Shakow and Mohammed El Ashry, among others.

48. World Bank, "Disclosure," BP 17.50, par. 5. The provision was originally drafted by Washington-based NGOs during the period in which they were commenting on drafts of the new information policy.

49. World Bank, "Disclosure," Bp 17.50.

50. PIDs reviewed include those for: Indonesia, Second Tree Crop Small Holder Development; Indonesia, Biodiversity Conservation; Indonesia, Fifth Kabupaten Roads; Indonesia, Dam Safety Project; India, MP Forestry; and all projects coming before the board before April 1995. PIDs are available from the World Bank's Public Information Center (www.worldbank.org).

51. An example of a problem PID is the one produced for the Arun III project. Despite the high level of public interest in the project both in Nepal and internationally, the PID for Arun was never updated. It was missing critical information—including a background description of the country, region, or Arun Valley; justification for Bank involvement; background on the sector and sector strategy; how the project fit into overall planning for the energy sector in Nepal; and information on Arun valley, its unique ecosystems or risks and benefits of the project. Another example is the PID for the Third Andhra Pradesh Irrigation Project in India. When compared with the initial executive project summary for the same project, the PID leaves out critical information such as background on the agricultural sector; sector strategy; reforms in the sector that the Bank is seeking from the state government; reason for failure in implementation of previous Bank-financed projects in the sector; economic rate of return; and risks associated with the project. Nor does the PID lay out the serious problems associated with past or present forcible resettlement of more than 150,000 people, especially in India, one of the governments with a poor record in following the Bank's resettlement policy.

52. Examples of bypassing democratic institutions range from approving projects before review by government agencies to ignoring court orders or ongoing law suits that have direct bearing on the project for which the Bank is lending.

For example, the World Bank approved the Sardar Sarovar Dam loan in 1985, despite the fact that the Indian Department of Environment and Forests had not given an environmental clearance.

53. For more detailed information on the Arun III project see World Bank, "Staff Appraisal Report" (draft) (Washington, D.C.: World Bank, August 1994); *Environmental Impact Assessment for Arun Access Road*, vols. 1 and 2 (Washington, D.C.: World Bank, 1994); *Environmental and Socio-Economic Impact Study*, vols. 1 and 2 (Washington, D.C.: World Bank, 1994).

54. Other donors included Japan, Germany, France Finland, Sweden, and the Asian Development Bank.

55. Environmental concerns included deforestation, loss of subsistence farm land caused by road construction and the impacts of an influx of up to ten thousand construction workers and their families.

56. Alliance for Energy is an organization in Nepal that develops community–based micro and small hydroelectric projects throughout the country. The Alliance has two engineers and two social scientists. Their international counterpart, Intermediate Technology Development Group (ITDG), is based in England and works in many developing countries to promote small-scale, community-based water projects. It also provides technical support and training to local NGOs.

57. Bikash Pandey to Lewis Preston, 1992.

58. See Intermediate Technology, "Hydropower Development in Nepal: Taking a Sector-Wide View," (July 1993); "Arun III: Cheaper Energy for Nepal" (April 1994).

59. W. A. Buehring and V. S. Koritarov, "Analysis of Options For the Nepal Electrical Generating System," (Argonne, IL: Argonne National Laboratory, May 1994).

60. *Ibid.*, p. 35.

61. Joe Wood, vice President for the South Asia Region also informed NGOs that it was "unlikely" that the Bank would take into account the evidence that NGOs were presenting because it had adequately investigated alternatives. Other Bank staff has told NGOs privately that they were sympathetic to the NGO view of alternatives. However, they were not prepared to defend that position inside the Bank.

62. Martin Karcher, division chief for Population and Human Resources, Country Department 1 in the South Asia Region at the World Bank, interview by Environmental Defense Fund, Washington, D.C., 9 September 1994. In the interview, Karcher's concern was that investments in health, family planning, and education would be crowded out by Arun III:

It's possible to construct scenarios where that will not happen, but they usually rest on rather optimistic assumptions. The latest country economic memorandum of the World Bank on Nepal, issued last March, describes a set of measures the Government of Nepal would have to implement in order to avoid the crowding out impact. I personally fear that those measures, which include revenue mobilization, strict recurrent expenditure control, investment prioritization and steep tariff increases, may be very difficult to implement and sustain over a long period of time.... I think there is significant risk that the government will have to cut back on its priority programs in the social sectors, as well as in some other important sectors. Prudence would argue in favor of less risky alternatives.

Karcher firmly believed that the economic analysis for Arun III needed to be reviewed, but he was concerned about timing of the revision:

The economic analysis had to be redone. The problem, however, with waiting for such a late stage in the project processing cycle before doing a proper economic analysis is the opportunity to use the results of the analysis to shape the design of the investment program is then lost. The analysis merely serves to justify the project after the fact.

In the interview, Karcher revealed that he received little support from the Nepal Country Department director when trying to promote a targeted poverty alleviation strategy and program in Nepal:

My feeling was that the project was not being handled in an objective and evenhanded manner. Since senior management seemed committed to the project, a serious and open debate was no longer possible, and even common sense questions were being dismissed. All the available energy went into building the case in favor of the project, rather then examining alternatives.

63. *Financial Times,* 22 October 1994.

64. Madhav Kumar Nepal to Lewis Preston, 18 October 1994.

65. At the heart of the debates in government and parliamentary circles was the fear that if the Nepali government did not accept the project as designed it would be forfeiting World Bank and bilateral funding for the energy sector because neither the World Bank nor other bilaterals had publicly committed to funding alternatives to Arun III. NGOs did not know to what extent the Bank pressured the government behind the scenes to accept Arun III or attached conditionality, but rumors of a January 8 deadline for giving final government approval surfaced in the press and were discussed in Parliament and by government officials. Heinz Vergin, acting vice president for the Division, denied that they had given a deadline to the government.

66. They proposed three changes: (1) access road alignment and road construction, (2) tariffs, and (3) construction of small and medium projects. The delegation sought to change the road alignment from the Arun Valley floor to the upper ridge. They also wanted the road to be built by Nepalese companies instead of international contractors, which they said would cut the cost by 45 percent. The delegation also requested flexibility in adjusting tariffs and for the government to build new projects above 10 megawatts.

67. In a memorandum entitled "Green Appeal: A Proposal for an Environmental Commission of Enquiry at the World Bank" (Washington, D.C.: NRDC, 1990)

NRDC argued that existing Bank mechanisms such as the environment department were not sufficient to deal with or prevent controversial projects with extensive environmental and social problems.

68. Lori Udall (EDF) and David Hunter (CIEL), "Draft Resolution for an Independent Appeals Commission for the World Bank," unpublished memo, Washington, D.C.: EDF and CIEL, August 1993. See also Peter Bosshard, David Hunter, and Lori Udall, "Creating an Independent Appeals Commission at the World Bank" unpublished report, Washington, D.C., August 1993; see also Udall, testimony at *Hearing before the House Subcommittee on International Development, Finance, Trade and Monetary Policy* (5 May 1993).

69. Also during this time, a proposal for a World Bank "ombudsman" was being put forward by Daniel Bradlow, an international lawyer at Washington College of Law, (*Opening Statement before the Subcommittee on International Financial Institutions of the Canadian House of Commons Standing Committee on Finance* (18 February 1993). Although more modest than the CIEL-EDF appeals proposal, the ombudsman proposal also contained some important principles, such as access to Bank information and independence from Bank management.

70. The proposal had been presented in testimony before the House Subcommittee on International Development, Trade, Finance, and Monetary Policy.

71. World Bank, *The World Bank Inspection Panel*. Resolution No. 93-10; Resolution No. IDA 93-6, September 22, 1993.

72. Ibid.

73. World Bank Inspection Panel, *Operating Procedures* (Washington, D.C.: World Bank, August 1994).

74. World Bank, *The World Bank Inspection Panel*.

75. Inspection Panel, *Operating Procedures*.

76. Members of the Morse Commission established their independence and credibility early on by negotiating key features, such as an independent budget, before they accepted the position. An independent budget of $1.2 million allowed the team an extensive travel budget and capability to hire consultants and researchers and support staff. Another important element was the fact that the report was published independently of the Bank and made available to the public at the same time it was made available to the executive directors and Bank management. The Morse Commission holds the copyright for the report. The commission also had access to all Bank information on the project and the cooperation of all parties concerned—including Bank staff, local NGOs, the Indian government, and international NGOs. The commission was not based inside the Bank, and after the terms of reference were agreed upon and a contract was signed, the commission did not take instructions from Bank management or executive directors.

77. Because the panel's job is to review Bank management's performance in regard to specific actions and omissions regarding violations of the Bank policy,

procedures, and loan agreements, it seems highly unusual that management so tightly controlled the nomination process or even had a stake in the final decision.

78. Arun Concerned Group, "Request for Inspection," submitted to the World Bank Inspection Panel, 24 October 1994.

79. World Bank Inspection Panel, *Inspection Panel Report on* "Request for Inspection," *Nepal: Proposed Arun III Hydroelectric Project and Restructuring of the Arun III Access Road* (Washington, D.C.: World Bank, December 1994).

80. D. Joseph Wood, "*Status Report on Arun III Project*, (Washington, D.C.: World Bank, 20 December 1994). See also Inspection Panel, *Arun III Project Updated* (Washington, D.C.: World Bank, December 1994); World Bank, "*Summary of Meeting of the Executive Directors*," (Washington, D.C.: World Bank, 20 December 1994, and 27 January, 1995).

81. Inspection Panel, memorandum to the executive directors, World Bank, Washington, D.C., 16 December 1994.

82. Inspection Panel, "Request for Inspection," p. 7.

83. *Ibid.*

84. *Ibid.*, pp. 1–15.

85. Arun Concerned Group to Lewis Preston, 5 January 1995.

86. The claimants did meet with Richard Bissell of the Inspection Panel, in Kathmandu.

87. When public inquiries were made, Bank management even told people that the claim itself was confidential!

88. Gopal Siwakoti to Lewis Preston, 5 January 1995. See also Lori Udall to Lewis Preston, 5 January 1995.

89. World Bank Inspection Panel, personal communication to the author, December 1994. Vice president and chief legal counsel Ibrahim Shihata also indirectly questioned the claimants' standing by issuing a legal opinion that contained a very restrictive view of "standing" immediately after the claim was filed. See Ibrahim Shihata, "Role of the Inspection Panel in the Preliminary Assessment of Whether to Recommend Inspection," memo, 3 January 1995.

90. World Bank, *Inspection Panel*, point 12.

91. Staff in U.S. executive director's office, personal communication to the author, February 1995.

92. Arun Concerned Group, "Request for Inspection."

93. Inspection Panel, "Report on Request for Inspection."

94. Inspection Panel, "The Inspection Panel Investigative Report–Nepal: Arun III Proposed Hydroelectric Project and Restructuring on IDA Credit 2029-NEP," Washington, D.C., World Bank, 22 June 1995.

95. *Ibid.*, p. 5.

96. For commentary on the World Bank Inspection Panel, see Ibrahim F. I. Shihata, *The World Bank Inspection Panel* (Oxford: Oxford University Press, 1994); Daniel D. Bradlow, "International Organizations and Private Complaints: The Case of the World Bank Inspection Panel," *Virginia Journal of International Law* 34 (spring 1994); David Hunter and Lori Udall, "The World Bank's New Inspection Panel: Will it Increase the Bank's Accountability?" *CIEL Brief* No. 1 (April 1994); and David Hunter and Lori Udall, "The World Bank's New Inspection Panel," *Environment* (November 1994), p. 2.

97. In all cases except the Arun III claim, management's initial written response questioned the eligibility of the claimants. Regarding the Arun III claim, management did not question the eligibility on paper, but in back-door communications with both board members and the Inspection Panel itself.

98. Lori Udall and David Hunter to James D. Wolfensohn, 11 February 1996.

99. For example, in the case of Planafloro, despite an extensive and detailed claim filed by affected people, the Bank argued that the claimants had to provide more specifics on the alleged damage, identify specific policies that were violated, and specifically link their allegations to harm caused. (See discussion in Keck's chapter, this volume.)

100. World Bank, *The World Bank Inspection Panel*. Resolution no. 93-10.

101. The best example of this is the Sardar Sarovar Dam project. Ibrahim F. I. Shihata, World Bank General Counsel, issued a memorandum in 1993, which stated that the legal agreements on the project between the World Bank and government of India were still in effect. However, the government of India was violating at least five sections of the loan agreement dealing with resettlement and rehabilitation of the affected people to be displaced by the project. As the waters rose behind the partially constructed dam, people in the villages closest to the dam site were threatened with losing their houses, crops and belongings as a result of the failure of the government to offer them rehabilitation consistent with the loan agreements. The World Bank had not attempted to enforce its loan agreements, and as a consequence, the affected people should have recourse to send a claim to the panel.

102. World Bank, *The World Bank Inspection Panel*, Resolution no. 93-10.

103. The ability to retain independent counsel for panel-related deliberations and review is comparable to the retention of outside counsel by directors of a corporation—a procedure that is relatively routine on matters where senior management may have interests that conflict with those of the corporation. Reliance on outside counsel does not suggest any impropriety on the part of management, but reflects the need to subject all management decisions, including legal matters, to independent review.

IV

Conclusions

12

Accountability within Transnational Coalitions

L. David Brown and Jonathan A. Fox

This chapter examines the "struggle for accountability" *within* transnational environment and development coalitions: the effort to construct balanced power relations among coalition members and particularly between grassroots groups and the international nongovernmental organizations (INGOs) that in theory represent them. Balanced partnerships within transnational coalitions are often very difficult to establish and preserve, given large differences in interests, power, wealth, ideology, culture, and geography that often separate grassroots groups, national NGOs and international advocacy organizations. This chapter focuses particularly on how intracoalition power relations affect and are affected by the evolution of coalition goals, strategy, and organization over the course of sustained advocacy campaigns.

Accountability has been defined as "holding individuals and organizations responsible for performance."[1] It is inherently relational, however, and therefore can be defined clearly only if the actors are specified. Accountability is often enforced by vertical power relations, but when actors are not subject to a common hierarchical authority, accountability may depend on their ability to "exit" to alternative sources of support or to mobilize political "voice" that influences others.

Previous chapters have examined the accountability of multilateral development banks with respect to standards set by their own projects and policies, and it is clear that the problems of institutional accountability are complex. Much of our current theory and practice for ensuring accountability[2] is designed for use within bureaucratic organizations, which assumes the existence of shared organizational structures and systems, mutually accepted goals and standards of performance, easily available information about performance, and agreements about divisions of labor and distributions of authority. Ensuring accountability among autonomous actors who do not share an organizational hierarchy can be problematic.

The problems of accountability may be particularly challenging within loose and fluid coalitions, such as the alliances of NGOs and social movement organizations involved in transnational campaigns. Given the differences in power and resources among coalition members, it is not surprising that some Southern observers have characterized relations between Northern and Southern NGOs as "emerging colonialism," in which "Third World NGOs have had to suit their agendas to the agendas of Northern NGOs."[3]

The problems of creating accountability are particularly difficult when the actors work across great power differences with little shared organization; when goals, values, ideologies, and interests are diverse, ambiguous, or conflicting; when crucial campaign information is often lost or distorted across great distances in geography, culture, and perspective; and when the actors differ about who is in charge or responsible for different tasks. Within NGO coalitions, such circumstances are the rule, not the exception. They begin with few shared goals or standards; information is difficult to acquire and not always shared; members have different political, cultural, economic, and social perspectives; and there is little explicit agreement about how decisions should be made or who should participate. Accountability through hierarchical control is clearly not applicable because ostensibly horizontal North-South NGO networks lack a shared hierarchy of authority. The threat of "exit" does not compel accountability when there are few alternative partners to join in a coalition. For many coalition members, having a voice in coalition decisions is essential to holding other members accountable. Balanced power relations and mutual influence processes are critical in the social and political construction of accountability within the coalition.

The importance of interorganizational relations for social problem solving has become increasingly obvious in the last decade, as analysts from a wide range of perspectives have assessed the impacts of such relations on communities and societies.[4] Interorganizational coalitions can be critical to many activities, within and across the boundaries of the market, state, and voluntary sectors.[5] They can bring diverse resources to bear on problems and at the same time reduce the risks of failure to any single organization. Transnational advocacy coalitions, for example, can combine the detailed information and the capacity for representing disenfranchised grassroots groups of social movements, with the institutional capacity, the societal perspective, and the political access of national NGOs, and the expertise on policy formulation and lobbying networks of international advocacy organizations.

The creation of such coalitions is not simple, however. Member differences frequently produce conflict rather than cooperation. The potential for conflict is particularly high in circumstances characterized by cultural differences, diverse ideologies, discrepancies in wealth, and power inequalities.[6] These differences may be further complicated when international NGOs seek to work with grassroots movements[7]—as was the case for many of the coalitions described in earlier chapters in this volume. Although NGOs in North and South may share some core political or ideological values, there is likely to be greater distance between international NGOs and grassroots movements in the South. NGO-grassroots coalitions are therefore more challenging and difficult to sustain, yet can also be more powerful in the long run because a consolidated grassroots base confers both national and international legitimacy to the coalition—even in the eyes of relevant factions within the multilateral development banks (MDBs). The resulting coalitions can be a powerful tool for dealing with transnational problems.

Mutual influence and accountability do not spring full-blown from initial meetings and shared targets, but rather evolve over time and interaction among coalition participants. In the analysis that follows here, we focus on three aspects of the evolving relations among coalition partners: (1) phases of coalition evolution, (2) organizing mechanisms that regulate actor behavior and promote mutual influence, and (3) interaction processes that construct or undermine coalition building. We use the conceptual windows of phases, organizing mechanisms, and interaction processes to examine the evolution of mutual influence and accountability within coalitions. We focus first on coalitions organized to affect specific projects and then on coalitions organized to influence more general policies. In the last section, we explore some implications of these coalitions and their interaction with the World Bank for transnational social capital and social learning.

Accountability in Project Campaign Coalitions

Transnational coalitions between Northern NGOs, Southern NGOs, and grassroots groups can bring together a wide range of resources for influencing specific development projects of multilateral development banks: grassroots groups have information about actual on-the-ground impacts and the legitimacy to challenge noxious consequences; national NGOs have knowledge of government policies and priorities, as well as understanding of national traditions relevant to a project; international advocacy NGOs understand broad international trends, have contacts for cross-national campaigns, and know the relevant pressure points within target institutions. Effective joint use of these resources, however, requires that the parties develop capacities for communication and mutual influence that span vast gaps in experience and resources. Mutual influence and accountability within the coalition may be complicated by a lack of shared goals and interests, diverse values and cultural expectations, difficult communications across long distances, and divergent perceptions of needed and actual power in the relationship.[8]

What is accountability in the context of these transnational coalitions, and how can we know when we see it? At a minimum, coalition members should be able to define their expectations for performance, monitor compliance with those expectations, and successfully press for changes in behavior that does not meet minimal standards. For coalitions organized around specific projects, agreements on these issues seldom exist at the outset, so accountability must evolve with the coalition. Shared goals, strategies, and responsibilities must be articulated before parties can implement coalition plans, let alone hold one another accountable for performance.

A central issue in the negotiation of coalition organization is the distribution of power among members. Some coalitions are initiated and initially largely controlled by highly organized grassroots groups, sometimes full-blown social movements, that are concerned about problems that directly affect their base. The challenge to the land reform law in

Ecuador, for example, was initiated by the Confederation of Indigenous Nationalities of Ecuador (CONAIE) as the representative of an indigenous peoples movement (see Treakle chapter in this volume). Other coalitions are launched by international NGOs primarily interested in shaping the programs and policies of institutions such as the World Bank. The initial concerns about the Planafloro project in Brazil were raised by the Environmental Defense Fund (EDF), an international advocacy NGO (see Keck chapter). Power and influence within the coalition may be unequal at the outset, and power distributions may change as the coalition evolves. We are particularly interested here in how the patterns of influence and accountability evolve over time within these coalitions. Also, how do coalitions come to terms with the differences between international NGOs and issue networks, on the one hand, and grassroots organizations and social movements, on the other? For example, are accountability relations within coalitions "path dependent," in the sense that their founding process leaves a strong imprint on subsequent power relations, or are they fluid and contingent on many factors, subject to revision and renegotiation in the course of a campaign whatever their origin?

Table 12.1 summarizes the major actors and key events in the four project campaigns documented in this volume. The last column summarizes the degree of mutual influence that appears to have evolved among grassroots organizations, national NGOs, and international advocacy NGOs. *High* mutual influence reflects much interaction and two-way influence. The Ecuador case, for example, describes high levels of interaction and influence between indigenous peoples' organizations and national NGOs such as Acción Ecológica. *Medium* mutual influence refers to less interaction—as in the fewer direct links between Ecuador grassroots groups and international NGOs—or to disagreement about mutual influence, as in the debates about the mission of the Philippine Development Forum between international and national NGOs (see Royo chapter). *Low* refers to little contact or mutual influence—as in the lack of links between local groups and the International NGO Group on Indonesia (INGI, see Rumansara chapter) or the lack of grassroots groups to influence other parties at the beginning of Brazil's Planafloro debate. In general, we expect the chain of influence to have stronger links between

Table 12.1
Accountability in project reform campaigns

Case	Key actors	Critical events in alliance evolution (by year)	Mutual influence
Kedung Ombo Dam	GRO: Oustee families, students	84: Resettlement plan claims 75% of oustees willing to transmigrate.	NGO-GRO: LOW to MEDIUM.
	NGO: Local legal aid; other local NGOs; National Legal Aid; poverty, environment NGOs	85: Construction started; 5,000 families will be ousted.	Oustees not organized; some oustees seek NGO legal help for better compensation.
		86: Farmers ask NGOs to sue for better land compensation.	INGO-NGO: MEDIUM.
	BO: International NGO Group on Indonesia (INGI)	87: Competition among local and national NGOs to represent oustees; military accuses resisters of being Communists.	National NGOs and INGOs plan and implement campaign;
	INGO: members of INGI from Northern countries	88: INGI presents Kedung Ombo at poverty conference; INGOs/NGOs send first letters to World Bank.	NGO-GRO: LOW.
		89: Dam completed; INGI protests impacts to World Bank; students, religious groups protest rights violations; Bank mission evaluates resettlement; Bank questions GoI about INGI campaign; GoI attacks INGI for "hanging out dirty linen."	Local NGOs criticize national NGOs as cause of military repression of local groups.
		90: INGI dicusses dam with Bank staff; GoI makes available other sites for resettlement.	

	93: Bank and others agree to avoid future "Kedung Ombos."	
	94: GoI rejects Supreme Court decision for better compensation.	
Mt. Apo Thermal Plant	87/88: Indigenous groups and farmers raise questions and begin organizing local, regional, and national networks to challenge project.	NGO-GRO: HIGH. IP rights accepted as priority by task forces.
GRO: Indigenous peoples groups; local farmers.	89: Bank missions reach conflicting conclusions; elders swear to defend Mt. Apo to "last drop of blood"; initial links started to INGOs and PDF.	INGO-NGO: MEDIUM. Differences over advocacy role, PDF accepts limits.
NGO: local, regional, national task forces; Legal Rights and Natural Resources Center.	90: PDF-US director meets elders and chooses Mt. Apo as focus; GoP certifies project for environmental compliance.	INGO-GRO: HIGH. PDF leader committed to no compromise position of indigenous elders.
BO: PDF (Philippine Development Forum); BIC (Bank Information Center).	92: LRC sues unsuccessfully to stop project; PDF advises Bank and arranges field visits; national NGOs debate PDF role; National Solidarity Conferences set strategy; Bank rejects EIA of GoP; GoP withdraws loan request.	
NGO: EDF; Columban Fathers for Justice and Peace; Greenpeace.	93: Further solidarity conferences set goals and strategy; alliance lobbies Export-Import Banks.	

Table 12.1 (continued)

Case	Key actors	Critical events in alliance evolution (by year)	Mutual influence
Planafloro Natural Resource Management	GRO: Rubber tappers, farmers organization. NGO: Local NGOs (some expatriate-led); national NGOs and networks (IAMA, CNS, IEA, and others). BO: Rondonia NGO Forum. INGO: Environmental Defense Fund, World Wildlife Federation, others.	89: INGOs (EDF) protest lack of local participation; project suspended for more local consultation. 90: Few existing GROs; national NGOs vie to organize GROs. 91: INGO (WWF) moderates conflicts among outside NGOs; Rondônia NGO Forum links local NGOs and INGOs, but excludes NNGOs; Forum supports loan given parity on deliberative council. 92: State agencies violate loan terms in spite of Forum role; Forum challenges violations to no avail. 94: Forum requests Bank to suspend disbursement. 95: Forum requests Inspection Panel review of project. Planafloro accepted as case by Bank Inspection Panel.	NGO-GRO: LOW to MEDIUM. No GROs at start; later GROs work with Rondônia NGO Forum. INGO-NGO: MEDIUM to HIGH. Expatriate-led NGOs provide information; Forum enables joint planning. INGO-GRO: LOW to MEDIUM. Some GROs have direct links to INGOs.

Ecuador Structural Adjustment			
GRO: Indigenous peoples (IP) groups; environmental groups.	86:	Federation (CONAIE) created; represents over 70% of indigenous peoples.	NGO-GRO: HIGH. IP groups participate in CONAIE, Accion Ecologica decision-making.
NNGO: CONAIE (IP federation); Acción Ecológica.	90:	First civic uprising for indigenous rights.	
BO: Ecuador Network, BIC (Bank Information Center).	92:	NGOs ally with INGOs to challenge WB loan to privatize oil business.	NGO-GRO: HIGH. IP groups participate in CONAIE, Accion Ecologica decision-making.
INGO: Rainforest Action Network; Oxfam; CAIA.	93:	Coalition debate with Bank moderates oil law.	NNGOs.
	94:	CONAIE negotiates IP role in oil leasing, EIAs; NGOs focus on country strategy role of MDBs; Ecuador Network (EN) arranges NGO Bank visits; New land law is pro-agribusiness, anti-IP; "Mobilization for Life" alliance demands repeal; Two-week demonstration paralyzes country; EN information to Mobilization and Banks; GoE, Mobilization, business leaders negotiate law; IDB President hosts CONAIE on Washington visit.	INGO to GRO: MEDIUM. IP groups influence INGOs through participation in NNGO federations and NGOs.

GRO = grassroots organizations; NGO = local and national NGO; BO = bridging organization; INGO = international NGO.

groups that are closer in social, cultural, political, and economic terms: the links between international and national NGOs or between national NGOs and grassroots organizations are expected to be stronger than those between international NGOs and grassroots groups.

Patterns of Influence and Accountability

The four cases described in table 12.1 represent quite different patterns of influence and accountability. They differ especially in the relative influence wielded by international NGOs and grassroots social movements. Some coalitions are initiated and shaped by grassroots groups and movements. The coalition to influence the Ecuador structural adjustment program, for example, was organized by the indigenous peoples' movement and their environmental NGO allies to challenge national policies about oil development and the privatization of land ownership. They began relationships with the Bank Information Center (BIC) and other international NGOs as resources in the struggle with the government of Ecuador and the MDBs. Similarly, the Philippines coalition on Mt. Apo was organized by the indigenous peoples movement and allied NGOs to preserve their rights to ancestral lands. These campaigns were launched around what might be called a *national problem coalition,* which sought allies among the international NGO community to seek MDB response to grassroots priorities.

In contrast, other coalitions were organized and strongly influenced by international NGOs (INGOs). The Indonesian and early Brazilian coalitions were launched by national and international NGOs with little grassroots participation at the outset. In Indonesia, a number of oustees protested compensation policies and human rights issues, but the primary impetus for the international campaign came from large, Jakarta-based NGOs that were less vulnerable to oppression by local authorities. The local stakeholders in the Brazil project were initially unorganized and unaware of the issues, though the Rondônia NGO Forum subsequently took an active monitoring and advocacy role. In essence, these coalitions were largely dominated by *international issues networks* composed of international and national NGOs concerned about reforming Bank policies and projects.[9] In the service of MDB reform, these networks used local information and witnesses to build legitimate cases against current MDB practices. Local concerns may be less influential than Bank priorities in shaping such networks' decisions and activities.

The Evolution of Project Coalitions

The case studies in this volume suggest that initial patterns do not necessarily determine the evolution of the coalition. Relations among parties evolve through interaction and response to contextual challenges, and new patterns of influence may merge from negotiations over strategy and implementation. In the Philippines and Brazilian coalitions, for example, the agendas of local stakeholders and international NGOs were combined to work simultaneously on international issues and local problems. When the concerns of indigenous peoples and environmental groups came into conflict in the campaign, national "solidarity conferences" were held to decide which interests and issues would take precedence. The indigenous peoples' federation, the national NGOs, and the international NGOs debated their respective concerns in the Philippine Development Forum (PDF), a preexisting coalition of national NGOs and INGOs, to create shared priorities and advocacy strategies. In Brazil, the Rondônia NGO Forum emerged as an arena for grassroots groups and local NGOs to engage international NGOs. Because the forum excluded national NGOs, however, it limited the possible range of mutual influence and trust building. Negotiations in these forums worked out agreements on goals and priorities and also provided opportunities for practicing mutual influence and for building trust among the parties. These forums acted as *bridging organizations,*[10] which provide arenas for contact and joint action among representatives of diverse actors.

Much of the research on interorganizational relations has focused on preconditions that support the emergence of coalitions or on formal structures that guide cooperation.[11] In recent years, research has begun to identify critical processes and steps in the evolution of interorganizational cooperation.[12] Investigators suggest that interorganizational collaborations often pass through several *evolutionary phases.*[13] We focus here on four phases for these coalitions: (1) problem defining, in which the parties agree on concerns around which the coalition can be organized; (2) direction setting, in which the parties agree on basic strategy and tactics; (3) implementing, in which the parties carry out problem-solving actions; and (4) revising, in which parties deal with new challenges generated by events.

The evolution of mutual influence and accountability can be traced through the phases of coalition development. In the *problem definition* phase, for example, the possibility of mutual influence and accountability is shaped by decisions about the nature of the problem and the composition of the coalition. The definition of the problem determines whose resources are relevant and who will be influential. The indigenous peoples in the Philippines gained the support of the national solidarity councils to define land rights of indigenous peoples as the central issue in the Mt. Apo case. This decision increased the influence of indigenous peoples and reduced the influence of the environmental protection activists on campaign goals and strategies. Problem definition is not always limited to early parts of a campaign; the emphasis on indigenous peoples' rights at Mt. Apo came out of councils called well after the initiation of the campaign, in response to possible concessions on environmental protection.

Coalition decisions about *direction setting* define campaign strategy and tactics, thus shaping the distribution of resources among coalition members. The choice of legal strategies in Indonesia and the Philippines enhanced the influence of legal assistance NGOs; the emphasis on directly lobbying the Bank early in the Brazil case empowered the international NGOs and their information-gathering colleagues in Rondônia; the choice of mass demonstrations in Ecuador reflected the strengths of CONAIE and the indigenous peoples' movement. Changes inside and outside coalitions may call for a redefinition of strategy, even though the problem definition remains the same. External forces, such as the delay and subsequent rejection of a favorable Supreme Court decision in Indonesia, undermined the influence of legal strategies and lawyers in that campaign. Developments within a coalition may also change strategies and related influence patterns. The creation of the Rondônia NGO Forum as an actor dealing with the Planafloro project provided a rationale for restarting the project by including NGOs in the deliberative council that was charged with making project policy.

Patterns of influence may also be affirmed or revised by experience stemming from campaign *implementation*. Implementation presses the parties to be specific about plans, responsibilities, and actions. The fact that CONAIE and Acción Ecológica mobilized thousands of activists for public demonstrations made them important players—important enough to sit down with Bank representatives, business leaders, and senior government officials to rewrite the proposed land law. The inability of court challenges to stop projects in Indonesia or the Philippines, on the other hand, altered the influence base of legal NGOs. Their capacity to bring suit became less valuable as a consequence. In implementing coalition plans, existing capacities can make an important difference. Bridging organizations, like the PDF in the Philippines and INGI in Indonesia, were already bringing national and international NGOs together for joint deliberation and action. The availability of such bridges, already linked by bonds of trust and solidarity, made joint action across great differences much easier.

Finally, as already suggested, internal and external events sometimes call for fundamental *revision* of coalition characteristics and activities— the need to redefine problems or reset directions in response to changed circumstances. These challenges also create opportunities for redistributing influence and redefining accountability within the coalition. The government decision to build the Mt. Apo plant without Bank loans, for example, forced the coalition to reorganize in order to challenge proposed loans from export-import banks. The creation of the Bank Inspection Panel offered a new tool—official complaints—to influence the Bank, which strengthened the hand of local and international NGOs in the Rondônia NGO Forum. In Ecuador, the land reform law was a new threat to CONAIE and its members that catalyzed a dramatic expansion of the coalition—creating the Mobilization for Life and the expanding links to the international Ecuador Network.

This outline of coalition problem definition, direction setting, implemention, and revision is deceptively linear. In practice, joint action often precedes agreement on problem definitions; agreements on common directions may emerge only after many discussions; and revisions may be required at many points in coalition evolution. But that evolution, however fragmented and circular, can produce increasing agreement on coalition goals and organization that is vital to mutual influence and accountability.

The Organization of Project Coalitions

Coalition evolution is reflected in the development of the *coalition organization* that guides its collective actions. Such organization may range from tightly organized federations or mergers with a centralized decision-making authority to loosely organized networks of largely autonomous members who are relatively unconstrained by their membership. Many coalitions have no coordinated organization at the outset, and the creation of shared assumptions and arrangements to support and enable joint action is a critical task for coalition effectiveness.[14] Two aspects of coalition organization are particularly important here: (1) *coalition definition,* or agreement on fundamental purposes and membership, and (2) *organizing mechanisms,* or arrangements that facilitate effective collective action. Although most coalitions start out as loose and informal networks, the demands of effective advocacy often require more clearly specified organization that guides action and provides a basis for holding each other accountable.[15]

Coalition definition, at a minimum, refers to coalition membership, but more importantly to coalition purposes and the rationale for its composition and membership. The coalition in the Planafloro case, for example, was initially an international issues network focused on reforming World Bank policies and practices. It needed NGOs in Rondônia to provide local information on compliance with Bank policies. With the emergence of the Rondônia NGO Forum as a more legitimate representative of grassroots interests, the coalition became more of a national problem coalition, concerned about the effects project implementation would have on local groups. This redefinition altered relations among the parties: local NGOs and grassroots groups became more active on the implementation council and later in calling for Bank action to change the project; Brazilian national NGOs were excluded from the forum; and international NGOs became participants rather than decision makers for the coalition.

Accountability and mutual influence within the coalition is shaped in part by who is included and at what point in the campaign. When the coalition is defined to require the resources of grassroots movements, it is more likely that their interests will be taken into account in decisions. Federations of indigenous peoples that could respond to the Ecuador land privatization law and the Mt. Apo thermal energy project were critical in defining problems, setting directions, and implementing advocacy strategies. In these cases, the indigenous peoples were already organized and could respond to the issue as part of a long-term effort to protect their interests and livelihoods. In the Indonesian dam case, in contrast, grassroots actors never became organized as a movement to participate in the coalition: the Kedung Ombo oustees were primarily interested in more compensation or better land, and neither the oustees nor the competing local NGOs became a sustained movement that could speak for local concerns.

The evolution of *organizing mechanisms* to support collective planning and action is also central to mutual influence and accountability. In these coalitions, organizing mechanisms have included strategies and plans that defined actors' resources and expectations; bridging organizations whose structures, norms, and expectations supported joint work; and leaders who could connect to other constituencies without losing credibility with their followers.

Strategies and plans offer guides to action—standards against which performance can be evaluated—and frames for recognizing the resources of diverse actors. The latter is perhaps particularly important in setting the stage for mutual influence because parties with no relevant resources are also likely to be accorded little influence. In Ecuador, for example, CONAIE's need for information about Bank policies and priorities led them to contact the Bank Information Center and other international NGO allies, who in turn were eager to gain access to CONAIE's local information and to gain support from its broad social base. In the Philippines, INGOs needed trustworthy information about indigenous peoples' groups and government actions, and local groups needed international contacts and their lobbying skills for influencing the Bank. Where issues were defined in terms that did not highlight the interdependence of coalition partners, mutual influence was less likely. The emphasis on improving compensation in Indonesia through court battles and legal expertise made the resources of local groups largely irrelevant, except as relatively passive plaintiffs. In Brazil, the strategies evolved: the initial challenge was mounted by INGOs and expatriate-led NGOs on the grounds that no local participation was solicited in project design, whereas subsequent challenges by the Rondônia NGO Forum were based on local contacts. Strategies, in short, offered contexts in which participant contributions could be recognized, demanded, and assessed by other members.

Organizing coalitions is often facilitated by preexisting organizational arrangements—usually in the forms of informal norms and relationships—that can support negotiation and cooperative problem solving. Several coalitions incorporated bridging organizations created by prior coalition building among national and international NGOs. PDF and INGI already had histories of cooperation among their national and international NGO members prior to their involvement in the coalitions. Some coalitions created new organizational arrangements to support debate and discussion. CONAIE and Acción Ecológica helped create the Ecuador Network with like-minded INGOs for future struggles. The Mt. Apo coalition in the Philippines created networks at the local, regional, and national levels, and then organized "national solidarity conferences" to set goals, articulate strategies, and manage disagreements. These forums enabled coalition participants to engage one another in face-to-face discussions of key issues and to negotiate agreement on strategies and action plans. They provided arenas where recurrent interaction built shared perspectives, goals, and trust on which to base joint action.[16]

The organization of these coalitions often depended, particularly at the outset, on the convergence of a few local, national, and international activists who were able to contact and work with each other without losing credibility with their own constituents, even when those constituents were quite suspicious of the other actors. These "bridging individuals" or "interlocutors" were willing to engage in the time-consuming and sometimes conflictive process of learning the perspectives of diverse actors and of spanning the social and political chasms among them. Once established as trustworthy intermediaries, they wielded much influence in building coalitions and creating the necessary bridging organizations. Some of these leaders were from international NGOs—such as representatives of EDF in Brazil and Friends of the Earth (FOE) in the Philippines, who played central roles in launching those coalitions. Some were leaders of bridging organizations—such as the executive directors of INGI and PDF, who acted as links between national and international NGOs and represented the coalition in discussions with World Bank officials. Some were leaders of NGOs or grassroots groups—such as the elders of the indigenous peoples' groups in the Philippines or the leader of CONAIE in Ecuador, who balanced their obligations to their constituents with the demands of negotiating with government and Bank officials. Although the coalitions often represented large numbers of constituents in different regions, coalition decisions were often made by a few individuals who could work with each other across large differences in background and constituency. Much of the effectiveness of these coalitions and their ability to remain organized in spite of tremendous centrifugal forces depended on the skills and commitment of a few key bridging individuals.

Interaction within Project Coalitions

The evolution of coalition organization and mutual influence turns in large part on interaction among its members. Although some analysts suggest that the patterns of organization can be largely predicted from considerations of efficiency or political power, others suggest that interorganizational cooperation builds trust and organization by means of social and political "construction" processes through which participants negotiate shared understandings and expectations in recurrent cycles of reciprocal commitment and action.[17]

The negotiation of shared goals and plans in itself can contribute to the building of social trust and shared community as the parties demonstrate their willingness to act in good faith.[18] This trust may be particularly important in bridging the gaps between grassroots social movements and international advocacy NGOs because both are organizational forms that are excruciatingly sensitive to differences in values and ideologies. Successful coalitions grow out of cycles of negotiation that bridge the gaps among their members—negotiations that deal with inevitable conflicts without destroying trust among the parties.[19]

In the coalitions described in this volume, direct, repeated, sustained contact among bridging individuals appears to have been an important factor in the development of solidarity, trust, and shared values among participants. Many coalition participants mentioned the importance of developing close relationships across great differences in culture, wealth, and experience. Extensive interaction, particularly face to face, seems important to the development of trust and solidarity. The opportunities for face-to-face discussion—as in the solidarity councils in the Philippines or the Mobilization for Life meetings in Ecuador—are important events in the construction of coalition goals, strategies, and organizations. Such interactions may be particularly important when participants come from very different worlds. The leader of the Philippine Development Forum noted that meeting the elders of the indigenous groups on Mt. Apo was critical to her commitment to the coalition because she thereafter felt personally accountable to the elders of the indigenous groups. International members of the Narmada Action Committee speak of their visits to "the Valley" as central to their solidarity with the grassroots movement. These personal commitments support information sharing and mutual influence among coalition members, even when power, wealth, and prestige differences might tilt decision-making power in favor of some at the expense of others.

Partnership Advocacy?

Are these coalitions in fact examples of "partnership advocacy" (see Wirth chapter), in which relatively powerful and wealthy actors share influence and accountability with their less fortunate partners? Has there been a shift in power from Northern to Southern partners in these coalitions, away from the inequalities implicit in the usual distribution of resources and wealth? The third column of table 12.1 suggests that these coalitions vary considerably in the extent to which grassroots, national, and international partners are accountable to one another. The coalitions formed to target the Philippines and Ecuador projects had relatively high levels of mutual influence among the various parties. In the Indonesia coalition, there was mutual influence between national and international NGOs, but little accountability to grassroots organizations. The Brazil coalition appeared to evolve from an INGO-dominated network into an coalition with more local participation, although it continued to exclude national NGOs.

In the histories of specific coalitions, some but not all show a trend toward increased accountability. Several observers see increasing Southern influence over Northern partners in general over the last decade (see

Wirth and Gray chapters). The legitimacy of Northern advocates without close ties to credible grassroots representatives is increasingly subject to question, so the value of grassroots participants in a project coalition is increasingly obvious. At the same time, it has become more and more clear that project influence is easier "early in the pipeline," so Southern parties need early access to information about projects, knowledge of Bank priorities, and links to sympathetic Bank staff—resources often held by INGOs. Project coalitions have not always been internally accountable or subject to mutual influence, but they will be more so in the future if the factors identified here—strong grassroots organizations, recognition of the resources of diverse coalition participants, wider availability of bridging organizations, and interactions that build solidarity and trust—become more common aspects of the institutional landscape.

Accountability in Policy Campaign Coalitions

Comparing *policy reform* campaign and *project* campaign dynamics in the case studies suggests that they pose different challenges to coalition accountability relations. This section examines coalitions to influence policies from the same perspectives applied to the coalitions to influence projects.[20]

Many chapters of this book emphasize the discrepancies between policies and practices. In at least one early case, public policy statements appear to be "figleaves" for the "real" policies that remain confidential. The World Bank policy on indigenous peoples published in 1982 was superseded by a confidential de facto policy that conceded much less indigenous influence over projects (see Gray chapter). Other policies exist on paper but seem to have little impact on practice. The Bank has had an information policy favoring disclosure since 1989, but Bank staff have routinely opted against disclosure even when host governments have been willing to share information (see Udall chapter). The resettlement review found that resettlement policies were actually followed in less than half the cases, though the review itself seemed to stimulate more internal attention and compliance (see Fox chapter). There are, in short, reasons to be skeptical about the utility of policy reform, by itself, as an engine for major improvements in Bank practices.

Nonetheless, changes in Bank policies are not irrelevant to its operations. Even though the institution of a new policy does not automatically change the practices of staff members and borrowing governments, many of these actors are guided by policies and many more follow them if compliance will be monitored. Bank policies also have effects beyond the immediate activities of Bank staff: they set international standards for audiences that see the Bank as a source of good development practice; they legitimate the commitment of Bank financial resources; they provide criteria for external stakeholders to hold the Bank accountable in the future; and in some cases they alter institutional arrangements to make future civil society influence possible. Thus, coalitions that contribute to the creation or revision of Bank policies can make a substantial contribution to changes in practice.

Are there patterns of evolution across policy reform campaigns, and do those patterns have implications for mutual influence and accountability within coalitions? Table 12.2 summarizes major actors, key events, and levels of mutual influence in the evolution of the policy campaigns described in this volume. As in table 12.1, a *High* rating refers to extensive interaction and mutual influence among the parties, as in the coalition that grew out of the Narmada and IDA 10 replenishment campaigns to press for the new information and Inspection Panel policies. A rating of *medium* reflects moderate interaction and mutual influence: international NGOs tried to mobilize Southern participation in the water resources management policy deliberations, but it was difficult to create and maintain the coalition without more immediate Southern interests in its outcomes. A *low* rating reflects little interaction or mutual influence: the Washington-centered review of the resettlement policy provided little opportunity or incentive for grassroots groups to influence the process.

Patterns of Policy Coalition

Should we expect the coalition dynamics identified in the project campaigns to apply to policy reform campaigns? There is likely to be some overlap: we know that some coalitions created to deal with projects have eventually influenced the development of such policies as the indigenous peoples policy and the information and inspection policy.

Table 12.2
Accountability in policy reform campaigns

Case	Key actors	Critical events in alliance evolution (by year)	Mutual influence
Indigenous peoples' policy	GROs: Indigenous peoples' and environment movements (Chico River, Polonoroeste, Narmada, Transmigration).	81: Chico River: first IP victory over Bank.	GRO-NGO: LOW. GROs focus more on specific projects than general policy.
	NGOs: Linked to project alliances.	82: Published policy stronger than "real" policy.	NGO-INGO: LOW.
	INGOs: International Work Group on Indigenous Affairs, Cultural Survival, International Survival.	83–87: Polonoroeste, Transmigration, other campaigns raise environment, IP issues.	NGOs have more interest in project than in policy. GRO-INGO: MEDIUM.
	WB: Social scientists; IP policy supporters.	87: Five-year review focuses Bank attention on environment.	GRO actions legitimate INGO policy advocacy.
		88–93: IP movements gain public visibility and shape projects around world;	GRO-WB: MEDIUM.
		90s: Political tensions among INGOs; increasing North/South NGO coordination.	GRO concern legitimates Bank reform proposals. NGO-WB: LOW:
		91: New IP policy (OD 4.20) enables more challenge to Bank IP practice.	NGOs focused on projects. INGO-WB: HIGH.
		92: Continuing challenges to new projects.	Interaction to assess policies.

Table 12.2 (continued)

Case	Key actors	Critical events in alliance evolution (by year)		Mutual influence
Resettlement policy	GROs: Oustee movements in many projects. NGOs: Allies of oustees like Narmada Bachao Andolan. INGOs: Narmada Action Committee, International NGO Group on Indonesia. WB: Resettlement Review Task Force.	86:	Policy requires review of resettlement plans.	GRO-NGO: LOW. GROs focus on projects.
		91:	Compliance at 30% in first five years; improves in response to external criticism.	NGO-INGO: LOW. NGOs focus on projects.
		92:	Morse report on policy violations inspires review to discover "other Narmadas."	GRO-INGO: LOW. GROs focus on projects. GRO-WB: LOW.
		93:	Task force "guerilla warfare" to get data from some reluctant Bank staff.	Little direct contact. NGO-WB: LOW.
		94:	Find 33% underestimate of oustees; policy compliance increases; internal bargaining over report content; early released to forestall report leaks.	Little direct contact. INGO-WB: HIGH. Internal review empowered by external interest.
Water resources management policy	GROs: Not involved. NNGOs: 50 volunteer to provide input to policy discussions. INGOs: International Rivers Network, Environmental Defense Fund. WB: Operations and engineering staff.	91:	Bank proposes to revise water policy; workshop for GOs on water policy; Bank decides against consulting NGOs.	GRO-NGO: LOW. Little consultation with GROs. NGO-INGO: MEDIUM.
		92:	NGOs invite NGO comments, fifty respond; Bank agrees to NGO consultations; Bank reneges on further consultation; Bank economists dominate policy drafting; INGOs lobby executive directors for policy changes.	INGOs invite but cannot support more consultation. GRO-INGO: LOW. Little consultation. GRO-WB: LOW. Little interaction.

Information and inspection Panel policy reform campaign	GROs: Sometimes consulted.	93:	Final policy includes some INGO ideas.
			NGO-WB: LOW. Limited interaction. INGO-WB: HIGH. Sustained interaction with sympathetic Bank staff.
	NGOs: Narmada, IDA 10 campaign activists.	89:	Policy presumption for disclosure.
		91:	Narmada campaign raises issues for executive directors.
			GRO-NGO: MEDIUM. Some GROs want access. NGO-INGO: HIGH. Bonds from past campaigns.
	INGOs: EDF, BIC, other Bank reform campaigners; Fifty Years Is Enough.	92:	Propose weak policy; minority disagrees; Morse Commission reinforces doubts; Wapenhans Report questions Bank performance; IDA 10 alliance demands new policies.
			GRO-INGO: MEDIUM. GRO campaigns legitimate INGO policy advocacy.
	WB: Allies of more open disclosure policy.	93:	Congress requires policies for funding; Fifty Years Is Enough demands reform as price of continued support.
			GRO-WB: LOW. Little contact on policy. NGO-WB: LOW. Little contact on policy. INGO-WB: HIGH.
	Others: Congressional Oversight Committee; U.S., Japan, Europe executive directors.	94:	Arun III files first claim to Inspection Panel; Republicans win control of U.S. Congress.
		95:	Panel reports flaws in Arun III; Arun III canceled by new Bank president.
			Constant interest, contact. INGO-Congress: HIGH. Close links enable pressure for new policy.

Key: GRO = grassroots organization; NGO = national NGO; INGO = international NGO; WB = World Bank staff; Other = other alliance participant.

But there are differences between projects and policies that affect the nature of coalitions organized to influence them. At least four differences are important. First, projects have tangible and sometimes immediate effects on specific people and places, whereas policies are by definition general and therefore somewhat intangible by comparison. So grassroots groups and local social movements have more at stake in influencing projects than in reforming policy. Second, even successful policy influence changes only future projects and then only when the policy is applied. Outcomes are delayed by three different lead times: time to change the policy, time to affect new project design, and time to implement the project. Policy coalitions thus need members with capacity for long-term engagement. Third, the links between policy reforms and project operations are embedded inside the Bank and in Bank–borrower government relations; both arenas, however, lack transparency to outsiders, so policy coalitions need insider participation.[21] Fourth, policy decisions are disproportionately influenced by donor governments because the Bank itself follows these policies. Projects, in contrast, are the shared responsibility of the Bank and borrowing governments, therefore raising the legitimacy of Southern voices much more directly. In policy reform campaigns, these differences tend to shift influence to coalition actors who have long-term, continuing contact with the policy issues and with Bank activities—such as international advocacy NGOs and sympathetic Bank staff.

We identified two patterns of coalition in the project cases: (1) *national problem coalitions,* in which active social movements and civil society organizations are concerned with solving local problems created by Bank projects, and (2) *international issue networks,* in which international NGOs use specific projects to reform Bank policies and practices. *National problem coalitions* are focused on solving the immediate problems produced by specific projects: they may provide cases that are considered in the development of new policies, but their members are less interested in policies whose relevance to national problems is debatable. The grassroots organizations (GROs) and national NGOs centrally involved in challenging the projects covered by the indigenous peoples and resettlement policies, for example, were less involved in the policy debate than international NGOs. *International issue networks,* in contrast, are well suited to participate in policy coalitions because they are often focused on policy reforms and can develop the intensive interaction with internal Bank constituencies required to understand existing policies, examine experience over many projects, develop feasible alternatives, and build linkages to formulate and implement policy changes.

The examination of policy cases in table 12.2 suggests the possibility of a third form of coalition important to policy changes. Although Bank staff usually played important but supporting roles in most policy coalitions, some policy changes were largely *initiated* by Bank staff, either joined with outsiders or empowered by the enabling environment created by external pressure. In these *internal reform initiatives,* actions to revise policies were taken by individuals and groups within the Bank to improve its operations. The resettlement and water resources policy reviews, for example, were led by insiders empowered by external criticism. The information disclosure and Inspection Panel policies, in contrast, were pushed more by external pressure from a variety of coalitions, although insider support was important in producing approved policy drafts in those cases as well. In most cases of policy reform, Bank staff were coalition allies, but in internal reform initiatives, they were initiators and leaders of the coalition.

The Evolution of Policy Coalitions

The policy cases can also be examined in terms of the evolution of mutual influence within coalitions over time. The *definition of policy problems,* for example, emerges from interactions among coalition members and Bank staff, in some cases over many years. The Bank's need for an indigenous peoples policy, for example, has been defined by challenges from indigenous movements to many Bank projects. A small group of international NGOs concerned with indigenous peoples, for example, has helped link national movements in order to maintain concern about the Bank's policy. In other cases, international NGOs have seized opportunities presented by project performance problems to build new coalitions. The review of the water resource management policy was a Bank initiative initially restricted to government representatives until international NGOs proposed a more participatory process. Still other problems have been defined by long-term coalitions committed to institutional reforms in the Bank. Thus, the Bank's information policies were defined as a problem by coalitions created for the Narmada campaign and the IDA 10 campaign. These problem definition processes set quite diverse precedents for within-coalition influence: Bank staff had more influence on the agenda of the resettlement and water resource management policy coalitions than on the information and inspection policy coalition.

In policy reform campaigns, *direction setting* articulates preferred policy alternatives and develops coalition strategies for achieving them. For many reasons, international NGOs are positioned for considerable influence in the direction-setting process because they have access to information about experience across many countries; contact and understanding of pressure points and constituencies sympathetic to reform within the Bank; connections to executive directors and donor country constituencies that might support policy reforms; and institutional interests in reforming Bank policies, which grassroots groups or national NGOs primarily concerned with national problems do not have. Where there are no compelling local concerns about policy, grassroots organizations and national NGOs may remain uninterested. The INGOs concerned with water resource management policy, for example, found it difficult to engage many local groups, even given some Bank interest in Southern perspectives. Even when strong grassroots movements exist, as among the indigenous peoples in many countries, they still may not put much pressure on INGOs to develop and maintain a coherent international strategy for influencing Bank policy. In the past few years, the INGOs concerned with indigenous peoples have adopted quite diverse strategies, from advocacy to cooperation. The abstract nature of policy campaigns may reduce INGO interaction with national and grassroots groups, although it may increase interaction with potential allies in the Bank and other Northern constituencies.

International NGOs may also play influential roles in *implementing* policy reform campaigns because they have access to and familiarity with Bank processes for policy formulation. Also, because policy decisions are ultimately made within the Bank, Bank staff also play central roles. In all of the cases recounted here, some Bank staff were sympathetic to alternative policies from the start. The first draft of an indigenous peoples policy, for example, offered more support to indigenous peoples' movements than any subsequent policy, and some Bank staff supported policies consistent with coalition proposals regarding both the water resources management and information policies. Building support within the Bank for new policies is critical to implementing reform, so coalition members with credibility inside the Bank become important resources. As part of coalition strategy it may be necessary to recruit other actors—such as donor country governments and Northern media—to support reform. In the information policy coalition, the INGOs' ability to expand their coalition to include key members of the U.S. Congress, for example, was central to pressing Bank management to produce a new policy.

Finally, the evolution of policy coalitions may also include substantial *revision* of coalition purposes and strategies as circumstances change, and those revisions may alter mutual influence and accountability within the coalition. The rise of indigenous peoples' movements around the world and the divergence of many INGOs' ideologies and strategies made it difficult to preserve the public INGO unity of the 1980s; what had been one transnational coalition became several, revolving around quite different perspectives and policy programs in the 1990s (see Gray chapter). The information and Inspection Panel policy coalition grew out of prior coalitions that emphasized projects and policies with direct local impacts; the new coalition shifted power and influence to international NGOs, Northern governments, and sympathetic Bank staff to promote policy reform. Policy coalitions, like project coalitions, must adapt to changing circumstances. Changes in Bank policy are strongly influenced by the forces that operate inside and close to the Bank, so the balance of coalition power can be expected to shift toward actors who are familiar with and able to utilize those forces.

The Organization of Policy Coalitions

The organization and internal accountability of policy coalitions are shaped by many of the same factors that affect project coalitions, but there are also some important differences. Characteristics of the policy influence process—such as the relevance of experience in many countries, the prospective application of new policies to future projects, and the Washington, Anglophone venue of primary discussions—affect *coalition definition*. The composition of policy coalition actors is quite different from project coalition actors: International NGOs take primary roles in pressing for alternate policies; Northern government actors and sympathetic Bank staff may serve as important resources as well. The definition of a policy coalition may range from clear to vague. Out of a series of project campaigns, the information and inspection policy coalition, for example, grew to become a well-defined coalition with a long history of coordinated action on Bank reform matters. The indigenous peoples policy coalition, in contrast, tended to fragment over differences in political and ideological orientation, so the policy was influenced by a series of coalitions made between INGOs and national movements organized around specific projects.[22] It is probably not coincidental that explicit policy changes never did materialize in this area, despite many successful project coalitions and worldwide concern for indigenous peoples. Regarding the water resources management policy review, it proved difficult to maintain a coalition with local and national organizations, although the international NGOs did seek to recruit Southern partners. Policy coalitions are often composed of international NGOs and Northern actors who can shape policies, even when for legitimacy they rely on organized grassroots voice from projects influenced by the policy.

The *organizing mechanisms* of policy coalitions also reflect their differences from project coalitions. Policy coalition strategies focus on assessing and influencing Bank policies, so they accord power and influence to actors with relevant resources—including allies within the Bank, influential members of donor governments, and international NGOs with long experience on the issues. In the water resources management policy process, for example, international NGOs were critical because they kept national and regional NGOs advised about the internal policy review and made Southern voices audible.

Organizational arrangements that bridge critical differences and enable wide participation in policy analysis can be central to policy influence. Some bridging arrangements are created as part of specific policy review processes, such as the NGO Conference on Water Resource Management proposed by international NGOs and partially funded by the Bank. Preexisting bridging organizations may be vital to mobilizing information and perspectives. The NGO Bank Information Center, for example, provides information and contacts on policy issues to concerned NGOs,

North and South. The World Bank–NGO Working Group, jointly created by the Bank and NGOs, provides an arena for "critical cooperation" to examine the implications of existing policies (see Covey chapter). Bridging organizations may be essential to the continuity required for policy reform coalitions, but all the organizations described above are largely located in the North near the source of the policy.

As in the project coalitions, bridging individuals in key positions have played critical roles in organizing and preserving the most successful policy coalitions. Key actors in the information and inspection policy coalition drew on relationships established in many years of joint work on the Narmada campaign. The indigenous peoples policy coalition, in contrast, was fragmented by the diverse political orientations of several NGO leaders. Moreover, because policy changes are essentially events within the Bank, they cannot be formulated or implemented without support within the Bank, so key Bank staff have often also become central actors, particularly in Bank-initiated policy reviews, such as those for resettlement and water resources. In some cases, such as the information and inspection policy processes, donor government officials took critical leadership roles as well.

Interaction within Policy Coalitions

The organization of a policy coalition at once shapes and is shaped by the interaction among its members. Interaction in policy coalitions is facilitated by the fact that the geography, wealth, and culture of their members are often the same. Because the members of international NGOs, Bank staff, and donor governments share common characteristics, communications among them are less subject to the chasms that separate members of some project coalitions.

On the other hand, these different parties may be more at odds in terms of political and ideological orientations and interests than the social movements, national NGOs, and international NGOs that participate in project coalitions. The construction of coalition goals and strategies across the institutional difference between NGOs, donor governments, and the Bank can be very challenging. The result may be splits among NGOs over how much to cooperate with Bank staff—as happened in the process of dealing with the indigenous peoples policy—or polarizations that alienate potential coalition members, as occurred in the information and inspection policy coalition.

The quality of interaction and the composition of the coalition can be affected by how much change is required by policy shifts. Incremental changes that involve few major departures from past experience are easier to accept and implement than transformational changes that call for large changes in existing policies and behavior.[23] Changes that build on long-term development of policies may include committed Bank staff in a process marked by debate and disagreement, but relatively little escalated conflict—as demonstrated in the water resources and resettlement reviews. More fundamental changes, such as the proposed changes in information and accountability policies, are more likely to promote intense and adversarial confrontations in which it is more difficult for insiders to challenge official Bank positions. It is doubtful whether such fundamental policy changes are possible without strong, external pressure. In the case of the information/accountability policies, strong resistance from Bank staff was overcome by a campaign that used the disillusionment with Bank management generated by the Narmada campaign and the Morse Commission report among executive directors, the U.S. Congress, and the general public to promote change. Transformational changes may require the convergence of strong pressures from inside *and* outside the organization.

Expanding Stakes

Policy coalitions bring together a mix of actors different from those involved in project coalitions and often include powerful actors within the Bank, in donor governments, and in larger publics. Some policy coalitions are initiated by the Bank itself, by staff sympathetic to values and concerns of NGOs and social movements. Policy coalitions depend on the long-term development of sympathetic constituencies within the Bank and donor governments, built throughout years of experience in many countries. Often their success depends on growing awareness within the Bank, among donor and borrower governments, among grassroots networks and movements, and in the general public of the problems in current policies.

Policy campaigns may contribute to a widening international concern about particular issues: the struggle of indigenous peoples around the world, the global consequences of rainforest devastation in Polonoroeste, and the threat of mass suicide in the Narmada Valley. They expand to involve a wider range of constituents—donor governments, Bank staff, international media—and promote more public awareness across international boundaries. At the same time, they may be beneficiaries of widening awareness that puts public pressure on the Bank. Thus, policy coalitions may be both a consequence and a cause of wider international trends and forces.

Accountability within Coalitions

The cases exemplified in this volume suggest that transnational coalitions vary substantially in their accountability to their members, in particular to members at the grassroots. Table 12.3 summarizes levels of accountability reported in previous tables across both policy and project cases. The rows describe accountability between actors separated by varying degrees of geographic and political distance; the columns indicate different degrees of mutual influence and accountability.

The distribution of cases in table 12.3 suggests several implications of this analysis. First, it seems clear that it is easier to exercise mutual influence across relatively small geographic and political distances between actors. High mutual influence is more common in project coalitions between actors that are relatively close—such as grassroots groups and NGOs, or national and international NGOs. Accountability is more difficult when the actors are separated by wide gulfs of physical, cultural, and political differences: high levels of accountability are less common (though not impossible) between grassroots groups and international NGOs or Bank staff.

Second, it also seems clear that grassroots influence on other actors is more common in project coalitions, where they have a direct interest. International NGOs are more likely to have high levels of mutual influence with Bank staff and national NGOs in policy coalitions, as might be expected from the nature of the issues and the long-term character of policy campaigns.

Table 12.3
Accountability and organizational distance in project and policy coalitions

Actors and organizational distance	Mutual influence and accountability		
	Low	Medium	High
Project coalitions			
NGO-INGO (1)		Brazil (early) Indonesia	Brazil (late)* Philippines* Ecuador*
GRO-NGO (1)	Indonesia (early) Brazil (early)	Indonesia (late) Brazil (late)*	Philippines* Ecuador*
GRO-INGO (2)	Brazil (early) Indonesia	Brazil (late)* Philippines* Ecuador*	
Policy coalitions			
INGO–World Bank (1)		Indigenous peoples*	Resettlement Water resources Information and Inspection Panel*
NGO-INGO (1)	Indigenous peoples* Resettlement	Water resources	Information and Inspection Panel*
GRO-NGO (1)	Indigenous peoples* Resettlement Water resources	Information and Inspection Panel*	
NGO–World Bank (2)	Indigenous peoples* Resettlement Water resources Information and Inspection Panel*		
GRO-INGO (2)	Resettlement Water resources	Indigenous peoples* Information and Inspection Panel*	
GRO–World Bank (3)	Resettlement Water resources Information and Inspection Panel*	Indigenous peoples*	

Key: GRO = grassroots organization; NGO = national NGO; INGO = international NGO.
An * indicates that a coalition was supported by a grassroots social movement.

Finally, table 12.3 suggests that different kinds of coalitions produce different levels of mutual influence. National problem coalitions (marked with an asterisk in the table) are much more likely to create high levels of accountability in project coalitions, but only somewhat more likely to foster accountability among international actors in policy coalitions. Issues networks may produce lower levels of accountability unless they work closely with Bank initiators of policy changes (as demonstrated by the resettlement and water resources coalitions) or unless the campaign leads to increased activism by grassroots actors (as, for example, happened regarding the Brazil project).

More generally, patterns visible across the last two decades suggest that these coalitions are learning that mutual influence and accountability have important benefits. In part, this change reflects changes in local capacities: local NGOs and organized grassroots groups are increasingly likely to be active in coalitions organized to affect Bank projects and, to a lesser extent, policies.[24] In part, the change reflects the need for coalition legitimacy: when the Bank and its executive directors question international advocacy NGOs about the credibility of their Southern partners, they create pressure for coalitions to include members who have real legitimacy as representatives of affected populations. In part, the increase in mutual influence reflects experience with advocacy and critical cooperation. As relationships are built from interaction among coalition partners, the possibilities of future joint action increase.

Creating Social Capital

There has been much recent interest in the concept of *social capital*—defined as relationships, organizations, norms, and social trust that enable horizontal social problem solving among actors.[25] Evidence indicates that stocks of social capital contribute to both a responsive government and rapid economic development, so creating social capital is potentially an important consequence of collective action.

How social capital comes into existence is a contested issue. Some analysts have suggested that it is the path-dependent product of long-term social processes that take decades or centuries.[26] In this view, creating social capital is not a short-term option. Others argue that cooperative relationships, organizations, and social norms can be constructed in relatively short periods of time.[27] Ongoing cycles of interaction construct and reconstruct the relationships, norms, organizational arrangements and levels of social trust that constitute social capital. Establishing transnational coalitions may be facilitated by existing stocks of social capital among actors. Do such coalitions also create new relationships, organizations, norms, and trust—thus increasing the stocks of social capital for future transnational problem solving and mutual influence?

The case reports suggest that transnational coalitions can generate social capital in several ways. First, such coalitions have in these cases supported the *projection of local organizations and movements* as national and international actors. Especially in the movement-dominated alliances, coalitions and federations of local organizations emerged to speak for poor and underrepresented populations to policymakers at the highest levels. In Ecuador, the struggle created a new voice for indigenous peoples and environmental NGOs in policy making with Bank officials, senior government officials, and business executives, thus strengthening the voice of indigenous peoples and other grassroots groups in the Mobilization for Life. In the Philippines, the organization of local, regional, and national networks increased the capacity of indigenous groups to work with other like-minded organizations. The emergence of these local organizations supported the transnational coalition, and the coalition in turn helped the local organizations exert their voice in negotiations with the Bank and their own governments. This pattern of local capacity building is particularly visible in struggles over projects that directly affect local interests and so involve clear stakes for grassroots populations. Such organizations are themselves a form of local and national social capital that will support grassroots participation in future problem solving.

A second pattern of social capital development is the *creation of bridging organizations* that enable mutual influence among coalition members. Some of these bridging arrangements make it possible for regional and national groups to negotiate shared goals and strategies. The task forces that brought together churches, unions, farmers, and indigenous peoples at regional and national levels in the Philippines, and the "solidarity councils" by which they articulated shared goals and strategies, are examples of bridging organizations within the country coalition.

Other bridging organizations link national and international NGOs concerned with projects and countries. The Ecuador Network, the Philippine Development Forum, the Rondônia NGO Forum, and the International NGO Group on Indonesia permitted collective reflection and campaign planning across national differences. Such bridges enable diverse parties to process their cultural, economic, political, and social differences; to negotiate the terms of joint action; and to build mutual trust and influence.

A third form of bridging organization links NGOs interested in long-term influence on international institutions, such as the NGO Working Group on the World Bank and the Bank Information Center. These organizations expand the social capital available to support problem solving around similar issues in the future, across national and international boundaries.

A fourth form of bridging organization connects coalition members to other actors with stakes in the project or policy. The Rondônia NGO Forum participated with government officials in the deliberative council charged with implementing the Planafloro program, and the leaders of CONAIE and Acción Ecológica negotiated with senior Ecuadoran government officials and agribusiness leaders to revise the land privatization law. Prior to the activities of the coalitions in these cases, there were no seats at the policy decision-making tables for NGOs or grassroots groups.

Finally, transnational coalitions contribute to social capital in some cases by *building relationships and trust* among bridging individuals or interlocutors. The interactions that produce mutual influence and coalition accountability can also produce trust that extends beyond the immediate issues of the campaign. The executive director of the Philippine Development Forum, for example, felt personally responsible to the elders who were committed to shedding "the last drop of blood" in defense of Mt Apo (see Royo chapter). The members of the Narmada Action Committee developed personal relationships that provided a basis for later campaigns. Abiding trust and relationships are not the outcome of all coalition interactions, of course, but when trust is developed, it provides a foundation for future joint action.

The social capital created by these coalitions does not necessarily accumulate. New relationships and trust may not survive the coalition itself, which may be terminated by either success or failure, and the trust developed in such new relationships is, at least initially, embodied in *individuals* rather than in groups or organizations, so the capital may not be transferable to their organizational successors.[28] There is evidence for the creation of social capital in these cases, but how it becomes institutionalized as a durable feature of an interorganizational relationship is less clear. These cases do suggest, however, that social capital can be created in the iterative negotiations that produce norms of cooperation, mutual influence, and social trust in coalitions—and that such newly created capital can persist through years of extended campaigns.

Social Learning
Have these coalitions and their interaction with the World Bank contributed to *social learning*—that is, articulating new paradigms that can alter the perspectives, goals, and behaviors of social systems larger than particular organizations? If organizational learning involves changes that go beyond individual learning to affect organizational perspectives and behaviors, then social learning must involve changes that go beyond organizational learning to alter perspectives and behaviors among broader social actors.[29] The contributions of transnational coalitions to learning may have wide effects on social perspectives and activities in three areas: (1) changing the World Bank and its policies as global symbols, (2) redefining development "accountability" as a multistakeholder concept, and (3) fostering global civic politics to influence international relations.

The World Bank is widely observed and emulated by other development agencies, and its seal of approval on a project is often enough to attract funding from other donors. Influence on Bank projects and policies can thus have wide multiplier effects. Perhaps more important, the Bank's efforts to deal with NGO communities, however limited, have been a model for other international actors. If transnational NGO coalitions can promote Bank learning from grassroots voices, they may catalyze similar changes in many other organizations. Changes in Bank policies and practices may foster social learning by providing new paradigms, by pressing governments to pay more attention to civil society, and by encouraging civil society organizations to raise their voices more often.

Transnational coalitions' challenges to Bank projects and policies have focused attention on the concept of accountability and the importance of local participation in defining, designing, implementing, monitoring, and evaluating project impacts. *Accountability* in this context refers to the responsibility for performance to beneficiaries and victims of Bank projects, as well as to donor and recipient country directors. To the extent the well-publicized debates between the Bank and its critics produce widely distributed new understanding of concepts such as accountability, what coalitions contribute to social learning may have impacts well beyond the activities and perspectives grounded in a specific campaign. Ironically, wider awareness of the multifaceted nature of accountability is also reshaping transnational coalitions. The legitimacy of coalitions without genuine roots among grassroots populations is subject to increasing challenge, not least by Bank staff, so the coalitions are forced to attend to their own accountability as a consequence of their challenge to the Bank.

Finally, the policy debates and dialogues between these coalitions and the Bank contribute to national and international publics' awareness of issues such as the fate of indigenous peoples, the problems of resettlement, or the importance of rainforests in the global ecology. By reframing problems in terms of widely held values and by publicly challenging the policies and actions of major actors, transnational networks organized around specific principles and values can reshape public awareness as well as the social and political contexts for implementing projects and policies.[30] The challenges raised by these coalitions, especially in controversial and drawn-out struggles like Mt. Apo and Narmada, have promoted social learning in the form of fundamental changes in public perspectives in both Southern and Northern countries. Institutions like the World Bank ignore such changes at their peril. The struggle over Narmada led directly to the IDA 10 replenishment campaign, which threatened the availability of funds to the Bank, and to the "50 Years is Enough" campaign, which conditioned continuing monetary support on fundamental institutional reforms. When campaigns frame the issues in terms of widely held values and concerns, they can catalyze social learning that makes external environments increasingly hostile to the targets.

A targeted institution's efforts to preserve its autonomy may also have perverse effects if escalated campaigns build more public activism and reveal further failures of the institution to live by its own policies.

The rise of transnational coalitions to influence the environment and development policies of multilateral development banks is not an isolated phenomenon. Transnational NGO coalitions have emerged to challenge corporate and government policies as well as the activities of multilateral development agencies.[31] The frequency of civil society network and coalition challenges to actions by international or multinational agencies seems to be increasing.

When do transnational networks and coalitions become transnational social movements? In the cases reported here, the transnational campaigns involved coalitions between social movements and NGOs rather than full-blown transnational social movements. Yet, in the long run, such coalitions may generate the social capital—reflected in the proliferation of bridging organizations and individuals capable of building relationships and trust among diverse actors—required to construct effective transnational movements. The coalitions can also help to catalyze the kinds of social learning—reflected in transnational recognition of shared values, common framing of problems, agreement on joint responsibilities, and widespread capacities for collective action—required for effective action on international problems in an increasingly interdependent world. Social learning is potentially enhanced by the diversity of perspectives, information, and resources brought to the table by the wide range of actors involved in these coalitions, and high stocks of social capital increase the likelihood that such diversity can produce joint learning rather than misunderstanding and conflict. Past experiences in the construction of problem-solving coalitions across differences in wealth, power, and culture may be replicated frequently throughout the coming decades, at once using and creating the social capital needed for future social learning and problem solving.

Notes

1. Samuel Paul, "Accountability in Public Services: Exit, Voice and Control," *World Development* 20, no. 7 (1992), p. 1047.

2. Paul has reviewed much of the literature on holding government agencies accountable for performance in developing countries. He argues that ordinary accountability to hierarchy utilized within bureaucratic organizations must be supplemented by mechanisms that promote consumer "voice" in the sense of political influence and opportunities for "exit" in the sense of wider market options in order to improve the performance of government agencies. See Paul, "Accountability in Public Services," and Albert Hirschman, *Exit, Voice and Loyalty* (Cambridge, Mass: Harvard University Press, 1970).

3. George Aditjondro, "A Reflection about a Decade of International Advocacy Efforts on Indonesian Environmental Issues," paper presented at the International NGO Group on Indonesia Conference, Bonn, Germany, April 1990, p. 16.

4. In community problem solving, for example, it appears that interorganizational connections are central to community responses to environmental changes and opportunities. See Edward P. Laumann, Joseph Galaskiewicz, and Peter V. Marsden, "Community Structure as Inter-organizational Linkages," *Annual Review of Sociology* 4 (1976), pp. 455–84. Eric Trist has argued that the interorganizational networks are critical mediators between macroinstitutions and their microcounterparts in modern societies. See his "Referent Organizations and the Development of Inter-Organizational Domains," *Human Relations* 36, no. 3 (1983), pp. 269–84.

5. In recent years, interorganizational cooperation in and between the private and public sectors has expanded dramatically. See Barbara Gray, *Collaborating: Finding Common Ground for Multiparty Problems* (San Francisco: Jossey-Bass, 1989); Walter W. Powell, "Neither Market nor Hierarchy: Network Forms of Organization," in B. Staw and L. L. Cummings, eds., *Research in Organizational Behavior* (Greenwich, Conn.: JAI Press, 1990), pp. 295–336; and Paul Lawrence and Russell Johnston, "Beyond Vertical Integration: The Rise of the Value-Adding Partnership, *Harvard Business Review* (July–August 1988), pp. 91–101. The expansion of international associations and interorganizational networks in the nongovernmental sector has also skyrocketed. See Julie Fisher, *The Road from Rio: Sustainable Development and the Nongovernmental Movement in the Third World* (New York: Praeger, 1993).

6. Power differences can lead to explosive escalations of conflict when they combine with cultural misunderstandings or ideological differences. See L. David Brown, "Interface Analysis and the Management of Unequal Conflict," in G. B. J. Bonners and Richard B. Peterson, eds., *Conflict Management and Industrial Relations* (Boston, Mass.: Kluwer-Nijhoff, 1982), pp. 60–78; L. David Brown, *Managing Conflict at Organizational Interfaces* (Reading, Mass.: Addison-Wesley, 1983); and Gray, *Collaborating*. When these kinds of conflicts take place within organizations that are closely linked to external constituencies, the resulting tensions can be highly volatile. See L. David Brown and Jane Covey Brown, "Organizational Microcosms and Ideological Negotiation," in Max H. Bazerman and Roy J. Lewicki, eds., *Negotiating in Organizations* (Newbury Park, Calif.: Sage Publications, 1983), pp. 227–47.

7. International advocacy NGOs are focused on promoting Bank reform and environmental responsibility, which are the main concerns of their Northern, middle-class members and staffs. Grassroots movements, on the other hand, are typically concerned with the welfare of their members. Because both organize around visions and values that are compelling to their members, their disagreements are often easily polarized into difficult-to-resolve ideological debates. See Budd L. Hall, "Building a Global Learning Network: The International Council for Adult Education," in B. Cassara, ed., *Adult Education in International Context* (New York: Krieger, 1993). As Sidney Tarrow has pointed out, the interorganizational linkages described here as "coalitions" are "contingent coalitions" between national movements and international NGOs that have temporarily shared objectives. See his "Fishnets, Internets, and Catnets: Globalization and Transnational Collective Action," *Occasional Paper,* Juan March Foundation, Center for Advanced Study in the Social Sciences (winter 1995).

8. In this section, the discussion draws heavily on earlier chapters about the Ecuador structural adjustment program, the Mt. Apo thermal power plant in the Philippines, the Planafloro natural resource management project in Brazil, and the Kedung Ombo Dam in Indonesia.

9. See Margaret Keck and Kathryn Sikkink, "International Issue Networks in the Environment and Human Rights," unpublished working paper, New Haven, Conn, Yale University, 1993; *Activists beyond Borders* (Ithaca, N.Y.: Cornell University Press, 1998).

10. See L. David Brown, "Bridging Organizations and Sustainable Development," *Human Relations* 44, no. 8 (1991), pp. 807–31, for a discussion of such organizations in the development process. Bridging organizations have been major actors in promoting cooperation among diverse actors in other settings as well. See L. David Brown and Darcy Ashman, "Participation, Social Capital and Intersectoral Problem-Solving," *World Development* 24, no. 9 (1996), pp. 1476–79.

11. See Peter S. Ring and Andrew H. Van de Ven, "Developmental Processes of Cooperative Interorganizational Relationships," *Academy of Management Review* 19, no. 1 (1994), pp. 90–118.

12. For examples see Gray, *Collaborating;* Barbara Gray and Donna J. Wood, "Collaborative Alliances: Moving from Practice to Theory," *Journal of Applied Behavioral Science* 22, no. 1 (1991), pp. 3–20; and David Chrislip and Carl Larson, *Collaborative Leadership* (San Francisco: Jossey-Bass, 1994).

13. For illustrations of these phases in industrialized countries, see Gray, *Collaborating,* and Sandra Waddock, "Lessons from the National Alliance of Business Compact Project: Business and Public Education Reform," *Human Relations* 46, no. 7 (1993), pp. 777–802. For description and analysis of revising coalitions in response to challenges encountered during implementation, see Cynthia Price-Taylor, *Theory of Evolution in Inter-Organizational Relationships: Exploration of the Implementation Phase,* DBA dissertation, Boston University, 1996.

14. Coalitions, like other forms of social organization, may vary from tight to loose in their internal organization; their external boundaries can vary from permeable to impermeable. See Brown, *Managing Conflict,* pp. 19–46. Interorganizational coalitions at the outset are likely to have relatively undefined and permeable boundaries and relatively loose organization (i.e., in regulation of member activity). In the evolution of coalitions, we would expect more clarity about boundary definition and permeability, and an internal organization appropriate to coalition strategies and tasks.

15. Although any organization needs some minimum level of shared expectations to coordinate behavior, it is particularly important for interorganizational coalitions—whose members are often very skittish about infringements on their autonomy—to develop shared understandings about decision making and conflict management. This is especially true of coalitions where power inequalities and cultural differences can combine with conflicts of interest to produce explosions of misunderstanding and conflict. See Brown, "Interface Analysis" and *Managing Conflict.*

16. Bridging organizations can be catalysts in development projects; see Brown, "Bridging Organizations." Organizations and individuals that bridge the North-South divide can be the "relays" that permit collective action and mutual trust between national social movements and Northern advocacy movements described in Tarrow, "Fishnets." Such bridging organizations can facilitate the discussions among INGOs and movement activists that are required for constructing common perspectives and goals. See Bert Klandermans, "The Social Construction of Protest and Multiorganizational Fields," in Aldon D. Morris and Carol M. Mueller, eds., *Frontiers in Social Movement Theory* (New Haven: Yale University Press, 1992), pp. 77–103; also David A. Snow and Robert D. Benford, "Master Frames and Cycles of Protest," in Morris and Mueller, pp. 133–55.

17. Relationships, norms, and social trust tend to be "socially constructed" in the course of interaction among actors, rather than established by agreement or fiat. See Peter Berger and Thomas Luckmann, *The Social Construction of Reality* (Harmondsworth, U.K.: Penguin, 1971); and Anselm Strauss, *Negotiations: Varieties, Contexts, Processes, and Social Order* (San Francisco: Jossey-Bass, 1979). Interaction among parties that have little experience with one another— members of a new coalition, for example—may create or destroy relationships and social capital. Acting reliably or in ways that imply respect and good will for others can create social capital: good relationships, shared trust, and norms of tolerance and cooperation. Acting opportunistically or in ways that maximize self-interest at others' expense may destroy trust and social capital. See Ring and Van de Ven, "Developmental Processes," and Robert D. Putnam, *Making Democracy Work: Civic Traditions in Modern Italy* (Princeton, N.J.: Princeton University Press, 1993).

18. The negotiation of interpersonal and interorganizational trust is central to building alliances, according to Ring and Van de Ven, "Developmental Processes," and Gray, *Collaborating.* The emergence of a "moral community"

among actors with histories of intense competition, by means of actions that demonstrated mutual commitment and trustworthiness, has been documented in Lawrence D. Browning, Janice M. Beyer, and J. C. Shetler, "Building Cooperation in a Competitive Industry: Sematech and the Semiconductor Industry," *Academy of Management Review* 38, no. 1 (1995), pp. 113–53. Similarly, Mark Leach has assessed in detail the evolution of expectations, trust, and shared metaphors for partnership between an international NGO and Indian NGOs. See *Organizing Images and the Structuring of Interorganizational Relations,* Ph.D. dissertation, Boston University, 1995. In the context of development projects, Judith Tendler and Sara Freedheim have described the development of social trust between communities and government health workers in Brazil. See "Trust in a Rent-Seeking World: Health and Government Transformed in Northeast Brazil," *World Development* 22, no. 12 (1994), pp. 1771–92.

19. The concept of trust as a critical variable in interorganizational relations has drawn increasing attention in the last few years. For political scientists' examination of the role of trust in the context of relations among states, see Robert O. Keohane and J. S. Nye, *Power and Interdependence,* 2d ed. (New York: Harper/Collins, 1989). Students of national economies have also argued that trust is crucial: see Francis Fukuyama, "Social Capital and the Global Economy," *Foreign Affairs* 74, no. 5 (1995), pp. 89–103. Others have looked at the impact of trust on problem solving and civic engagement: see Putnam, *Making Democracy Work.* Economists and organization theorists have also begun to examine the relevance of trust to economic performance: see William G. Ouchi, "Markets, Bureaucracies and Clans," *Administrative Science Quarterly* 25 (1980), pp. 129–41; and Oliver E. Williamson, *Markets and Hierarchies: Analysis and Antitrust Implications* (New York: Free Press, 1975). Peter S. Ring and Andrew H. Van de Ven have argued that the existence of trust permits more effective (in some situations) organizational alternatives to markets and hierarchies. See "Structuring Cooperative Relationships between Organizations," *Strategic Management Journal* 13 (1992), pp. 483–98.

20. The discussion in this section draws on earlier chapters on the indigenous peoples policy, the resettlement policy, the water resource management policy, and the information and Inspection Panel policy.

21. It may be easier for advocacy groups to form de facto coalitions with insider Bank reformers on policy reforms than on project reforms because fewer tangible interests within the Bank and borrowing governments are directly challenged. It is easier to extract promises to be good in the future than to change misdirected projects in midstream, once they have developed momentum and entrenched interests.

22. Tarrow has characterized many "transnational social movements" as being in fact "contingent coalitions" between national social movements and international NGOs ("Fishnets"). Such coalitions are more tactical and less strategic, more focused on instrumental goals than shared social visions, and often more loosely organized than movements.

23. Organizational learning that reframes fundamental assumptions typically requires more resources, energy, and basic revision of organizational structures, operations, tasks, and cultures. See Chris Argyris and Donald Schon, *Organizational Learning: A Theory of Action Perspective* (Reading, Mass.: Addison-Wesley, 1978; and J. Swieringa and A. Wierdsma, *Becoming a Learning Organization: Beyond the Learning Curve* (Reading, Mass.: Addison-Wesley, 1992. Experience with efforts aimed at major organizational change suggest that "frame-breaking" changes that call for basic shifts in organizational paradigms are much more difficult than "framebending" incremental changes. See David A. Nadler and Michael L. Tushman, "Organizational Frame Bending," *Academy of Management Executive* 3, no. 3 (1989), pp. 194–204.

24. The rise of NGOs as major actors in national and international development matters has been noted by many investigators. See, for example, Lester M. Salomon, "The Rise of the Nonprofit Sector," *Foreign Affairs* 73, no. 4 (1994), pp. 111–24. Some have also emphasized the important role being played by the rapid development of networks, coalitions, and alliances within regions and countries. See Fisher, *The Road from Rio*.

25. We focus here on the forms of social capital identified by Robert D. Putnam as contributors to civic cooperation and horizontal problem solving. Putnam emphasizes the social features—such as associations, norms, and social trust—that enable horizontal (rather than vertical) cooperation (*Making Democracy Work*).

26. Putnam suggests that the social capital that shaped political and economic activity in the Italian states is grounded in hundreds of years of history (*Making Democracy Work*). Similarly, as Fukuyama's analysis suggests, social capital that shapes economic behavior is grounded in national cultures, which change only over the long term ("Social Capital").

On the other hand, Putnam's more recent work has focused on the rapid decline of social capital in the United States, a phenomenon he believes to be closely linked to the introduction of television in the fifties. See "The Strange Disappearance of Civic America," *American Prospect* no. 24 (1996), pp. 34–48. This relatively rapid decline suggests that social capital may be more volatile than is implied in the earlier analysis. See Nicholas Leeman, "Kicking in Groups," *Atlantic Monthly* (April 1996), pp. 22–6.

27. Jonathan Fox has argued that civil society is "thickened" and social capital is constructed by iterative cycles of negotiations and conflict between state and the civil society actors over time. See "How Does Civil Society Thicken? The Political Construction of Social Capital in Rural Mexico," *World Development* 24, no. 6 (1996). Brown suggests that cooperative interactions across sectors among grassroots groups, NGOs, and government agencies can create social capital. See L. David Brown, "Creating Social Capital: Nongovernmental Development Organizations and Intersectoral Problem Solving," in Walter W. Powell and Elizabeth Clemens, *Private Action and the Public Good* (New Haven, Conn.: Yale University Press, 1998). See also Norman Uphoff's analysis of the Gal Oya

irrigation project, where invoking values on cooperation, new organizational ideas for implementing cooperation, and new friendship ties combined to foster extensive and rapid improvements in a seriously debilitated irrigation system: *Learning from Gal Oya: Possibilities for Participatory Development and Post-Newtonian Social Science* (Ithaca, N.Y.: Cornell University Press, 1992). Similarly, Tendler and Freedheim describe a health program in Brazil that built "trust in a rent-seeking world" and enabled program success even in a situation where bureaucratic mismanagement and corruption might be expected ("Trust in a Rent-Seeking World", p. 1771). For a collection of papers on the creation of social capital in state-society relations, see Peter B. Evans, ed., *State-Society Synergy: Government and Social Capital in Development* (Berkeley, Calif.: International and Area Studies Research Series, U.C. Berkeley, 1997).

28. The combination of coalition and individual mortality may undermine the accumulation of social capital that has not been translated into more durable and institutionalized forms. There is some evidence, however, that social capital may be more durable than it first appears. For a discussion of how past experiences, even failures, may shape the social resources available for future action, see Albert Hirschman, *Getting Ahead Collectively: Grassroots Experiences in Latin America* (New York: Pergamon Press, 1984).

29. Although the concept of "social learning" has recently been recommended as a strategy for dealing with the large-scale problems facing modern societies, little attempt has been made to define the concept very systematically. See Matthias Finger and Philomene Verlaan, "Learning Our Way Out: A Conceptual Framework for Social-Environmental Learning," *World Development* 23, no. 3 (1995), pp. 503–14; Lester Milbrath, *Envisioning a Sustainable Society: Learning Our Way Out* (Albany, N.Y.: State University of New York Press, 1989); and Thomas Princen, Matthias Finger, and J. P. Manno, "Translational Linkages," in Thomas Princen and Matthias Finger, eds., *Environmental NGOs in World Politics: Linking the Local and the Global* (London: Routledge, 1994), pp. 217–36. L. David Brown and Darcy Ashman have explored the possibility that differences may be a source of social learning in intersectoral problem solving when the social capital that enables effective cooperation is present. See "Social Capital, Mutual Influence and Social Learning in Intersectoral Problem-Solving," in David Cooperrider and Jane Dutton, eds., *Organizational Dimensions of Global Change* (Beverly Hills, Calif.: Sage, 1998).

30. NGOs and NGO networks may play central roles in social learning as sources of social innovation, as advocates for policy changes, as articulators of new perspectives and frames for understanding problems, and as promoters of citizen action on newly recognized problems. See Keck and Sikkink, *Activists Beyond Borders*; Princen, Finger, and Manno, "Transnational Linkages," pp. 220–28; and Paul Wapner, "Politics beyond the State: Environmental Activism and World Civic Politics," *World Politics* 47, no. 3 (1995), pp. 311–40.

31. Perhaps the best-known example of a challenge to corporate decision making is the International Babyfood Campaign, which sought to regulate the marketing

of infant formula, especially by the Nestle Corporation, to mothers in developing countries. See, for example, Douglas A. Johnson, "Confronting Corporate Power: Strategies and Phases of the Nestle Boycott," in L. Preston and J. Post, eds., *Research in Corporate Social Performance and Policy* (Greenwich, Conn.: JAI Press, 1986), pp. 323–44. Other social movements—such as the human rights, environmental, and women's movements—have also developed transnational coalitions to influence governments as well as international agencies. See, for example, Fisher, *The Road from Rio;* Keck and Sikkink, *Activists beyond Borders;* and Wapner, "Politics beyond the State."

13

Assessing the Impact of NGO Advocacy Campaigns on World Bank Projects and Policies

Jonathan A. Fox and L. David Brown

Has the World Bank really begun to reform its social and environmental actions? If so, what role did nongovernmental advocacy groups and grassroots protest play in the process? This process has been fought out in the media, in legislatures, and with mass citizen action for a decade and a half—more a "war of position" than a "war of movement." To the degree change has occurred, it has been slow, inconsistent, and ambiguous, with few clear-cut breakthroughs as far as people directly affected by projects are concerned. This observation should not be surprising; the World Bank is a bank, after all. Any change would come in degrees and would be combined with business as usual.

This concluding chapter focuses on the key patterns that emerge from the volume's diverse efforts to assess the role of external pressures in effecting change. Three analytical dilemmas have to be faced in the attempt to discern such patterns: (1) how can we assess changes in the Bank's *portfolio*; (2) what is the "set" of Bank *projects* that have been tangibly influenced, at least in part by public protest and (3) to what degree are social and environmental reform *policies* actually put into practice? One major conclusion is that many protest campaigns manage to influence subsequent World Bank policies, but have limited impact on the projects that provoked the protest in the first place.

This overview begins by illustrating such analytical challenges with a notable case of clear-cut "NGO impact" on World Bank decision making: its withdrawal of support for Nepal's proposed Arun III Dam after seven years in the design process. The chapter draws out the broad patterns of World Bank reform and concludes by proposing the specific conditions under which its "sustainable development" reforms could actually be implemented. Future comparative studies will have to address the related issue of precisely which kinds of NGO/grassroots strategies and tactics have had the greatest impact.

Lessons from the Arun III Dam Campaign

The World Bank's eleventh-hour 1995 decision to cancel its planned funding of the Arun III Hydroelectric Dam surprised both the project's critics and supporters. Arun was a classic Bank "megaproject," a $1 billion investment involving $175 million from the World Bank, influential private sector interests, and a Nepali government interested in cheap electricity. After many years in the design process, however, it had provoked widespread criticism from international and Nepali public interest advocacy groups on social, environmental, and economic grounds—a campaign that utilized the momentum of the early 1990s international campaign to stop India's hotly contested Narmada River dam project.[1] The August 1995 cancellation of the proposed loan showed that an emerging international NGO "alarm system" had gained a growing capacity to block questionable development projects *before* they were built.

At first glance, Nepal's Arun III was not an obvious candidate to follow India's Narmada Dam as the second-most conflictive World Bank project of the 1990s. Unlike many other Bank-funded dams, Arun was a run-of-the-river operation that involved relatively little flooding—the kind of large dam usually considered to be as environmentally friendly as they get. Arun III Dam also promised little of the forced resettlement that made the Narmada Dam into such a contentious human rights issue. It did threaten isolated indigenous communities and forests with a massive influx of outsiders, however, and for one of the poorest countries in the world, the dam's huge cost made it seem to be a "paradigm case" for challenging the World Bank's longstanding commitment to large infrastructure projects.[2] An international NGO network convinced the U.S. executive director of the Bank to join the opposition, supported by a U.S. Agency for International Development (USAID) study favoring a smaller-scale alternative energy strategy. Under pressure from a vocal NGO campaign, the German government also pulled back after its federal audit office questioned the project's "economic viability, sustainability and the minimization of risk."[3] Japan's expected aid contribution hung by a thread.

The international NGO campaign was further legitimated when a ranking World Bank dissenter came forward, taking early retirement and making public a powerful critique of the project's questionable economic assumptions and social opportunity costs. He observed,

Since senior management seemed committed to the project, a serious and open debate was no longer possible, and even common sense questions were being dismissed. All the available energy went into building the case in favor of the project, rather than examining alternatives.... The project is not in conformity with the Bank's poverty alleviation strategy for Nepal.... It is an unbalanced use of Bank funds with an overemphasis on energy which will crowd out investments in the social sector such as rural infrastructure and agriculture."[4]

Then the Bank's new Inspection Panel, the semiautonomous official accountability channel set up in response to the Narmada controversy, accepted Arun as its first case. The panel questioned whether Bank officials had followed their own indigenous peoples and resettlement policies.[5] Arun's backers in the Bank had so clearly ignored social, environmental, and economic concerns that radical NGO campaigners were eventually joined by quite moderate project critics—those who felt that project risks could have been acceptably mitigated if only the Bank's policies had been followed. Indeed, the project had been so widely questioned inside as well as outside the Bank that the Inspection Panel's credibility depended on accepting the case (even though very few actual residents of the Arun Valley were involved in the complaint).[6]

In spite of this convergence of internal and external criticism, Bank management remained committed to the project, distorting the Inspection Panel's findings and moving the project toward board approval. The Bank's board of directors has never rejected a project proposed by the Bank's management. Having lost his battle to continue funding India's discredited Narmada project, the highest official responsible for projects in South Asia dug himself in and declared that approval of Arun was crucial for the Bank's "credibility," thus raising the stakes beyond the project itself (a logic reminiscent of the U.S. government's justification for not withdrawing from the Vietnam War).

To the surprise of many Bank critics, the new Bank president, James Wolfensohn chose to overrule his own top management: "the risks to

Nepal were too great to justify proceeding with the project." The Bank's official explanation claimed that the proposed social and environmental mitigation measures were satisfactory, but suggested that these measures, in addition to increased power rates, "would have imposed requirements which the Bank now judges to be beyond what Nepal could realistically have achieved at present."[7] This public recognition of unviable assumptions about proposed mitigation measures was new; prior internal recognition of such problems had not stopped classic project disasters in the past, as demonstrated by the Polonoroeste Amazon road project of the early 1980s.[8] As an investment banker, Wolfensohn also looked more carefully at the project's shaky economic assumptions—an action clearly in contrast to the approach usually taken by long-time Bank managers accustomed to the entrenched official "culture of approval."[9] This "culture of approval" problem had been highlighted by a major internal evaluation (the Wapenhans Report) of rising rates of "unsatisfactory" project performance.[10]

Environmental NGO advocates reacted to Wolfensohn's rejection of Arun III with cautious optimism, and some hoped that the decision would send a strong message to World Bank staff that future large dam projects would receive much more scrutiny in the future. Indeed, the World Bank is currently preparing fewer large dam projects than it funded in the mid-1980s (outside of China, that is, where independent advocacy groups are not tolerated).[11] Gopal Siwakoti of the Katmandu-based Arun Concerned Group concluded: "This is a victory for Nepal and stunning defeat for the Bank.... This is [also] a very smart face-saving measure on the part of the Bank. It is trying to wriggle out of admitting that it violated its own policies by approving the Inspection Panel's critical report."[12] After the cancellation, however, Nepalese NGOs continued discussions with Bank staff and executive directors, lobbying in support of Bank funding for an alternative, smaller-scale hydroelectric development project for their country.[13]

Is the Arun III cancellation evidence that the World Bank is finally "greening" itself, after more than a decade of sustained environmental and human rights protest? Or was this an isolated concession to environmental critics? Although it is still too early to draw firm conclusions, it should be noted that the new Bank president was not personally invested in the project and that because Nepal was a small borrower, it was easier for the Bank to pull out. One could therefore argue that canceling the Arun project was a politically low-cost way to allay NGO critics, to show his commitment to improving project effectiveness, and to show management that—unlike his predecessors—he was not going to be a rubber-stamp president. The project was especially vulnerable because it was to be funded by the Bank's concessional, low-interest aid window, the International Development Association (IDA), whose budget was under sharp attack in the Republican-controlled U.S. Congress.[14] By itself, then, environmental and indigenous rights protest alone does not explain the Bank's cancellation decision, but it does explain why the controversy reached the top of the new president's agenda. He was faced with a decision about whether to begin to invest his own fresh political capital at a time when the Inspection Panel's criticisms had already legitimized what was sure to be a prolonged protest campaign.

As suggested in this volume's introduction, the Arun III cancellation was one dramatic instance of a much broader process of political conflict between the World Bank, NGOs, and national governments over how to allocate resources in the name of development. The outcome of the conflict was mediated by a set of World Bank policies and institutions created in response to previous waves of protest—notably the Inspection Panel's capacity to give political "teeth" to the Bank's new set of social and environmental reform policies involving environmental impact assessment, public information access, "involuntary resettlement," and indigenous peoples. The Arun case also suggests that there are limits to the "organizational-learning" explanation of Bank environmental reform because seven years of project preparation and internal debate did not prevent it from almost being sent to the board for approval.[15] Rather, the case illustrates a broader pattern of institutional change: neither advocates of environmental and social issues within the Bank nor external criticism alone were sufficient to defeat the project; each reinforced the other, with the external critique tipping the balance in an internally divided Bank. At the same time, much of Bank activity remains "more of the same"—as evidenced by the strong management support for Arun III until the very end, based on the very same information that led the new president to cancel it.

Assessing Institutional Change: Analytical Dilemmas

How does one take stock of the impact of transnational advocacy campaigns on an institution as large, opaque, and slow moving as the World Bank? Four methodological problems come up. The first is the question of the counterfactual: would World Bank projects have been even *more* socially and environmentally costly—or would a larger number of them have been destructive—in the absence of external scrutiny and protest? Or, as World Bank officials often claim, would national governments have been *less* environmentally responsible in the absence of World Bank "greening?" Second, where the Bank does appear to have responded to public pressure, how does one open up the black box of official decision making to disentangle the relative weights of the various different factors—internal and external, ideological and interest driven—that come into play? Third, how does one avoid the conflation of normative and analytical criteria for assessing change? This problem is most serious for projects that were redesigned in response to pressure, but that remain socially and environmentally costly. For example, because of protest against Thailand's Bank-funded Pak Mun Dam, the dam was lowered and resited, thus reducing the estimated number of people affected from twenty thousand to approximately five thousand displaced.[16] These five thousand still lost homes or livelihoods, but some NGO critics neglect to mention the significant changes in the project along the way.[17] Fourth is the problem of the time frame for assessing change. Many World Bank defenders have long argued that changing its institutional direction is like turning around an ocean liner—a necessarily slow and arduous process.[18] The long lead time between changes in top-level decision-making processes and outcomes on the ground creates the "pipeline effect": at the front end of the pipe, policymakers claim new projects will be different, but at the receiving end citizens groups continue to experience the results of past decisions. As discussed below, the pipeline effect creates an ongoing political dissonance problem between the Bank and its critics because reform promises can never be "definitively" assessed until an ever-moving point in the indefinite future.

Although Bank officials emphatically affirm that they have changed, many Bank campaign critics argue that the institution does not comply with its own reforms. The Arun III project cancellation would be a sign of the times for the former, but an isolated exception for the latter. In terms of assessing change, both sets of actors should be considered interested parties. It is obvious why Bank officials would claim to have "learned their lesson," but it is less obvious why Bank critics might underestimate possible change. There is a tension between short-term political "campaign logics" and a more distanced, longer-term assessment of protest impact. In the short term, many advocacy groups see the recognition of partial reform as undermining the case for further change. Declaring victory before the job is done risks falling into the trap of accepting small concessions "instead of" more significant change. Some partial concessions can coopt critics, though others serve as wedges that make deeper changes possible. Whether and when small changes are accepted "instead of" or considered "steps towards" bigger changes is an open empirical question that will vary from case to case. Either way, documenting partial changes is nevertheless significant, both in terms of informing future "pro-accountability" strategies and for potential "development refugees" whose lives are not disrupted as a result.

Assessing degrees of change can be framed in terms of three major analytical dilemmas, each associated with a different level of analysis of World Bank actions:

1. *Portfolio trends.* What criteria does one use to assess whether social and environmental reform has occurred? At the broadest macrolevel, one can look at the distribution of funds across the portfolio in terms of positive and negative categories.

2. *Impact on projects.* How does one determine the "relative weight" of a particular set of public interest advocacy pressures as distinct from other factors? The broader the level of change, the more difficult it is to trace external impact. The clearest examples are in cases of specific projects.

3. *Impact on policies.* Policies clearly have changed, but to what degree are they reflected in the actual projects? Policies ostensibly mediate between broad changes in institutional direction and the final impact of

specific project-lending decisions. What do the available studies tell us about the degree to which reform policies are actually implemented?[19]

Portfolio Trends: "The Good, the Bad, and the Ugly"

Perhaps the broadest indicator of the Bank's actions is its diverse portfolio. Rather than go into a detailed assessment of changing trends in terms of types of actual loans—a daunting empirical task—we can frame the issue in hypothetical terms. For the sake of discussion, let us say that projects can be divided into different categories in terms of their social and environmental impact. Some projects clearly have worse impacts than others; indeed, international NGO networks are constantly on the lookout for especially devastating projects. For the sake of discussion, projects in the portfolio can be categorized as "the good, the bad, and the ugly." These categories refer not to conventional sectors, such as energy versus education, but rather to the degree to which loans are consistent with some hypothetical minimum sustainable development criteria. If the Bank's changes in policies and discourse managed to influence its actual lending patterns, the *relative weights* of "the good, the bad, and the ugly" within the portfolio would change.

"Good" Projects   One set of projects might be "good" from a sustainable development point of view because they offer more access to basic education for girls, primary health care, reproductive choice, safe drinking water for poor people, titling and demarcation of smallholder and ancestral indigenous lands, biodiversity protection, industrial pollution control, microenterprise loans for low-income women, AIDS prevention, funding for worker-managed extractive reserves in the Amazon forest, and so forth.[20] Some analysts assert that a sizeable fraction of this set of projects is no doubt much less enlightened in practice than in theory, as can be seen in the case of social service projects that require onerous "cost recovery" charges for poor people, or in the case of decentralized projects that risk strengthening authoritarian local governments.[21] Other analysts make a broader critique of the Bank's entire "market-led plus targeting the poorest" approach to social policy, arguing that it is replacing the principle of universal access to basic health care as a right with limited

welfare for the destitute.[22] This critique of official Bank human resource investment policies has yet to be followed up by systematic empirical studies that link projects with actual social impact on the ground.

Some Bank critics might argue that the "reformed" category of lending merely serves to legitimate the rest of what the institution does, which is much more negative on balance, or that such funding serves to divide and conquer NGO critics. For the most severely indebted low-income countries, the nominal purpose of "good-sounding" projects may well be moot because in practice they must devote most incoming loan funds to debt service. For countries that pay a large fraction of their national income back to international financial institutions, fresh development funds end up flowing from one building in downtown Washington, D.C., through national capitals, and back to another building in downtown Washington—from the World Bank to the International Monetary Fund (IMF) (or to another window of the World Bank itself).[23] It would still be important to know whether these hypothetically pro–sustainable development projects really are what they claim to be and whether this category of "good" projects is growing.[24]

The Bank announced the shift toward "poverty reduction" as its overarching objective in its 1990 *World Development Report*, which described a three-prong strategy to attain this goal: export-oriented, labor-intensive growth; investment in the poor via the development of human capital (mainly health and education); and the promotion of safety nets and targeted social programs to support those who fall through the cracks.[25] According to a top Bank social scientist, these changes came about "partly in response to NGO concern."[26] The most rapid increase in lending occurred in "human resource" investments (education, health, population, and nutrition). In the 1980s, human resource lending averaged about 5 percent of operations; this amount tripled to more than 15 percent in the fiscal 1993–1995 period.[27] Nominally poverty-targeted projects accounted for 24 percent of total 1995 lending and 43 percent of IDA lending. The World Bank has also officially encouraged more projects to involve "stakeholder" and NGO participation through the "project cycle." At least until recently, NGO involvement was largely limited to "retail" service delivery, but according to World Bank NGO

Table 13.1
World Bank lending since Rio (FY 1993–1995)

Types of lending	Billions of dollars	Percentages
Total lending	67	100
Environmental projects	6	9
"Win-win" projects	20	30
Investments with potentially significant and harmful impacts on the environment	13	19
All other lending	28	42

Source: Environment Department, *Mainstreaming the Environment: The World Bank Group and the Environment since the Rio Earth Summit, Fiscal 1995* (Washington: The World Bank, 1995), p. 13.

liaison staff, this pattern is beginning to change, with some NGO involvement in project design and evaluation.[28]

On the environmental side, the World Bank reports that its mix of projects with either positive or negative environmental effects is changing. Environment Department analysts divide up the "post-Rio" World Bank lending portfolio into four categories: (1) positive environmental impact; (2) "win-win" projects, whose economic focus is on producing positive environmental benefits; (3) investments with significant environmental risk; (Category A projects, which require full environmental assessment; and (4) projects without direct environmental impact (see table 13.1). Many advocacy groups would differ over how to categorize specific projects, but the fraction of lending *claimed* to be pro-environmental is clearly growing, whether in *green* (natural resource) or *brown* (pollution control) categories. The degree to which Category A projects—that is, those acknowledged to be environmentally risky—are actually mitigated in practice varies widely, as does the quality of the environmental assessment (see below for further discussion of the Bank's environmental assessment policy implementation).

The reliability of the Bank's environmental and social project categories is contested by NGO watchdog groups, in part because the Bank's ratings are usually based exclusively on its own internal or borrowing government sources, or on official intentions expressed in project documents rather than on actual implementation experience. At the same time, however, NGO critics also lack independent, field-based assessments of most ostensibly "good" projects because their monitoring energies are focused on more overtly threatening projects. For the sake of this conceptual exercise, let us assume that some undetermined subset of the broader categories of anti-poverty and pro-environmental projects might actually fit those descriptions, pending future field-based studies of outcomes.

"Bad" Projects From the point of view of some hypothetical minimal sustainable development criteria, another category of projects might be considered simply "bad." This hypothetical category includes projects or policy-based loans that contribute to ongoing environmental degradation and social inequity, or are largely wasted through corruption, patronage, and/or support for local elites and international contractors. These projects often promote unsustainable paths of energy and natural resource use, as in the case of the Bank's preference for huge investments in thermal power plants rather than energy conservation. One recent study found that the World Bank, including its private sector arm, now lends *more than one hundred times as much on fossil fuel investments than it spends on the entire Global Environmental Facility (GEF) budget for projects that avert greenhouse gas emissions.*[29]

Some structural adjustment packages would fit in this category as well, depending on their varied social impact and the degree to which regimes were going to carry them out anyway to please private financiers and the IMF (or the U.S. government, as in the case of the Mexican financial crisis). Indeed, assessing the impact of structural adjustment packages is especially problematic because governments tend to accept them only once they are already in economic crisis. Bank defenders can therefore claim that, as bad as poverty got, the alternative of "not adjusting" would have been even worse (based on the assumption that its particular adjustment path was the only one possible).[30] The main point here is that a large fraction of World Bank projects can be considered "more of the same" in that they do not encourage change toward more equitable and environmentally sustainable development paths. This category is considered here to be "bad" rather than "ugly" because it mainly reinforces *existing* trends in borrowing government development policies (leaving the problematic issue of the counterfactual aside for the sake of this exercise).

"Ugly" Projects   The effects of the third hypothetical category of projects are even worse in terms of sustainable development criteria: "ugly" projects *directly* immiserate large numbers of low-income people, endanger fragile indigenous cultures, encourage the dangerous spread of toxics, promote irreversible biodiversity loss, and prop up dictatorships that might otherwise fall. The conceptual distinction between "bad" and "ugly" would be quite difficult to operationalize empirically, but it serves to illustrate the analytical dilemma of how one needs to "unpack" the diverse set of Bank activities to determine the relative weights of different kinds of project impacts. It matters how much money goes to each category. For example, it is possible that the relative share of "ugly" projects is growing smaller as a result of the cumulative impact of international protests and related institutional reforms. This possible decrease does not in any way make the "bad" projects less "bad," nor does the growing set of "good" projects compensate for the combined set of "bad" and "ugly" projects. Another possibility is that the weight of the "ugly" projects is changing in terms of the scale of their social and environmental costs. As suggested above, the classic controversial project of the late 1980s and early 1990s—the Narmada Dam—threatened direct devastation to many thousands more people than the proposed Arun III Dam did. Moreover, by the mid-1990s, it was more likely that projects as contentious as Narmada would be blocked long before approval.

Damage Control: Assessing Protest Impact on Projects

Project cycles can last a decade or more, from design through approval and implementation. The Bank usually plays an intensive role in design at the beginning of the process, but the responsibility gradually shifts more to borrowing governments in the course of implementation. This cycle affects the opportunities for external influence. The first question is: to what degree did controversial projects change? Change is necessary but not sufficient to demonstrate advocacy impact, however. Bank staff can always claim that project improvements, such as mitigation of environmental and social impacts, were due exclusively to the effectiveness of their own autonomous implementation of their "reformed" policies. As senior Bank social scientist Michael Cernea put it early on, however, "Even though their [NGOs'] criticism of certain Bank-assisted projects has sometimes been harsh ... it has helped the Bank and some of its borrowing agencies become more keenly aware of some projects' implications on vulnerable groups, on resettlement, on nonrenewable resources."[31] The point that external critics had influence is not controversial in itself; the difficulty is in specifying the relative weight of that influence and under what conditions it had the most impact.

The influence of public interest advocacy on the World Bank varies greatly—over time, from country to country, across regime types and sectors, and through the project cycle—but this volume's studies suggests that we consider two factors in particular. First, the studies found no direct link between the intensity of grassroots mobilization and impact on projects. Even though the most intense and sustained Bank protest so far led the World Bank to withdraw support from the Narmada Dam, Indian authorities moved ahead until domestic judicial challenges suspended construction. Conversely, a relatively small amount of strategic lobbying, combined with the potential *threat* of overt mobilization, can be sufficient to block a project in its early stages. The preemptive avoidance of controversial projects is known in Bank jargon as "negative selection."

Second, the studies suggest that additional factors are required to explain the outcome of most cases of protest impact. As noted in the introduction to this volume, a simple two-actor model—the Bank versus the local communities and their international NGO allies—is not enough to explain the broader question of who is the winner and who is the loser in development projects. National states, local governments, divisions within local communities, national and international private sector interests, and distinct policy currents within the Bank all play important roles as well. As illustrated by figure 13.1, international pressure is mediated in important ways by the Bank's governance structure of executive directors—representatives of national governments who can in turn be pressured both by private business and by public interest groups in democratic member countries.

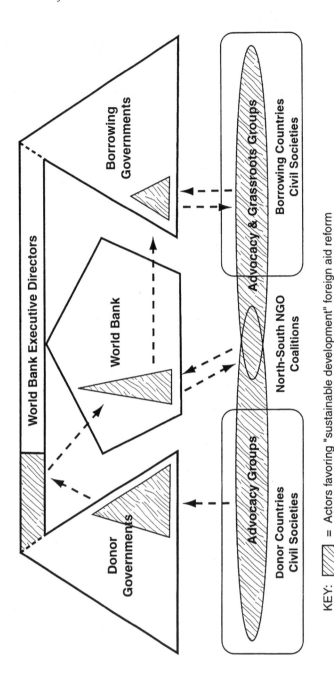

KEY: ▨ = Actors favoring "sustainable development" foreign aid reform

- ◂ = Political support

Figure 13.1
The political dynamics of World Bank "sustainable development" reforms

Protest still matters, however, and it is not a coincidence that many of the more influential movements have focused on rainforests and dams. Rainforest destruction provokes high levels of concern even among the most moderate of environmental groups in donor countries, while dams provoke especially high levels of grassroots resistance. More than most development projects, the social costs of dams are especially clear to those who are displaced. Dam reservoirs create intensely shared interests and identities among those affected, potentially facilitating mobilization, but many protests gather momentum only once construction is well under way and is therefore especially difficult to stop.

Table 13.2 lists the thirty-six project campaigns found to have had some impact. Of this set, twelve involved dams, fourteen involved forests or other community natural resources, five involved mines or industrial energy projects, and only two involved urban infrastructure. At least twenty-three involved indigenous peoples' rights—an important factor in all of these sectors. Most project campaigns involving macroeconomic structural adjustment projects, urban transportation and water infrastructure projects, and social sector investments had little impact.

In terms of the geographic distribution of cases that have involved some protest impact, most have been concentrated in a handful of countries. Just more than half were located in Brazil (9); the Philippines (4); India (4); Indonesia (4); and Mexico (3). Notably, the impact cases did not occur only in countries with democratic political regimes. They did have in common relatively strong pro-democracy *movements* in civil society, however. More detailed analysis of the cross-national distribution of transnational advocacy coalitions and their impact would require much more field-based analysis.

Table 13.2 outlines most of the World Bank projects that appear to have been significantly influenced by advocacy coalitions. Some campaigns involved transnational partnerships, while others were either largely nationally driven or strictly local. Ideally, this data set would include the full set of advocacy campaigns to permit systematic comparison of successes and failures. It does include, however, those campaigns that failed to influence their projects but had spillover effects on subsequent projects and policies. The table is organized to show campaigns both in chronological sequence and in terms of the different kinds of impact.

Table 13.2
Conflictive World Bank projects influenced by NGO/grassroots protest

| Peak Conflict Periods | Projects canceled or suspended at different points in the project cycle | | | |
	Project loan blocked before Bank approval	Ongoing project loan canceled	Loan disbursements temporarily suspended	Subproject within ongoing loan blocked before approval
1975				
late 70s	Philippines: Chico River dams			
1979				
1981		India: Bastar Forestry		
early 80s			Brazil: Polonoroeste	
mid/late 80s				
1986				
1987				
1987				
1987				Brazil: Machadinho Dam (Power Sector I)
mid/late 80s				
late 80s				
1989	Brazil: Xingu Dam (Power Sector II)			
1989				
1990–91				
late 80s/ early 90s		India: Narmada Dam (Sardar Sarovar)		

Social/environmental impact mitigated	New project designed in response to past protest	Spillover impact of project campaigns on subsequent World Bank policies and projects
Philippines: Tondo Foreshore		
		Philippines: Chico River dams
		Brazil: Sobradinho dam
		Brazil: Polonoroeste
		Indonesia: Transmigration
		Brazil: Power Sector
Ethiopia Forestry		India: Karnataka Social Forestry
Brazil: Ita Dam (Power Sector I)		
	Brazil: Itaparica Resettlement and Irrigation	
		Indonesia: Kedung Ombo Dam
		Guinea Forestry
		Ivory Coast Forestry
		India: Narmada Dam

Table 13.2 (continued)

	Projects canceled or suspended at different points in the project cycle			
Peak Conflict Periods	Project loan blocked before Bank approval	Ongoing project loan canceled	Loan disbursements temporarily suspended	Subproject within ongoing loan blocked before approval
1990–91	Indonesia: Irian Jaya Area Development			
1990–94				
1990–96				
1991–92				Mexico: San Juan Tetelcingo Dam (Mexico Hydro)
1991–92	Cameroon Forestry			
1992		Northern Mexican Forestry		
1992				Philippines: Mt. Apo Geothermal (Energy Sector)
1992				
early 90s	China: Three Gorges Dam			
1992–96				
1993–94	Congo Natural Resources			
1993				

Social/environmental impact mitigated	New project designed in response to past protest	Spillover impact of project campaigns on subsequent World Bank policies and projects
Brazil: Rio de Janeiro Flood Reconstruction		
	Brazil: Planafloro (Rondônia Natural Resources Mgt.)	Brazil: Planafloro
Thailand: Pak Mun Dam (Power Sector III)		
	Pilot Program for Brazilian Rainforest (G-7)	
		India: Singrauli Thermal Power

Table 13.2 (continued)

	Projects canceled or suspended at different points in the project cycle			
Peak Conflict Periods	Project loan blocked before Bank approval	Ongoing project loan canceled	Loan disbursements temporarily suspended	Subproject within ongoing loan blocked before approval
1994				
1995	Nepal: Arun III Dam			
1995				
1995				
1995				
1995	Nigeria: LNG Pipeline and Production Facilities (IFC)			
1995–96				
1995–96			Mexico Aquaculture	
1995–96				

Social/environmental impact mitigated	New project designed in response to past protest	Spillover impact of project campaigns on subsequent World Bank policies and projects
Philippines: Integrated Protected Areas		
		Nepal: Arun III
Lesotho Highlands Dam		
		Chile: Pangue Dam (IFC)
	Ecuador: Indigenous Peoples Development Project	
Coal India Environmental and Social Mitigation Project		
		Indonesia (Irian Jaya): Freeport McMoran Gold & Copper Company (MIGA)

From the top to the bottom of Table 13.2, the campaign cases are listed in chronological order based on the peak years of conflict. From left to right, they are categorized in terms of where they influenced the projects in their respective cycles in order to show the range of different points at which project critics had impact. Project campaign impact falls into four main categories. The first category includes blocking, canceling, or temporarily suspending a project by means of advocacy and/or protest. This type of impact could be felt at different points in the project cycle—for example, blocking projects before they are even approved or forcing their cancellation after they are already under way. The second broad category of impact is perhaps the most common—when protest or advocacy leads to a significant degree of mitigation of a project's social or environmental costs. In cases where movements develop in response to projects already under implementation, this type of impact is often the most that can be expected. The third category of impact is small and new because it involves projects that are designed specifically to *respond* to protest and controversy, which was often provoked by previous projects in the same sector or region. The fourth column's category details campaign impact on the broader policies that guide *subsequent* World Bank projects. Many other projects have been changed, restructured, or canceled for reasons unrelated to civil society concerns, but they are not the subject of this analysis.

Loans Blocked before Approval   Table 13.2 begins with the category of protest impact in which projects in preparation were prevented from being approved. This data set is certainly missing cases that were internally vetoed very early in their cycles and are therefore little known outside the Bank. The first grassroots movement that succeeded in blocking a proposed Bank project was led by the indigenous Igorot peoples of the Philippines Cordillera region, who prevented a hydro dam from flooding their ancestral domain. As Gray's chapter shows, this conflict contributed to the Bank's first tribal peoples policy in 1982.[32] India's Bastar Plan to turn a primary forest into a plantation was also blocked by tribal protest at a very early stage: the preliminary technical assistance loan was canceled.[33] Gray also notes how in 1989 the indigenous-international NGO alliance against an Amazon dam contributed to blocking an entire energy sector loan to Brazil, as worldwide video coverage of the Altamira tribal summit added greatly to the World Bank's public relations problems. The transnational/Indonesian NGO coalition INFID successfully questioned a regional development project for conflictive Irian Jaya. In the early 1990s, NGOs also contributed to blocking a potentially destructive forestry project in Cameroon and a nominally "green" project that threatened rainforests in the Congo.[34]

The proposed support for Nigeria's 1995 private sector liquid natural gas project would have been channeled through the Bank's International Finance Corporation (IFC). Though not a focus of this volume, the World Bank's rapidly growing IFC investments have provoked increased public scrutiny and protest. The Nigerian project would have invested in an industry widely known for extensive pollution and systematic violations of the rights of the minority Ogoni people. When the Nigerian government executed a leader of the minority rights movement in 1995, the IFC responded to an international NGO protest campaign and withdrew its proposed investment.[35]

Loans Canceled   Protest rarely manages to cancel an ongoing project, largely because governments must formally agree to cancellation. Partial concessions, if any, are much more likely, as the size of the "impact mitigation" category suggests. The Narmada Dam project was the first large loan to be canceled on environmental and social grounds, nominally by the Indian government (see chapter 11 by Udall and chapter 9 by Fox). At first it appeared that construction would continue in spite of the Bank's pullout, but the government's loss of international support contributed to increased domestic momentum against the project. Because of financial problems and challenges in the court system, construction was largely suspended in 1995.

The 1989 Mexican Forestry Project, a pilot developed for indigenous and peasant-owned northern temperate forests, provoked a new cross-border NGO coalition between a local Mexican human rights group and U.S. environmentalists that gained leverage in the context of the debate over environmental concerns surrounding the North American Free

Trade Agreement.[36] The project's cancellation appeared to be driven by NGO pressure, and the project was already vulnerable because the government's trade opening undermined its economic logic, making it cheaper to import U.S. lumber than to take it from indigenous peoples' forests.

Loans Suspended   Canceling a loan is difficult for bureaucrats because it implies an official admission of failure. Instead, if Bank officials want to pressure governments to comply with loan conditionalities, they are more likely to suspend disbursements. The first suspension on environmental grounds occurred in 1985 in response to more than two years of international criticism of Brazil's Polonoroeste. This campaign case had very significant spillover effects; it contributed directly to the 1987 strengthening of the Bank's Environment Department and eventually led to the ostensibly "green" Planafloro project detailed in Keck's chapter in this volume. Note that Polonoroeste is not listed in table 13.2 as a case that resulted in mitigation, in spite of its "founding" status in the MDB campaign. Few analysts would argue that the ongoing rainforest destruction and invasions of indigenous lands changed significantly on the ground in the 1980s.

The Mexico Aquaculture project was suspended in part due to civil society concerns just before the loan was to be signed. In response to concerns expressed by independent, low-income fishers' groups and supportive Mexican NGOs about the social and environmental impact of proposed industrial aquaculture parks, the World Bank had encouraged the Mexican government to redefine the project to include more emphasis on environmental regulation in the sector, less support for the large-scale installations, site-specific social and environmental impact assessments, targeted funds for small-scale indigenous people's fishing projects, and a commitment to make project performance indicators public on an annual basis. By 1996, Mexico's Environment Ministry (responsible for the fishing sector) was willing to make these concessions, which the NGOs and the fishers saw as a victory, but then the Finance Ministry blocked the project for a year because of concern about setting precedents in favor of public participation and environmental assessments.[37]

Funds for Proposed Subprojects Blocked   The next impact category encompasses impacts on sector loans that pay for diverse subprojects. Because many of the actual projects and sites are not specified at the point when the loan is approved, sector loans are much more difficult for advocacy groups to monitor. Nevertheless, environmental and social impact criteria can be used to block specific subproject investments without changing the terms of the overall loan.

In Brazil, people affected by dams to be built on the Rio Uruguai had to face government power companies that were backed by a sectorwide Bank loan, but they managed to block dam construction. Brazil's diverse dam protest movements won concessions that varied from dam to dam, depending on the local and regional political balance (winning more in the southern states, where social movements were stronger and democratic governance was more consolidated).[38] Similarly, Mexico's electric power company reportedly hoped to fund a planned hydroelectric dam under an already approved 1989 sectorwide loan for hydroelectric dams. The San Juan Tetelcingo Dam would have evicted at least thirty thousand indigenous peasants. Sustained grassroots mobilization in 1991 and 1992, however, put the project on indefinite hold.[39] Because of the Mexican government's heavy-handed resettlement track record, the scale of displacement, and the degree of resistance, the World Bank did not even come close to approving the project.[40] As discussed in Royo's chapter in this volume, the Philippines' indigenous peoples and environmentalist alliance against the proposed Mt. Apo Geothermal Plant blocked the use of already approved World Bank sector funds, thus also empowering the Bank's environmental staff to veto a project supported by its own Energy Department.[41]

Social and Environmental Impact Mitigated   Further down the project cycle, the next kind of impact is probably the most common: tangible mitigation of a project's social or environmental costs. This mitigation can take the form of a range of actions—from redesigning a project to reduce its impact, as in the Thai dam mentioned above, to creating channels for negotiating solutions with grassroots groups and NGOs, as in the case of Brazil's Planafloro project, to the recognition of the legitimacy of Ecuador's indigenous organizations. NGO/grassroots protest has also been able to buffer the impact of involuntary resettlement driven by urban infrastructure projects. Even under the Marcos dictatorship, Philippine poor peoples' neighborhood movements were able to win some important concessions in the context of the Bank-funded Tondo Foreshore Urban Renewal Project.[42]

Isolated local resistance can also block project implementation, as in the case of the Ethiopian forestry project.[43] Broad-based, local mass movements of Brazil's *atingidos* ("affected people") called for "land for land"—a compensation strategy approved by the World Bank's resettlement policy but rarely fully respected. In the Ita Dam case, the movement's strong social base, together with Brazil's democratization and a strengthening of civil society more generally, helped the movement to win unprecedented concessions, which included even farmland for affected landless workers.[44] In another Brazilian project campaign that lacked an international advocacy wing—targeting a large-scale urban drainage infrastructure project—grassroots movements managed to improve the terms of resettlement significantly.[45]

Local protest against displacement by Thailand's Pak Mun Dam came to worldwide attention in the context of the World Bank's annual meetings, which were held in Bangkok in 1992. As mentioned earlier, the Pak Mun case is one in which the scale of displacement was significant, but significantly reduced by the resiting of the dam location. In the Philippines, where civil society is highly politicized and long aware of the impact of international financial institutions, sensitivity to impending problems was high as the World Bank began to plan an Integrated Protected Areas Project. The project began with a top-down approach to biodiversity conservation that ignored the indigenous peoples whose ancestral lands were to become "protected areas." After lengthy debates through the early 1990s between different NGO networks, the Philippine government, and the World Bank, the project was redesigned to create more space for NGO and indigenous community participation.[46] Large development NGOs ended up with significant control over resource allocation, though indigenous leaders were concerned about continued lack of government respect for their ancestral land claims.[47]

In the case of Brazil's Planafloro Natural Resource Management Project, NGOs and grassroots organizations concerned about more accountable project implementation took a formal complaint to the Inspection Panel, as Keck's chapter in this volume explains. In order to avoid a formal review, project authorities in the government and the Bank rushed to speed up implementation of its environmental provisions. Although Bank management decided not to pursue the complaint officially, the external pressure clearly provoked a response, accelerating implementation of key land demarcation measures for rubber-tappers and indigenous peoples.[48]

The Lesotho Highlands Water Project is another case in which heightened local and international NGO scrutiny bolstered the implementation of mitigation measures. According to a senior World Bank environmental analyst, NGO concern "created space" for greater Bank and government attention to mitigation measures. One World Bank resettlement specialist noted that they were already beginning to deal with the project's problems by relying on information from local NGOs before international NGOs began to scrutinize the project in 1995. The subsequent international attention did, however, "sharpen our sense of urgency," and "we went to the Lesotho NGOs very directly" in an effort to bolster their leverage vis-à-vis the government.[49] This example reflects a broader pattern in which public pressure from Northern NGOs encourages World Bank officials to grant more legitimacy to local NGOs as alternative interlocutors. This *backwards triangulation* process to create political space for local NGOs is an important outcome, beyond specific impacts on projects, of international pressure on the World Bank.

The 1996 Coal India Environmental and Social Mitigation Project is an example of an effort to respond to years of criticism leveled at one of the most controversial sectors of the Bank's India investments. In the new project, resettlement and indigenous development plans have been prepared for each of twenty-five coal mines stated for economic investment, and for several in the state of Bihar the plans improved significantly as the result of grassroots pressure and advocacy.[50] Because of NGO concern, especially in Europe, about the social and environmental cost of India's past coal projects, the World Bank's board of directors approved the project on the condition that management submit a progress report on mitigation measures before economic expansion begins.[51]

Several candidates among potential "impact mitigation" cases did not make the list. Thousands of villagers displaced by the Kedung Ombo

Dam found international allies late in the construction process, as discussed in Rumansara's chapter. The local protests and international scrutiny did manage to save villagers from forcible relocation to distant islands, but this project campaign is not considered here to be a case of significant mitigation because most villagers appear to have ended up economically worse off than before.[52]

New Projects Designed in Response to Past Protests   A few World Bank projects are actually designed to help governments respond to mass mobilization driven by negative experiences with previous projects. The first such project was Brazil's Itaparica Resettlement and Irrigation Project. The World Bank did not fund the hydroelectric dam itself, which displaced fifty thousand people, but instead supported the creation of alternative livelihoods for those people who were relocated. This project was the first attempt in Latin America at comprehensive resettlement of an entire displaced population. The World Bank had previously funded the Sobradinho Dam in 1979, in the same region, which displaced more than 120,000 people and left half without any compensation. That project was imposed during the military dictatorship, and dissent was limited to vocal clergy, but by the mid-1980s Brazil had begun its return to democracy, and the rural poor mobilized to defend their rights. By 1985, grassroots protest against displacement by the Itaparica Dam grew, gaining international support from the Environmental Defense Fund (EDF) and Oxfam (United Kingdom). The World Bank made approval of the second and third tranches of its $500 million national electric power sector loan conditional on improved resettlement terms for those affected by the dam. Stepped-up grassroots direct action combined with international pressure led the Bank to encourage the government to begin to negotiate with the network of independent rural trade unions (Pólo Sindical). By 1987, the World Bank approved a separate Itaparica Resettlement and Irrigation Project to fund the investment in irrigation to permit high-value agriculture for formerly rainfed or landless peasants.[53] Only years of subsequent mass mobilization convinced the government actually to install the irrigation works, however. By 1996 less than half of the displaced population had access to functioning irrigation. The rest of the population continues to pressure the government for access to irrigation.[54]

The next two projects designed specifically to respond to social, political, and environmental problems created by past projects were also Brazilian. As Keck's chapter shows, the Planafloro project began in the late 1980s mired in controversy, but combined local and international NGO pressure led to changes in project design that created notable opportunities for civil society participation. The third Bank project designed in response to grassroots/international advocacy followed on the heels of the 1992 global environmental summit in Rio. Mainly European countries agreed to provide $290 million for sustainable development projects under the "G-7 Pilot Project to Conserve the Brazilian Rain Forest." Though slow to get off the ground, the project was closely supervised by an influential, region-wide NGO network and a representative grassroots organization: the Atlantic Forest Network and the Amazon Working Group, chaired by the National Rubber Tappers Council.[55]

The fourth project in this unusual category emerged in response to Ecuador's massive grassroots indigenous protest. As Treakle's chapter shows, what began as a debate over land rights and planned oil investments evolved into a broader civil society questioning of the role of multilateral development banks. Treakle shows how two national civic strikes won indigenous leaders the right to negotiate directly with both the World Bank and the Inter-American Development Bank (IDB), leading to the World Bank's proposed Ecuadorian Indigenous Peoples Development Project. Its design was the result of an unusually pluralistic tripartite negotiation between the government, indigenous leaders, and the World Bank.[56] If approved and implemented, this project would represent the most comprehensive example thus far of the implementation of the World Bank's indigenous peoples policy (discussed in Gray's chapter).

Spillover Effects on World Bank Policies   One of the most important categories of project campaign impact takes the form of spillover effects: *influence on policies beyond the project that is the immediate focus of the campaign.* Although some project campaigns have influenced other projects, this category is limited to campaigns that have had policy impact. The process begins with the Bank's first social/environmental policy —the directive on resettlement. In the design of Brazil's Sobradinho Dam, no plans were made to resettle the population in the catchment area, and the military evicted them as the flood waters rose. Together with the more successful Philippine Chico River protest, the Sobradinho experience encouraged the Bank to listen to its social scientists' recommendation that a policy was needed.[57] Similarly, as mentioned, Brazil's Polonoroeste campaign contributed to the Bank's 1987 creation of the Environment Department, as did the U.S. executive director's precedent-setting environmental vote against Brazil's 1986 power sector loan (which was nevertheless approved).[58] The international campaign against Indonesia's transmigration program led the World Bank to cease funding settler-colonization schemes in rainforests in general.[59] As Rumansara's chapter notes, the local Indonesian and international Kedung Ombo campaign for compensation of displaced villagers raised the broader concern regarding noncompliance with the Bank's resettlement policy in general and encouraged bilateral donors, such as the Japanese, to carry out their projects more scrupulously.

Also in the late 1980s, grassroots protest against bureaucratic and commercial conceptions of "social forestry" raised the political cost of imposing top-down approaches on low-income, rural, natural resource–dependent communities. In 1987, the enclosure of poor peoples' commonlands in India to plant commercial eucalyptus plantations provoked widespread resistance to the Karnataka Social Forestry Project.[60] Villagers managed to limit the imposition of eucalyptus, and the conflict influenced subsequent social forestry projects in India and elsewhere by underscoring the importance of genuine community participation in natural resource management. In 1989 and 1990, the Guinea and Ivory Coast forestry projects were approved, but NGO criticism convinced the U.S. executive director to encourage the Bank to revise its forest policy, which later led to a ban on support for rainforest logging.[61]

International NGO criticism of the massive impoverishment and displacement created by India's Singrauli power investment, together with concern from Bank resettlement specialists, encouraged India's National Thermal Power Company to develop a resettlement policy and created the background for the new Coal Mitigation Project discussed above.

More recently, several project campaigns have set precedents involving the World Bank's new Inspection Panel. As noted, the Inspection Panel's critique of Nepal's Arun III project legitimated the NGO campaign and led directly to both the withdrawal of the project and the strengthening of the Inspection Panel more generally. The campaign targeting Brazil's Planafloro had more mixed results. Because the complainants had a strong case, the Inspection Panel was willing to accept it, but the board decided to avoid offending Brazil and found an indirect way to monitor the project.[62] Thus, even in an especially difficult case, the panel influenced the outcome. Local environmentalists also pursued an Inspection Panel claim in the case of Chile's Pangue Dam, a privately owned hydro dam that displaced indigenous peoples and was funded by the World Bank's International Finance Corporation (IFC). Officially, however, the Inspection Panel's mandate does not cover either of the World Bank's private sector arms, the IFC or the Multilateral Investment Guarantee Agency (MIGA, a political risk insurance enterprise). The board did not agree to extend the Inspection Panel's mandate in the Pangue Dam case, but it did agree to commission an independent assessment of whether World Bank environmental and social standards were being met, which set a precedent for the IFC.[63] Similarly, in the case of the Freeport McMoran Mine in Irian Jaya, Indonesia, an international-local coalition sought to apply World Bank social and environmental standards to the Bank's private insurance arm for the first time. Again, the board did extend the Inspection Panel's mandate but also agreed to an independent assessment of compliance with World Bank environmental and indigenous rights policies.[64]

To sum up this diverse set of project campaigns that appear to have had some tangible impact on the World Bank, most either mitigated impact or influenced subsequent policies. But to what degree do reform policies in turn influence projects?

The Challenge of Reform Implementation: Policies Versus Projects

The World Bank's social and environmental policies spell out the procedures by which Bank staff must carry out environmental impact assessments, consider alternative investments, minimize involuntary resettlement, prevent immiseration of those resettled, buffer the impact of projects on indigenous peoples and encourage their "informed participation," and encourage NGO and beneficiary collaboration in project design and implementation (see table 13.3).[65] Reading the Bank's policy

Table 13.3
Principal World Bank social and environmental policies

		Dates issued and main revisions
OP 4.01	Environmental Assessments	1989, 1991, under revision
OP 4.02	Environmental Action Plans	1994
OP 4.04	Natural Habitats	1995
OP 4.07	Water Resources Management	1993
OP 4.09	Pest Management (formerly OP 4.03)	1985, 1992, 1996
OP 4.10	Indigenous Peoples (formerly OD 4.20)	1982, 1991, under revision
OP 4.12	Involuntary Resettlement (formerly OD 4.30)	1980, 1986, 1990, under revision
OP 4.15	Poverty Reduction	1993, under revision
OP 4.20	Gender Dimensions of Development	1994
OP 4.36	Forestry	1993
GP 4.46	Energy Efficiency and Conservation	1992, under revision
OP 4.76	Tobacco	1994
OP 8.60	Adjustment Lending	1981, 1987, 1990, under revision
OP 10.20	Investments under the Global Environment Facility	1993, 1995
OP 10.21	Multilateral Fund of the Montreal Protocol	1993
GP 14.70	Involving Nongovernmental Organizations in Bank-Support Activities	1989
OP 17.50	Disclosure of Operational Information	1993, 1997, under revision
BP 17.55	Inspection Panel	1993, 1997

Note: Policies listed here as "under revision" were still being "reformatted" as of the end of 1997 (they are listed officially as "to be issued"). In these cases, the most recent policies ostensibly remain in effect. OPs refer to mandatory Operational Policies (formerly known as Operational Directives). BPs refer to Bank Procedures, and GPs are Good Practices. Most OPs are also backed up by more detailed BPs and GPs. See also the NGO descriptions of Bank policies in: Bank Information Center, "A Citizen's Guide to World Bank Environmental Assessment Procedures," (Washington, D.C.: Bank Information Center, 1992), Cindy Buhl, *A Citizen's Guide to the Multilateral Development Banks and Indigenous*

Table 13.3 (continued)

Peoples, (Washington, D.C.: Bank Information Center, 1994) and Lori Udall, *The World Bank Inspection Panel: A Three Year Review* (Washington, D.C.: Bank Information Center, October, 1997).
Source: World Bank, "Operational Manual: Operational Policies, Bank Procedures," Operations Policy Department, Sept. 28, 1995, updated in 1997 with the list of World Bank policies available from its Public Information Center [www.worldbank.org].

reforms creates a strong sense of dissonance between what the official policies say and what NGO critics claim is actual Bank practice. There are many reasons for such different assessment; some are due to differences over basic goals and conflicting visions of sustainable development, but many disputes are also over "the facts"—debates over what is actually happening on the ground. Where one stands often depends on where one sits. To a Bank staffer sitting in his or her office, reading official documents, the social costs of a development project may seem to have been dealt with, but at the same time local villagers might well see the floodwaters rising around them long before government project managers have offered them alternative homes and livelihoods.

Few comprehensive, field-based assessments of Bank and borrowing government compliance with these reform policies have been carried out. Most field-based assessments of actual project implementation cover specific projects rather than entire sectors or country portfolios. Moreover, most critiques of the World Bank cover a wide range of projects and policies, and only a few have isolated those projects that were approved *after* the reform policies were issued, in part because of the long lead time involved in project cycles. Most projects implemented in the mid-1990s were designed either before many of the reform policies or in the first few years of the policies' implementation. Most projects conceptualized since the reform policies of the early 1990s are just beginning to be approved and implemented on the ground. Because the policies did not apply retroactively, the fact that disastrous *"prereform"* projects are ongoing is not an adequate test of the degree to which later reforms are being complied with. The following discussion reviews available studies of *postreform* implementation in several of the areas where the World Bank has been most vigorously criticized by local and international

NGOs—including energy, water resources, indigenous peoples, involuntary resettlement, poverty targeting, gender, forestry, pest management, and environmental impact assessment.

So far, three kinds of Bankwide assessments of reform policy compliance have been carried out: *independent/external, internal/autonomous, and internal.* Independent assessments of the degree to which postreform projects have changed include studies of policies involving energy, indigenous peoples, and water resources. The second kind of assessment is carried out by the Bank itself, but with relative autonomy from the operational staff responsible for the projects and with the capacity to cross-check their information—for example, the field work of the Operations Evaluation Department (OED). Unlike most internal assessments, the Bank's resettlement review staff did not rely exclusively on information from project task managers and carried out their own field work and consultations with NGOs. An Operations Evaluation Department (OED) study of gender-related actions can also be characterized as an internal/autonomous assessment, though it did not include any independent field assessments. The third kind of assesment is limited to a "desk review," based on official Bank documentation and supplementary interviews with project task managers. These reviews are not as autonomous when they are carried out by departments that share responsibility for policy compliance because the departments are thus interested parties to some degree. Desk reviews of the implementation of policies guiding poverty reduction (1990), forest policy (1991), agricultural pest management (1985), and environmental assessment (1989/91) are discussed in the next few sections.

External Review: Energy Policy

The Bank's 1992 energy policy encouraged increased attention to more integrated resource planning and energy efficiency rather than just to production. The EDF and National Resources Defense Council (NRDC) report assessed all power loans under consideration during the first half of 1993 ($7 billion), based on the official project information. The report finds that the Bank's energy loans, as designed, "do not comply with the Bank's stated policy of increasing its comprehensive support for end-use energy efficiency and conservation." Except for the area of pricing, the

Bank "failed to incorporate its own policies into the loan preparation process.... Bank staff have no requirement or incentive to operationalize the policy, and they have only applied it selectively." The report rates the forty-six loans covered and finds that only two complied with the Bank's policy, and only three contain comprehensive support for improved end-use efficiency.[66] Bank reformers might contend that the 1992 energy policy was advisory, without the ostensibly mandatory status of an operational directive. However, a much more recent study—based on a survey of Bank summaries of projects either in preparation or just approved in May 1995—found the same dominant pattern. Of fifty-six projects, three were found to be fully compliant, seventeen were "partially compliant," and thirty-six "did not comply."[67]

External Review: Indigenous Peoples
Though the Bank's indigenous peoples policy was one of its first reform mandates, its implementation record is still among the weakest. The World Bank's own internal review of implementation of the policy in its first five years found little evidence of progress.[68] As Andrew Gray's chapter and the discussion of table 13.3 suggest, violations of indigenous peoples' rights remain among the most frequent causes of conflict over Bank projects. A fully comprehensive assessment of the policy directive would have to go beyond the infrastructure projects that provoke most resistance; it would have to include the natural resource management projects that often involve pro forma consultation rather than substantive participation, as well as the large number of rural social service and agriculture projects that ostensibly benefit indigenous peoples. Such an assessment is beginning in Mexico, home to the largest indigenous population in the Americas. The preliminary findings suggest that only a tiny fraction of the Bank's vast indigenous-related Mexico portfolio can be considered to have even nominally applied the key operational directive 4.20 mandate for "informed participation" by ostensible beneficiaries in all phases of the project cycle.[69]

External Review: Water Resources
The water resources policy is much more recent and as a "policy paper" also lacks the obligatory quality that operational directives imply. Like

the energy sector reviews summarized above, however, Moore and Sklar's chapter offers a comprehensive assessment of the degree to which the policy's recommendations influenced projects in the design phase during its first three years. Based on the Bank's own summary project descriptions, Moore and Sklar find a small degree of responsiveness to the policy recommendations at the level of different kinds of water projects, with more for water supply and less for irrigation. They show small increases in the funding shares for infrastructure rehabilitation and institutional capacity. Although recognizing these changes, they also note that alternative-style projects, involving watershed management or smaller-scale initiatives, receive little funding, whereas privatization lending grew much more quickly. The number of large-scale infrastructure projects in the pipeline, with significant potential for negative environmental and social impact, appears to be decreasing, but many continue to provoke controversy.

Internal/Autonomous Review: Involuntary Resettlement

The 1994 resettlement review is still the most comprehensive analysis of Bank reform-policy compliance available.[70] Its findings have been interpreted in varying ways: the Bank sees the glass half-full, whereas NGO critics see it as half-empty, as Fox's chapter notes. Critics nevertheless recognize that the report offers an unusually frank and comprehensive assessment. The report indicates that involuntary resettlement was part of 192 projects active between 1986 and 1993, displacing approximately 2.5 million people, and that "projects appear often not to have succeeded in reestablishing resettlers at a better or equal living standard and that unsatisfactory performance still persists on a wide scale" (p. x). Most responsibility for these results went to national governments (notably India and Indonesia) and their lack of commitment to compensate those evicted, but the report also documents a long list of procedures often ignored by Bank staff (lack of resettlement plans, inadequate funding, no base-line data, weak institutions, etc.). It shows that some improvement was made starting in 1991, but acknowledges that apparent compliance may be superficial because "Bank appraisals tended to overestimate likely performance" (p. 140). Moreover, the study recognizes that internal reform and external criticism are mutually reinforcing.

Internal/Autonomous Review: Gender

The World Bank has long engaged in public relations claims involving "women in development" and, more recently, gender. Bank presidents have been proclaiming the importance of educating girls and improving the status of women (at least in order to reduce fertility), since Mac-Namara's 1977 *Population Address,* but the World Bank had no gender policy until 1994. The OED's 1995 comprehensive review of the Bank's explicitly gender-related lending indicates that substantive attention to gender issues began in the late 1980s. Other Bank researchers claim that 24 percent of projects approved between fiscal years 1986 and 1993 included some form of gender-related action. The OED study notes, however, that "the rating standards were applied loosely for FY88 to FY93, and projects were classified as having some action specifically designed to address gender issues on the basis of very minimal action."[71] A more recent report on World Bank gender actions, published as a follow-up to commitments made at the 1995 Beijing Conference on Women, "estimates" that in fiscal year 1995 "28 percent of World Bank operations contained gender-specific actions, and a further 9 percent contained a discussion of gender issues."[72] This optimistic rating system appears to have the same limitations pointed out in the 1995 OED evaluation.

The vast majority of projects considered to be "gender-related" involve education and health, with a small but growing number of microcredit projects open to women. Although these projects certainly do try to reach women, their success in addressing gender *roles* varies widely and can only be determined by field-based assessments.[73] The rating system also does not address the issue of supposedly "nongender" projects that turn out to have gender implications, which is part of the broader problem of the lack of field-based assessments of the relationship between project goals and actions. Overall, as one independent assessment concludes,

The Bank (like other donor agencies) has ... done significantly more on behalf of women, but as *mothers* rather than as *workers.* There is intellectual consensus within the institution on the importance of addressing gender in population, health, and education, especially in relationship to women's reproductive roles. This consensus does not exist in the productive sectors; despite recent Bank statements to the contrary ... the staff ... has yet to be convinced of the direct impact on development and on the Bank's own portfolio performance of boosting women's home and market productivity.[74]

Internal Review: Poverty-Targeted Lending

The World Bank's 1990 *World Development Report* on poverty describes a three-prong strategy to attain its goal of poverty reduction: export-oriented, labor-intensive growth; investment in the poor via the development of human capital (mainly health and education); and the promotion of safety nets and targeted social programs.[75] The Bank also emphasizes the importance of examining the actual composition of public spending and recommends that social spending be directed more toward the poor. Bank analysts now ask, for example, whether health spending is targeted at providing primary or tertiary care and if education monies fund primary schools in poor regions or college educations for urban middle classes.

Market-led growth remains the mantra of the Bank's poverty-reduction strategy, but some internal analyses are beginning to recognize that the *pattern* of growth (i.e., which social groups are gaining) matters as well and that specific targeting is necessary to influence this pattern. The 1996 Agricultural Action Plan, for example, acknowledges that smallholder-led growth should be the priority for growth that would lead to poverty reduction.[76] Although Bank research now recognizes that the social distribution of assets is crucial for determining the distribution of the *benefits* of growth, very little Bank lending promotes pro-poor asset redistribution.

Since 1990, the World Bank has developed an indicator to document its claim that it is increasingly targeting anti-poverty lending activities: the Program of Targeted Interventions (PTI). As table 13.4 indicates, PTI projects make up a greater percentage of the World Bank's low-interest lending window for the poorest countries (IDA) than of its regular lending, which is at near-commercial rates. This difference is related to the reasons for the creation of the PTI in the first place. The PTI indicator was designed as part of the effort to persuade skeptical donor governments that money directed to the IDA was directly helping to reduce poverty.[77] Most PTI projects are in agriculture and rural development, education, population, health, and nutrition. Table 13.4 shows that PTI projects now account for a growing and significant fraction of World Bank lending, especially to the poorest countries, reaching 54 percent of IDA investment lending in 1995. This figure is the basis for the claim that

Table 13.4
"Poverty-targeted" lending, 1993–1995

Fiscal years	1993	1994	1995
Total WB PTI lending (U.S.$ millions)	4,674	4,441	5,437
Share of WB investment lending (%)	27	25	32
Share of all Bank lending (%)	20	21	24
IDA's PTI lending (U.S.$ millions)	2,137	1,853	2,423
Share of IDA investment lending (%)	41	43	54
Share of all IDA lending (%)	32	28	43

Note: "Investment lending" excludes adjustment, debt and debt-service reduction operations, and emergency reconstruction operations.
Source: World Bank, *World Bank Annual Report, 1995* (Washington, D.C.: World Bank), p. 21. See also Poverty and Social Policy Department, World Bank, *Poverty Reduction and the World Bank: Progress and Challenges in the 1990s* (Washington, D.C.: World Bank Human Capital Development Division, 1996).

the World Bank is more often "directly targeting the poor"; however, several serious weaknesses in the indicator make it difficult to draw strong conclusions from this data. A project is included in the PTI "if it has a specific mechanism for targeting the poor and if the proportion of poor people among project beneficiaries is significantly larger than the proportion of the poor in the total population."[78] This definition permits Bank data to overestimate the relative weight of "directly targeted" lending operations in the portfolio. Even though only a small fraction of a given loan may be allocated to a targeted program, the Bank considers the *entire* loan to be poverty targeted. Assuming that targeting mechanisms actually target as designed, then aggregating these *subcomponents* would provide a more accurate sense of the degree to which the portfolio is poverty targeted. The second part of the official PTI definition is also very limited because it focuses on the percentage of the beneficiaries who are poor, rather than the percentage of the *loan benefits* actually received by the poor. For example, take a $100 million loan in a country where 50 percent of the population is poor. If 80 percent of the project beneficiaries are poor, and they receive 10 percent of the benefits, the *entire* loan is counted as a PTI project.[79] In sum, the volume of poverty-targeted lending may be growing, but it is also systematically exaggerated.

Internal Review: Forest Policy

Although the resettlement review team had the autonomy to commission their own field investigations to cross-check operational staff claims, the Bank's forest sector policy review did not. Like the NGO review of the energy sector portfolio, this review is based exclusively on official project documents and is therefore limited to information provided by interested parties: the staff responsible for the projects themselves. Similarly, it is also limited to projects still in their very early stages, though some had entered the implementation phase.

The forest sector review acknowledges that it is limited to the "*intentions* of work done since the issuance of the new forest policy" (p. iv, emphasis added).[80] In 1991, the Bank issued a forest policy that promised to: take into account the impact of nonforest projects on forests; promote sustainable forest development and conservation; strengthen institutions; rectify market failures; expand public participation; promote plantations (outside of intact natural forests); take a "precautionary approach" to logging in temporal and boreal forests; and no longer directly finance commercial rainforest logging (although most Bank impact on rainforests was not through direct logging projects). In a comparison between the 1984–1991 and 1991–1994 periods, lending for "protective and restorative activities increased from 7 to 27%, alternative livelihood support (including a few extractive reserves) rose from 1 to 14%, plantations fell from 32 to 23% and road construction fell from 10 to 0.4%." The report lauds the more extensive use of environmental assessments to gauge the forest impact of nonforestry projects (p. 38).

Although the report acknowledges that many nonforest policies affect deforestation, it does not consider the crucial issues of land tenure and agrarian reform outside of forest areas to be relevant.[81] It also alleges that log bans are ineffective at improving environmental stewardship (p. 10). It hints at possible negative impact from agricultural sector adjustment operations. The report admits a potential "sequencing problem" if extensive farming grows more quickly than intensive, employment-generating effects, but it doesn't actually examine any sectoral project effects on forests (p. 38). It acknowledges that social assessment and community participation are limited.

The report also indicates that forestry "governance" is a problem: "officials of the agencies responsible for reform often have strong personal motives for resisting change because of the rent-seeking opportunities created by distorted policies" (p. 13), and "powerful social classes ... can dominate decision-making" (p. 44). Not surprisingly, then, it concludes that forest management "institutions are failing" (p. 43), but there is little discussion of whether the Bank's effort to strengthen institutions might be bolstering the wrong institutions. The potential for business as usual persisting under the guise of new-style, green-sounding projects is evidenced by the experience of the Planafloro project. On paper, it appeared to be an example of organizational learning in terms of resource management and the protection of indigenous peoples' rights, but in practice it ended up repeating remarkably familiar problems—to the point of assigning the project to one of the same task managers responsible for the original Polonoroeste disaster.

Internal Review: Agricultural Pest Management

The World Bank first issued guidelines to regulate pesticide use in projects in response to a 1984 petition from more than two hundred NGOs. The 1985 *Pesticide Guidelines* announced that it would be policy to support integrated pest management (IPM) and to "aim to reduce dependence on chemical pesticides." An independent review of official descriptions of twenty-four World Bank projects funded from the beginning of the policy to 1988 found that the policy was ignored. IPM did not receive support, and all nine projects examined in detail actually continued or increased pesticide use.[82] Perhaps coincidentally, a major internal World Bank desk review begins where this study leaves off. Covering the 1988–1995 period, the Bank review analyzes ninety-five projects that involved pest management.[83] Forty-two of these projects involved pesticide purchases, and forty-eight claimed an integrated pest management component, but only twenty-two actually planned to implement an IPM approach. Eleven projects planned to use both approaches. Within this set of projects, $361 million was used to purchase conventional pesticides, whereas only $81 million funded on-farm integrated pest management. The vast majority of IPM funds ($51 million) were allocated to only two projects. A majority of the projects were not subjected to even partial environmental assessments. These findings suggest systematic noncompliance with both the Bank's pest management and environmental assessment policy directives, more than a decade after these directives were issued.

Internal Review: Environmental Impact Assessment

In 1995, Bank environmental staff conducted a desk-based review of compliance with a much broader policy—the linchpin of the "greening" of the Bank: its post–1989 environmental assessment (EA) policy. An earlier review covers the policy's first three years, indicating significant inconsistencies and little impact on projects. Because this period included the initiation of the policy, however, uneven implementation of such a new (for the Bank) methodology was certainly to be expected. The second review is more revealing of the limits and possibilities for reform because it covers the second three years of policy implementation (projects approved from fiscal years 1992 through 1995).[84]

The public summary of this review states that the Bank's EA "unfinished agenda" includes "implementing the portfolio of environmental projects ... [m]oving beyond project-specific environmental assessments ... [i]mproving the monitoring of on-the-ground impacts ... [and] [a]ddressing the social dimensions of environmental management."[85] The full, internal version of the review states that

EA is now a firmly rooted part of the Bank's normal business activity, effectively reducing the adverse environmental impacts of Bank-financed projects.... However, certain questions persist concerning the Bank's capacity to further improve the quality and effectiveness of EAs. In particular, there are questions about how to ensure adequate supervision of EA-related measures during project implementation, especially in light of the rapidly growing number of Category A [high environmental risk] projects that will enter the active portfolio over the next few years.[86]

The review indicates that EA quality improved, borrower government capacity to do the EAs increased, and progress was made in mitigation of direct impacts. However, "[t]he weakest aspects of EA work continue to be public consultation and analysis of alternatives" (p. ii). Supervision is also considered weak, which suggests that the Bank has limited information about the degree to which mitigation plans are actually implemented. A senior Bank environmental analyst confirmed that the quality of EAs has improved, but noted that the review downplays the systematic lack of implementation of recommended measures by borrowing governments.[87]

Overall, the 1992–1995 EA review found significant progress at the "end of the pipe," with much weaker performance in terms of key "upstream" EA processes—such as seriously considering alternatives, broader sectoral and regional EAs, as well as public transparency and participation. The EA compliance review is quite frank, but is limited by its reliance on internal Bank records rather than independent, field-based assessments. On the whole, the results show a significant degree of progress toward greater institutionalization of EAs but limited evidence of their actual impact. By 1995, however, a new trend emerged, as Bank management began planning to "reformat" mandatory operational directives into shorter "operation policies" and various nonbinding guidelines. In the view of Washington-based environmental NGOs and of U.S. Environmental Protection Agency policy analysts, however, this reformatting threatens to weaken the EA policy in important ways.[88]

Most of these reviews of reform policy compliance are limited to official Bank sources, however, so independent assessments of the *degree* to which policies were actually carried out by country, sector, and over time remain quite limited. The problems with desk-based reviews of compliance with reform policy are not limited to potential biases from World Bank sources, it turns out that these internal reports often lack reliable information on project outcomes. The Bank's Operations Evaluation Department carried out a study of the degree to which the twenty-year-old policy required projects to include monitoring and evaluation (M&E). The results were "disappointing.... The history of M&E in the Bank is characterized by non-compliance."[89] It found that basic project outcome information is systematically lacking or of poor quality. With such an inadequate information base, most desk-based reviews of policy implementation are inherently flawed.

Preventing Problems: Trip Wires, Back Channels, and Watchdogs

The three broad analytical dilemmas sketched out earlier in the chapter (i.e., the "good, bad and the ugly" in the portfolio, reviewing project campaign impact and Bankwide policy compliance, respectively) suggest that the outcome of civil society advocacy and protest is largely mediated by its impact on the shifting balance of forces within the both World Bank and borrowing governments. This concluding section reviews three kinds of institutional change mechanisms within the Bank in the context of their interaction with external critics and discusses the role of borrowing governments.

If the Bank's policies actually worked as written, its projects would be much less controversial. On paper, they are designed to prevent, or at least to channel, controversy. For example, one of the main potential results of social and environmental assessments is to fix or block projects early in the design process, before they gather bureaucratic and economic momentum. This potential for internal vetting brings up the dilemma of how to assess the changing portfolio mix. What are the factors that influence the degree to which impact assessments veto the "ugliest" projects early on, perhaps even before external critics manage to mobilize? The Bank's environmental staff are supposed to serve as internal "trip wires," alerting the institution to potential public relations disasters before they happen.[90] Although this function may serve the interests of the institution as a whole, it often conflicts with the interests of specific project task managers and their superiors, who are professionally rewarded for moving money quickly through the system. The capacity of environmental assessment staff to veto projects is therefore limited. In many of the most controversial projects, Bank staff alerted management to the social and environmental risks, but they were ignored. Because the trip wires do not always trip, they are necessary but not sufficient to encourage institutional change.

One of the main results of international advocacy campaigns is the increased number of World Bank staff dedicated to environmental and, to a lesser degree, social issues. Some of these staff are *reassigned* from other professional backgrounds and/or do not share NGO concerns. Others do, however, though to varying degrees. The number of institutional development specialists, who may be more likely than economists to encourage public participation and governmental accountability, remains very small. Although the social and environmental staff create at least a *potential* internal constituency for reform, their numbers alone do not necessarily give them influence over funding flows.

The Bank's internal institutional structure limits the power of insider reformists in several ways. The choice of environmental impact analyst is in the hands of an interested party—the official responsible for getting the project designed and approved. In turn, the budgets available for the in-house environmental impact analysts (in the technical departments) depend on the demand for their services. These analysts may therefore be discouraged from "biting the hand that feeds them." Project managers can contract outside consultants if in-house evaluators develop a reputation as too socially and environmentally rigorous. Independent minded environmental and social impact analysts are then unable to carry out direct field research, which in turn undermines the credibility of their critique, therefore making the implementation of promised mitigation measures a major problem. The structure of the environmental assessment process creates the public impression that the reviewers must formally approve or reject high-impact projects for the projects to proceed, but in practice the reviewers' role is usually limited to commenting on project design. Internal project reviewers rarely attempt to block projects, especially because formal veto power remains vested higher up, in the same managers who oversee those persons responsible for originally designing the projects (the regional vice presidents).[91]

When the "proper channels" for dealing with problem projects failed, environmental and social staff sometimes resorted to civil society "back channels." For example, confidential project information sometimes fell into the hands of concerned local and international NGOs, especially before the Bank's 1994 information disclosure reform. Indeed, it turns out that many early NGO campaigns against Bank projects were based initially on an insider tip-off. Valuable information flowed both ways. Such discreet information-sharing networks require high levels of personal trust between insiders and outsiders, who are to all appearances on opposite sides of highly contentious debates. Just as inside information empowers NGO critics, external pressure can empower insider reformists, as detailed in Fox's chapter. Nevertheless, many efforts to block or mitigate socially or environmentally destructive projects still fail, indicating powerful limits to both the institutionalization of trip wires and informal insider-outsider coalitions.

In-house Watchdogs: The Inspection Panel

Although social and environmental critics have pointed out weaknessess in reform policies as written, much of the recent public debate has focused on their lagging implementation (which varies greatly by policy, sector, and country, and over time). The reform policies have reshaped much of the institutional terrain on which NGO/grassroots protest unfolds, allowing critics to combine their own holistic criticisms of the basic logic of problematic projects with more technical "internal" critiques based on lack of compliance with the Bank's own standards. Noncompliance with reform policies turns out to be widespread in part because there are no systematic internal rules within the Bank for ensuring that staff consistently follow these reform policies. These directives are often time intensive and diplomatically challenging for technocrats used to dealing only with high-level government counterparts (who are usually at least as reluctant to encourage environmental impact assessments and informed local participation, if not more so). As the Bank's watershed Wapenhans Report suggested, most staff career incentives favor moving as much money as quickly as possible, thus encouraging merely pro forma application of these policies. The array of career carrots and sticks changed somewhat in 1993, however, with the creation of the Inspection Panel, which was designed to encourage Bank officials to meet the standards set by Bank policies.

The mandate of the Inspection Panel is to investigate when parties directly affected by projects submit complaints that official Bank policies were not followed. As Udall's chapter shows, the Inspection Panel's initial mandate was weaker and less independent of the Bank than its ad hoc predecessor, the Narmada case's Morse Commission. Yet the handling of the Arun III case suggests that the Panel's degree of autonomy is not predetermined. What remains to be seen is the degree to which the existence of a complaint channel can send an effective signal to Bank staff that they will be held accountable for not abiding by reform procedures that may conflict with more immediate career incentives. After all, the

Inspection Panel lacks the authority to choose its cases, to impose sanctions, or to provide compensation. It is clear, however, that most top Bank managers (and many board members) perceive the Inspection Panel as a threat, which suggests that it is having some effect. After its first year of activity, members of the Inspection Panel wondered whether the Bank's board would allow them to continue in the future. The board renewed the Inspection Panel's mandate in 1996, but seriously questioned its future in 1997.

In spite of its mission to increase World Bank accountability, however, the Inspection Panel has had a contradictory effect on efforts to encourage compliance with environmental and social policies. It was based on the premise that the reforms of the 1980s and 1990s set the standards against which the Bank could now be held accountable. Following the panel's inception, management argued that these policies were too detailed and unwieldy, and staff were therefore largely unfamiliar with many of their key provisions. They claimed that the policies needed to be "reformatted"—that is, separated into very brief mandatory sections (two pages)—and the "recommended" good practice section would then be much more extensive. As one senior manager recognized in an internal memo, "it has been hard for staff and managers to define clearly what is policy and what is advisory or good practice. *Our experiences with the Inspection Panel are teaching us that we have to be increasingly careful in setting policy that we are able to implement in practice*" (emphasis added).[92] As of 1997, it appears that the existence of an accountability mechanism provoked a powerful backlash in favor of watering down the Bank's own social and environmental policy standards.

Conclusions: Sustainable Development, Accountability, and the Pipeline Effect

Theoreticians often dismiss the concept of "sustainable development" as an oxymoron. Indeed, sustainable development is becoming all things to all people: a battle flag raised both by the technocrats defending the parapets of the besieged development institutions *and* by citizens' groups fighting on the front lines from civil society's trenches. Ideological contestation of the concept's legitimacy certainly continues, but now that the dominant international policy discourse has accepted sustainable development as a goal, the terrain of political conflict over environment and development issues has shifted significantly. In some countries and in some parts of the World Bank apparatus, citizen advocacy groups are now legitimate participants in the debate over *what counts* as an acceptable environmental assessment, reasonable access to project information, and appropriate grassroots participation.

In the 1980s, the dominant advocacy strategy of publicizing devastation after the fact held the moral high ground but won remarkably few tangible victories. Development disasters proved very difficult to stop. By the time the international alarm bells rang, rampant deforestation was well under way, or people forcibly evicted had already been immiserated and dispersed. Once launched, large projects inherently generate huge economic, political, and bureaucratic momentum. But the same accumulation of critical forces—the advocacy war of position that hammered the Bank via lobbying, mass media, legislatures, and grassroots direct action—was able to extract promises that the criteria and processes for making future lending decisions would change (without reversing project decisions already made). These promises took the form of new policies. The main impact of protest was therefore *indirect*—embedding new constraints, allies, and pressure points within the institution that were supposed to make *future* development disasters less likely. With increased access to project information earlier in the project cycle, and with the creation of an incipient mechanism through which the Bank can be held accountable for flouting its own policies, public interest groups now have greater leverage with which to try to prevent development disasters before they happen.

This chapter's review of studies of compliance with reform policies suggests that the impact of the Bank's reform policies is still quite limited. Therefore, the impact of protest remains limited. Before the creation of the Inspection Panel, there were no internal career incentives for Bank staff to follow the reform policies. Those staff members who tried to comply often encountered conflicts between their individual convictions and a powerful array of institutional disincentives and constraints. With the Inspection Panel, however, the implementation of the Bank's sustainable development reforms no longer depends exclusively on the good will of individual staffers. The Bank's dominant career incentive structure still encourages most staff to continue doing business as usual but the existence of a new public accountability mechanism creates an important potential counterweight.

The prospect that discontented locals could file a complaint about noncompliance with Bank policies provoked a wide range of reactions among staff. Some may rest assured that by the time a complaint makes its way through the system, they are likely to have been transferred to a different division or continent. There are still no mechanisms for holding accountable individual staff who flout social and environmental policies. Some staff have already invented the term *panel proofing* to describe preemptive measures such as pro forma consultations to create a defensible paper trail—just in case. Others take policy implementation more seriously—some merely adapting, others actually learning. The internal diversity within the Bank detailed in this volume's cases underscores the analytical importance of unpacking the institution to discover its distinct factions, interests, and ideologies. Most critics portray the Bank as monolithic, but this volume shows that one cannot explain the impact of protest without taking into account how external pressure is mediated by the Bank's internal policies, structure, and factions.[93]

The prospect that even limited policy reforms may make it possible to begin to hold the Bank accountable for its social and environmental damage creates new challenges for external critics. The Bank has hundreds of skilled professionals paid to design the new terrain of conflict, and many critics are understandably wary of hidden traps and dead ends. Yet the Inspection Panel's first case, Arun III, suggests that even limited policy reforms can lead to unexpected outcomes, especially when internal Bank conflicts create opportunities for greater public interest leverage. This leverage is likely to be greatest at the earliest stages of projects in the pipeline, before they generate bureaucratic and economic momentum. Both external critics and insider reformists are likely to have much less influence over projects well under way. This discrepency creates another challenge for public interest groups: how can they assess their own influence—not to mention the reliability of insider reformists—when tangibly devastating projects march on?

This question brings us back to the "pipeline effect"—the process whereby the Bank is *simultaneously* supporting ongoing projects created and implemented under an earlier set of rules and incentives, while designing future projects under a different set of rules. If the new set of rules has any impact at all, the new set of projects will be different, but to what degree? Studies of reform policy implementation suggest that the differences so far are minor. But if one returns to "the good, the bad, and the ugly" framework for assessing the Bank's portfolio, the differences may not be trivial. As a result of the social and environmental trip wires embedded in the institution as a result of public interest protest, "ugly" projects are more likely to get vetoed at such an early stage of the pipeline that they may barely reach NGO computer screens. The small category of "good" projects may grow, though most projects may still be considered "bad" from a sustainable development point of view, as the review of energy sector projects suggests.

If such a shift is under way, it poses a challenge for future efforts to hold the World Bank accountable for its actions. The most environmentally and socially outrageous projects of the 1980s gave the public interest groups their greatest leverage, but as the unambiguously "ugly" projects are increasingly vetoed or become less "ugly" (e.g., the contrast between Narmada and Arun), this strategy may become less effective. If insider-outsider reform coalitions are as important as this volume suggests, then more work needs to be done to strengthen the kind of critical cooperation that Covey outlines in her chapter because it would fill in the space on the political spectrum in between confrontational advocacy and uncritical collaboration.

The Arun experience also suggests that critics need to move beyond the strictly demarcated environmental and social arena and begin to assess the economic rationale of project decisions more systematically. The issue of the possible viability of alternative energy sector investments was crucial to the Arun cancellation, and the Bank's own analysis of its environmental assessment process recognizes that the evaluation of alternatives remains one of its major weaknesses. Because it is a bank, its lending decisions are primarily economic decisions. Sustainable development advocates both inside and outside the Bank have made some progress toward damage control and marginal "greening"; now they face the challenge of bringing social and environmental concerns to the center of the Bank's loan decision making.[94] Otherwise, the Bank may well succeed in avoiding the more politically costly projects, but still lend mainly for "more of the same."

To sum up, so far transnational advocacy coalitions have had more impact on policies than on projects. When policies are disregarded, they may be widely dismissed as window dressing, but they do set a standard to which the Bank can be held accountable. Accountability is determined more by bargaining power than by formal rules; but by setting minimum standards, however, those rules can empower challengers in potentially unexpected ways—as the Arun III Dam cancellation showed. Even the issue of who gets to participate in Bank project decisions is now on the table. The Bank's 1994 *Annual Report*—not one of its many publications designed primarily to mollify NGO critics—goes so far as to claim that "involving beneficiaries in project preparation is now beginning to become normal Bank procedure." Like its new recognition that good governance, transparency, and accountability are legitimate, this nominal acceptance of grassroots participation may open a Pandora's box for the Bank—as many staff undoubtedly fear. Some staff members argue that community participation should be limited to deciding where to lay the sewer pipes and then digging the trenches for free. The legitimization of grassroots participation in project decision making—like the acceptance of minimum standards for sustainable development—could lead to unexpected outcomes, however. Because of the pipeline effect, it is still difficult to tell.

Epilogue: What Drives the Implementation of Sustainable Development Reforms?

This volume has stressed the gap between official discourse and practice in the arena of environmental and social reforms at the World Bank. This gap is politically contingent, however, in the sense that its depth and breadth are determined by the relative balance of power between actors that support or oppose the implementation of the Bank's package of sustainable development reforms. This concluding discussion outlines an analytical framework for explaining the conditions under which the

World Bank can be held accountable to the minimum standards that it has recognized as legitimate.

The outcomes of interactions between the World Bank and civil societies are mediated largely by two other key sets of actors—government economic policymakers in donor and borrowing countries. The project and policy cases show how NGOs in donor countries use political access to their own nation-states as key levers over the World Bank and how the room to manuever for grassroots groups and NGOs in developing countries is conditioned by their respective national regimes. These cases offer diverse examples of transnational bargaining over resources within and between three intersecting arenas: the world's leading international development agency, diverse nation-states, and increasingly transnational civil societies. The cases show that the World Bank, nation-states, and civil societies (local, national, and international) *are all internally divided* over how to deal with the challenge of how and whether to promote sustainable development and public accountability. The main conceptual proposition here is that variation in project outcomes will be driven by bargaining processes that cut across state, civil society, and international actors. The degree to which reformists within states will be able to carry out reforms that increase institutional accountability will depend largely on their degree of support from outside allies (i.e., their mutually reinforcing interaction with pro-reform actors in organizations in other nations, internationally and within civil society). Similarly, the degree to which reformists within civil societies can reform their states will depend largely on their capacity to form broader transnational and national alliances. Internationally, the degree to which pro-accountability World Bank officials can implement their own reforms will depend on their capacity to bolster pro-reform interlocutors in both national states and civil societies. The specific coalitions needed to mitigate destructive action may be different, however, from those needed to promote positive environment and development policy.

This interactive approach informs the following attempt to depict the political dynamics that determine the nature and pace of implementation of sustainable development reforms. Figure 13.1 (p. 498) presents a stylized version of the North-South coalitions that drive both the formulation and implementation of sustainable development policies. Although this chapter has shown precisely how highly uneven and inconsistent the re-form process has been, this chart attempts to capture the political process that will determine the degree to which they might be implemented in the future.

Pro–sustainable development actors are defined in this chart in the very limited way: those actors from each arena who are committed to social and environmental reforms compatible with those promised by the Bank. The process begins in the two lower rectangles, as North-South NGO/grassroots coalitions begin to put the social and environmental costs of World Bank projects on the political agenda. Especially in the 1980s, most local organizations in borrowing countries had little lever-age over their governments, but their mobilization, authenticity, and credible alternative information bolstered their Northern NGO partners' efforts to encourage donor governments to pressure the World Bank for reform. Note that the shaded areas are not depicted to scale, but simply suggest that these transnational advocacy coalitions represent subgroups rather than entire societies, and that their relationships are often rooted in interlocking transnational wings of largely local or national movements.

Once North-South coalitions managed to put sustainable development reforms on donor government agendas, at best they managed to win over policy makers within the executive and legislative branches of their national governments—hence the shaded triangle on the left side of the chart. To the degree that they were able in turn to influence the World Bank, such impact was achieved mainly through their governmental representation on the board of directors, depicted as a horizontal bar above the World Bank itself. This body includes representation from both donor and borrowing governments, but is organizationally distinct from both individual governments and the World Bank apparatus itself. Pro–sustainable development reform supporters on the board of direc-tors rarely dominate votes; hence, they are depicted as a minority by the shaded area on the left-hand side of the horizontal bar. Because pro–sustainable development policymakers and NGOs in developing coun-tries rarely manage to influence their countries' representatives on the board of directors, this channel of influence is largely limited to Northern governments (hence, no pro-reform "support" arrow coming toward the board from borrowing governments on the right-hand side of the chart).

When reformists on the board do manage to exercise influence over the World Bank apparatus, it is largely by bolstering the power of pro-reform policy currents within the Bank itself, both through increased resources for potentially "good" projects and by reinforcing their authority over the operational staff (i.e., by strengthening mandatory reform policies). Insider reformists are depicted by the narrow triangle inside the Bank itself, sustained by the arrow of support from the upper left. They also often engage in mutual support relationships, overtly or implicitly, with transnational advocacy coalitions (relationships suggested by the two-way arrows in the center of the chart).

If and when Bank reformers manage to gain control over lending decisions and project design, they are well positioned to assign legitimacy and resources to pro-reform counterparts within borrowing governments (if there are any). Pro-reform national policymakers, depicted by the small shaded triangle on the right-hand side of the chart, in turn often have mutual support relationships with grassroots movements and NGOs in their countries, as suggested by the two-way arrows on the right-hand side. Each arrow depicting "political support" is implicitly accompanied by conflictive relationships—within civil societies, between civil societies and states, between states and the World Bank, and within the World Bank itself. The main thrust of this stylized picture is to underscore the importance of the contested balance of forces *within* as well as across diverse political arenas.

Against this backdrop, under what conditions will the Bank's growing category of potentially pro–sustainable development loans actually be able to meet reformists' ambitious goals? The framework depicted in figure 13.1 suggests a specific hypothesis that will hopefully be tested by future field-based research. The outcome of international sustainable development projects will depend on three conditions. To meet minimum sustainable development goals, projects must: (1) at the international level, be supported and controlled by committed reform elements within the international funding agency; (2) at the governmental level, be designed to target support specifically to agencies already controlled by reformist, pro-accountability elements within the state; and (3) within civil society, include informed participation by representative social organizations from the beginning of the design process. If *any one* of these three conditions is missing from the constellation of forces involved in the project, then it will likely fall short of even the World Bank's minimum sustainable development criteria. In other words, the impact of international sustainable development funding depends as much on the democratization of states and the mobilization of the underrepresented as on the intentions or interests of the World Bank.

Acknowledgements

Thanks very much for thoughtful comments from Peter Bosshard, Barbara Bramble, John Clark, John Gershman, Jo-Marie Griesgraber, Scott Guggenheim, Margaret Keck, Juliette Majot, Jane Pratt, Bruce Rich, Kay Treakle, Peter Van Tuijl, Warren Van Wicklin, and Robert Wade.

Notes

1. See chapter 11 by Udall and chapter 9 by Fox.
2. For an overview, see Lori Udall, "Arun III Hydroelectric Project in Nepal: Another World Bank Debacle?" Washington, D.C., International Rivers Network, March 1995. The project's disruption of a very precarious sub-subsistence economy would have put the very survival of villagers in the project area at direct risk. In spite of the World Bank's funding of social safety net projects elsewhere, the Arun project design was based on the gamble that market forces alone would be able to provide adequate income and food supplies to villagers. (Interview by Fox with long-time Arun observer Jane Pratt, former senior World Bank environmental advisor and director of the Mountain Institute, an NGO heavily involved in the Arun watershed, Washington, D.C., 31 October 1995).
3. Paul Lewis, "World Bank Cancels Nepal Project Loan," *New York Times*, 16 August 1995. The original Arun campaign in Germany was quite small, but it mobilized very influential actors, ranging from the mass media to trekker networks and the German-Nepal Friendship Societies (interviews with Heffa Scheuking, Urgewald (Germany), October 1995 and Bruce Rich, Environmental Defense Fund, Washington, D.C., December 1995)
4. See transcript, Martin Karcher, division chief for Population and Human Resources, Country Department I in the South Asia Region, World Bank, "Nepal's Arun Dam," interview by the Environmental Defense Fund Washington, D.C., 9 September 1994. See also Udall chapter, this volume.
5. See the preliminary conclusions by the Inspection Panel's in its December 16 memo to the World Bank's executive directors. The World Bank's board formally agreed that the Inspection Panel "should conduct an investigation into [the World

Bank's] ... adherence to its policies and procedures relating to Environmental Assessment, Involuntary Resettlement and Indigenous Peoples," but, in an odd catch-22, only allowed the panel to begin field research if the Nepali government decided to go ahead with the project (World Bank, "World Bank Board Authorizes Inspection of Nepalese Project," press release, Washington, D.C., 2 February 1995).

6. Inspection Panel member, interviews by Fox, Washington, D.C., November 1995.

7. World Bank, "Arun III Power Project," press release, Washington, D.C., 4 August 1995.

8. See David Price, *Before the Bulldozer* (Cabin John: Seven Locks, 1989), and Bruce Rich, *Mortgaging the Earth* (Boston: Beacon, 1994).

9. Wolfensohn consulted on Arun with Maurice Strong, who, in addition to his long service as an international environmental policymaker, was also a director of Ontario Hydro and therefore qualified to assess the project's economic viability (interview with Jane Pratt, Washington, D.C., 31 October 1995). For extensive background on Wolfensohn and his first months at the World Bank, see Garry Evans, "The World According to Wolfensohn," *Euromoney*, IMF/World Bank issue (September 1995), and Bruce Stokes, "The Banker's Hour," *National Journal*, 7 October 1995. See also Oxfam International, "Report Card on James Wolfensohn, June 1995–August 1996," Washington D.C.; Oxfam, August 1996. Their overall grade was a B– ("Promising starter, but World Bank action lags behind Wolfensohn rhetoric. Needs to focus more on equity and poverty").

10. The Wapenhans Report details how the Bank's project cycle and institutional priorities stress project approval more than successful implementation. See Portfolio Management Task Force, World Bank, *Effective Implementation: Key to Development Impact* (Washington, D.C.: World Bank, October 1992).

11. Patrick McCully, International Rivers Network, interview by Fox, Berkeley, CA, 16 August 1995. See also Paul Lewis, "World Bank Ends Heyday of the Big Project Loan," *International Herald Tribune*, 17 August 1995. For comprehensive critiques of the World Bank's role in dam building, see Leonard Sklar and Patrick McCully, *Damming the Rivers: The World Bank's Lending for Large Dams*, International Rivers Network Working Paper 5 Berkeley, CA, November 1994). and Patrick McCully, *Silenced Rivers: The Ecology and Politics of Large Dams* (London: Zed Books, 1996). Few international environmental groups have strong links in China. For one critique of World Bank funding of dams there, see "Spotlight on China," *Probe Alert* (September 1993).

12. Cited in Pratap Chatterjee, "Greens Laud World Bank Move to Scrap Nepal Dam Loan," InterPress Service, 8 August 1995.

13. Gopal Siwatoki, interview by Fox, Washington, D.C., 13 October 1995. Nepal is known for its emerging small-scale hydroelectricity industry. The NGO critique had stressed the proposition that a megaproject would benefit foreign contractors and would crush their own nascent hydroengineering capacity. Nepali

advocacy NGOs were nevertheless put on the defensive by pro-project nationalist critics (see Gopal Siwakoti, "No Foreign Hand in ARUN Campaign," and Rajendra Dahal, "To See a Foreign Hand is Nonsense," *Spotlight,* 24 February 1995).

14. Right-wing U.S. Bank critics might also have raised eyebrows at making a controversial loan to the recently elected (and short-lived) Communist-led minority parliamentary coalition government. It should be noted that right-wing critics of the World Bank are giving it a taste of its own medicine, calling for it to be privatized. See Nicholas Eberstadt and Clifford M. Lewis, "Privatizing the World Bank," *The National Interest* 40 (summer 1995). This potential threat creates an incentive for Bank management to fund the growing category of "green" and social projects because such projects distinguish the Bank from profit-maximizing private banks and therefore potentially broaden the constituency against privatization. See Adrian Wooldrige, "James Wolfensohn: Finding a New Role for the World Bank in the Global Economy," *Los Angeles Times,* 16 November, 1997.

15. See chapter 9 by Fox for further discussion of this issue.

16. World Bank resettlement officials, interviews by Fox, 18 January 1995. As one staffer commented, "How can you build dams and not move anyone? Pak Mun was much improved, and close to what we consider best practice." The dam destroyed fishing grounds, reportedly with the result that 2,500 fishers demanded compensation. According to one report, 50 percent of affected fishing families in fifty-three villages moved to the cities since construction (Rani Derasary, "Pak Mun Dam Destroys Fishing Communities," *World Rivers Review* 11, no. 1 (1996), p. 3.

17. See, for example, Carol Sherman, *Thailand's Energy Tentacles: Power Plants, Dams an Disaster Fuelling "Development" in Indochina* (Woollahra, Australia: Aid/Watch, May 1995).

18. Robert Ayres first applied a version of this metaphor to the Bank's anti-poverty shift in the 1970s in *Banking on the Poor* (Cambridge: MIT Press, 1983), pp. 110–111.

19. Peter Van Tuijl, formerly of the International NGO Forum on Indonesian Development (INFID), suggested that a fourth approach would be to assess whether a country portfolio has changed in response to civil society scrutiny, lobbying, and protest. Indeed, the Indonesian experience is by far the most advanced in this regard. Since 1985, INFID has been linking grassroots groups and national and international NGOs in an effort to influence multilateral development bank activity in Indonesia. In addition to its role in the Kedung Ombo Dam campaign discussed in Rumansara's chapter in this volume, INFID also managed to convince the World Bank not to fund future transmigration programs and to block a proposed "area development project" that would have threatened indigenous communities on Irian Jaya; it also made official population control programs "less coercive," according to Van Tuijl (interview by Fox, Washington, D.C., June 1996). Similar efforts in other countries are much less consolidated than INFID's, including those made by the Philippine Development Forum,

(discussed in Royo's chapter in this volume), whose activity peaked in the early 1990s, or by the Brazilian MDB Network, which (Rede Brasil) gained increasing strategic capacity by 1996.

20. This volume does not address the Global Environment Facility (GEF), in part because their projects are much smaller and usually less controversial than regular World Bank projects. The GEF is comanaged by the World Bank, the United Nations Environment Program (UNEP), and the United Nations Development Program (UNDP), and many critics suggest that it was "captured" by the Bank. Environmental NGOs have expressed concerns about the GEF, and local protests blocked GEF projects in Ecuador and Kenya in 1993. The main NGO criticism is that "unless the World Bank's annual lending of about $23 billion is consistent with protecting the environment, the GEF's planned yearly budget of $1 billion will barely be a Band-Aid" (Korinna Horta, "Environmental Band-Aid," *Boston Globe*, 23 May 1993). See, among other assessments: Christopher Plavin, "Banking Against Warming" *World Watch*, November-December 1997; UNEP, UNDP, and World Bank, *Report of the Independent Evaluation of the Global Environment Facility Pilot Phase* (Washington, D.C.: UNEP, ENDP, and World Bank, November 1993); Ian Bowles and Glenn Prickett, *Reframing the Green Window: An Analysis of the GEF Pilot Phase Approach to Biodiversity and Global Warming and Recommendations for the Operational Phase* (Washington, D.C.: Conservation International/Natural Resources Defense Council, 1994); and David Fairman, "Increments for the Earth," in Robert Keohane, ed., *Institutions for Environmental Aid: Pitfalls and Promise* (Cambridge: MIT Press, 1996). Analysts differ over the track record of a related multiagency, multilateral environmental aid mechanism. Compare Elizabeth DeSombre and Jeanne Kaufman, "The Montreal Protocol Multilateral Fund: A Partial Success Story," in Keohane, *Institutions,* with Steve Krezmann, *Money to Burn: The World Bank, Chemical Companies and Ozone Depletion* (Washington, D.C.: Greenpeace, September 1994).

21. For an analysis of the implications of turning poverty reduction projects over to local government, see Jonathan Fox and Josefina Aranda, *Decentralization and Rural Development in Mexico: Community Participation in Oaxaca's Municipal Funds Program* (La Jolla, Calif.: Center for U.S.-Mexican Studies, University of California, San Diego, 1996).

22. See Asa Cristina Laurell and Olivia López Arellano, "Market Commodities and Poor Relief: The World Bank's Proposal for Health", *International Journal of Health Services* 26, no. 1 (1996); Susanne Paul and James Paul, "The World Bank, Pensions and Income (In)Security in the Global South," *International Journal of Health Services* 25, no. 4 (1995).

23. See Oxfam International, *Multilateral Debt: The Human Costs,* position paper (Washington, D.C.: Oxfam, February 1996).

24. Frances Korten (Ford Foundation) has argued that this entire category of environmental (and by implication social sector) lending is fundamentally problematic because the loans must eventually be repaid in foreign exchange,

"creating pressures for exports that often involve damaging exploitation of natural resources.... The ... multilateral banks [are therefore] ... ill-suited to solving environmental problems" ("Questioning the Call for Environmental Loans: A Critical Examination of Forestry Lending in the Philippines," *World Development* 22, no. 7. [1994], p. 979). This argument is more relevant for loans from the World Bank's International Bank for Reconstruction and Development (IBRD) window, which are at near-commercial rates, than for IDA's very low-interest, long-term loans.

25. This is not the first time the Bank has made poverty a focus. During the presidency of Robert McNamara (1968–1981), the Bank also had an anti-poverty agenda (see chapter 1, note 29). Two main elements distinguish the current approach. First, more attention is now paid to targeting resources directly to the poor and underserved, mainly through providing public goods that benefit poor people rather than investing in agriculture per se (local elites find it easier to capture credit and fertilizer than rural clinics and schools). Second, there is currently an emphasis on examining the composition of national-level public spending in order to direct social spending more toward the poor (as part of the Bank's effort to influence national policy).

26. Michael Cernea, *Nongovernmental Organizations and Local Development*, World Bank Discussion Papers 40 (1988), p. 38.

27. The Bank announced in a press release at the World Summit on Social Development that it would increase social spending by 50 percent over three years (*World Bank News* 14, (no. 10, 9 March 1995). See also World Bank, *Annual Report* (Washington, D.C.: World Bank, 1995); *Advancing Social Development* (Washington, D.C.: World Bank, 1995) and *Investing in People* (Washington, D.C.: World Bank, 1995). According to the *Lancet*, the World Bank's 1993 "World Development Report," *Investing in Health*, represented "a shift in leadership on world health from the World Health Organization to the World Bank" (*Laucet* 342 10 July 1993).

28. On Bank-NGO collaboration, see Covey's chapter in this volume; Paul Nelson, *The World Bank and NGOs* (New York: St. Martins, 1995); the World Bank Operations Policy Group's annual progress reports (1995, 1996) (*Cooperation between the World Bank and NGOs* [Washington, D.C.: World Bank,]; and Tom Carroll, Mary Schmidt, and Tony Bebbington, *Participation through Intermediary NGOs*, Environment Department Papers, Participation Series 31 (Washington, D.C.: February 1996). More generally, see Bhuvan Bhatnagar and Aubrey Williams, eds., *Participatory Development and the World Bank* (Washington, D.C.: World Bank, 1992); Operations Policy Department, World Bank, *The World Bank and Participation* (Washington, D.C.: World Bank, September 1994), and the Department of Environmentally Sustainable Development, World Bank, *Participation Sourcebook* (Washington, D.C.: World Bank, February 1996).

29. See Daphne Wysham et al., "The World Bank and the G-7: Changing the Earth's Climate for Business" (Washington, D.C.: Institute for Policy Studies and

Sustainable Energy and Economy Network, 1997). According to Christopher Flavin of the World Watch Institute: "Bank officials believe that the IPS figures are overstated, and that fossil fuel project outweigh the greener ones by *six*-to-one. Even by that estimate, the Bank's energy loan portfolio is in no present danger of being labeled environmentally friendly." (Christopher Flavin, "Banking Against Warming," *World Watch*, November-December, 1997, p. 31.) See also Michael Philips, *The Least Cost Energy Path for Developing Countries: Energy Efficiency Investments for the Multilateral Development Banks* (Washington, D.C.: International Institute for Energy Conservation, 1991), and R. Govinda Rao, Gautam Dutt, and Michael Philips, *The Least Cost Energy Path for India: Energy Efficiency Investments for the Multilateral Development Banks* (Washington, D.C.: International Institute for Energy Conservation, 1991).

30. For contrasting views, see, for example, Development Gap, *The Other Side of the Story: The Real Impact of World Bank and IMF Structural Adjustment Programs* (Washington, D.C.: Development Gap, 1993; Nelson, *The World Bank and NGOs;* and Carl Jayarajah and William Branson, *Structural and Sectoral Adjustment: World Bank Experience, 1980–1992* (Washington, D.C.: World Bank Operations Evaluation, 1995) and note 9 for chapter 1 of this volume.

31. Cernea, *Nongovernmental Organizations*, p. 39.

32. See also Walden Bello, David Kinley, and Elaine Elinson, *Development Debacle: The World Bank and the Philippines* (San Francisco: Institute for Food and Development Policy, 1982), pp. 56–7, 85–6), and Operations Evaluation Department, World Bank, *Project Performance Audit Report: Chico River Irrigation Project,* unpublished report, (Washington, D.C.: World Bank, June 1989. Almost two decades later, these Igorot communities are again threatened with hydroelectric dam displacement, but this time led by private investors without apparent multilateral development bank funding.

33. See Robert S. Anderson and Walter Huber, *The Hour of the Fox: Tropical Forests, the World Bank and Indigenous People in Central India* (Seattle: University of Washington Press, 1988).

34. Korinna Horta, Environmental Defense Fund, interview by Fox, Washington, D.C., April 1996. See also Horta's "The Last Big Rush for the Green Gold: The Plundering of Cameroon's Rainforest," *The Ecologist* 21, no. 3 (May/June 1991).

35. See Andrea Durbin, "IFC Pulls Out of Shell Deal in Nigeria," *Bankcheck Quarterly* 13 (February 1996). See also Richard Richardson and Jonas Haralz, *Moving to the Market: The World Bank in Transition,* policy essay no. 17 (Washington, D.C.: Overseas Development Council, 1995), and Andrea Durbin, *The World Bank Group's Role in Private Sector Lending,* unpublished discussion paper (Washington, D.C.: Friends of the Earth, 1995).

36. For an overview of the environmental campaign against this project, see Richard Lowerre, *Evaluation of the Forestry Development Project of the World*

Bank in the Sierra Madre Occidental in Chihuahua and Durango, Mexico (Austin: Texas Center for Policy Studies, May 1994 [updated edition]).

37. Interviews, World Bank staff, Washington, D.C., May 1996, and Hilda Salazar, Desarrollo, Ambiente y Sociedad (a Mexican NGO), Mexico City May 1996). See Project Information Document, World Bank, *Mexico Aquaculture Project,* draft (Washington, D.C.: World Bank, March, 1996).

38. On Brazil's diverse anti-dam movements, see references listed in note 32, chapter 9.

39. See Catherine Good, "'Making the Struggle, One Big One,': Nahuatl Resistance to the San Juan Dam," paper presented to the Agrarian Studies Colloquium, Yale University, New Haven, Conn., 30 October 1992, and Jonathan Amith, *La tradición del Amate: Innovación y protesta en el arte mexicano/The Tradition of Amate: Innovation and Dissent in Mexican Art* (Albuquerque: University of New Mexico, 1996).

40. World Bank dam specialist, interview by Fox, Washington, D.C., October 1995.

41. To put the World Bank's leverage in context, the Mt. Apo project was important enough to the government's effort to provide cheap power for industry that it continued to drill without multilateral funding (in spite of the electricity surplus on the island of Mindanao).

42. See Bello et al., *Development Debacle,* pp. 108–18; Maria Anna de Rosas-Ignacio, *Collaborative Effort in Development: The Case of the Tondo Foreshoreland/Dagat-Dagatan Development Project,* case study (Boston: Institute for Development Research, 1991), and Ayres, *Banking on the Poor.*

43. In 1987, villagers resisted to the point where, according to an internal Bank assessment in a memo, "further resettlement was no longer tenable."

44. See sources listed in note 32, chapter 9.

45. Riverbank-dwellers won the right to be resettled near their own neighborhoods, for larger housing units and affordable terms of repayment (five years, at 15 percent of the minimum wage). These terms set a precedent for later urban infrastructure projects in Rio (Orlando Alves dos Santos Jr., director, Public Policy Unit, Federaçao de Orgaos Para Asistencia Social [FASE], interview by Fox, Managua, Nicaragua, June 1996). See also Jorge Florencio, Helio Ricardo Porto and Orlando Alves dos Santos Jr., *Saneamento ambiental na baixada: Cidadania e gestao democratica* (Rio de Janeiro: FASE/IAF, 1995).

46. See Thomas Wiens, "Philippines Integrated Protected Areas Project," in Environmentally Sustainable Development Department, World Bank, *Participation Sourcebook* (Washington, D.C.: World Bank, 1996).

47. Interviews, Victoria Corpuz, Asian Indigenous Women's Network; Joji Cariño, International Alliance of Indigenous and Tribal Peoples of the Tropical Forests; and Tony La Vina, Legal Rights and Natural Resources Center, Harper's Ferry, West Virginia, October 1995.

48. Interviews by Fox with the Inspection Panel, Washington, D.C., November 1995, and Stephan Schwartzman, Environmental Defense Fund, Washington, D.C., March 1996.

49. World Bank staff, interviews, Washington, D.C., April–May 1996. See Korinna Horta, "The Mountain Kingdom's White Oil: The Lesotho Highlands Water Project," *The Ecologist* 25, no. 6 (1995), and "Making the Earth Rumble," *Multinational Monitor* (May 1996).

50. Cindy Buhl, Indigenous Peoples Program, Bank Information Center, interview by Fox, Washington, D.C., June 1996.

51. Peter Bosshard, Berne Declaration (Swiss NGO), personal communication, May 1996.

52. See chapter 4 by Rumansara and World Bank, *Project Completion Report: Indonesia—Kedung Ombo Multipurpose Dam and Irrigation Project*, Ln. 2543-IND, unpublished (Washington, D.C.: World Bank, 1995).

53. See Anthony Hall's (a former Oxfam-U.K. staffer) detailed account, "From Victims to Victors: NGOs and the Politics of Empowerment at Itaparica," in Michael Edwards and David Hulme, eds., *Making a Difference: NGOs and Development in a Changing World* (London: Earthscan, 1992). See also R. Parry Scott's less optimistic account, "Dams, Forced Resettlement, and the Transformation of Peasant Economy in the Sao Francisco Valley, Brazil," unpublished ms. (1992). Aurélio Vianna and Lais Menezes, *O Pólo Sindical e a luta dos atingidos pela barragem de Itaparica* (Rio de Janeiro: CEDI/Koinonia - Pólo Sindical de Submédio São Francisco, 1994). The resettlers' difficult situation led them to file a complaint with the Inspection Panel in mid-1997, just before the loan was to close out.

54. Aurélio Vianna, Brazilian MDB Network (Rede Brasil), interview by Fox, Managua, June 1996 (former advisor to Itaparica rural workers movement).

55. According to Aurélio Vianna of the Brazilian MDB Network, civil society actors achieved a high degree of representation in project decision making (interview by Fox, Managua, June 1996). See also John Garrison, "The World Bank and Grassroots Participation," *Rain Forest Pilot Program* 4, no. 2 (1996).

56. See Public Information Document, World Bank, *Ecuador: Indigenous Peoples Development Project* (Washington, D.C.: World Bank Public Information Center, 1996).

57. See Michael Cernea, "Social Science Research and the Crafting of Policy on Population Resettlement," *Knowledge and Power* 6, no. 3–4 (1993), and "Social Integration and Population Displacement: The Contribution of Social Science," *International Social Science Journal* 143, no. 1 (1995).

58. See Rich, *Mortgaging the Earth*, pp. 136–137

59. *Ibid.*, pp. 34–38.

60. On the local and international critique of the project, see the joint letter to the president of the World Bank sent by the Environmental Defense Fund, 21 August 1987; and D. M Chandrashekhar et al., "Social Forestry in Karnátaka: An Impact Analysis," *Economic and Political Weekly* 22, no. 24, 13 June 1987. According to an internal World Bank evaluation, in spite of its nominal "satisfactory" rating, the project failed to meet its anti-poverty goals. The only positive outcomes "resulted largely from beneficiaries' independent responses to market conditions and external pressure from environmental groups on the Bank and the Borrower. Project sustainability is rated as unlikely." Operations Evaluation Department, World Bank, *Project Completion Report: India, Karnataka Social Forestry Project*, credit no. 1432-IN, unpublished (Washington, D.C.: World Bank, June 1993).

61. Korinna Horta, Environmental Defense Fund, interview by Fox, Washington D.C., April, 1996. As of mid-1997, the World Bank was actively considering lifting its ban on rainforest logging, in the name of "sustainable forestry."

62. According to one of the campaign strategists, "the [Board's] decision to seek an intermediate solution involving the Inspection Panel (instead of simply rejecting the inspection request) could bring significant improvements for the implementation of Planafloro." Brent Millikan, "Planafloro em Rondônia: Desafios do desenvolvimento sustentavel," *Políticas Ambientales* 11 (April 1996), p. 9. [translation by Fox]

63. Glenn Switkes, "Bank Rejects Chilean Citizens' Claim," *Banckcheck Quarterly* 13 (February 1996). See Jay Hair, et al., "Pangue Hydroelectric Project (Chile): An Independent Review of the International Finance Corporation's Compliance with Applicable World Bank Group Environmental and Social Requirements," (Santiago, Chile: Pangue Audit Team, 4 April, 1997 [the IFC released an edited version to the public]).

64. See Lori Udall, "Bank to Review Freeport Operations," *Bankcheck Quarterly* 13 (February 1996). To avoid international scrutiny, Freeport subsequently ended its MIGA insurance.

65. The official operational directives are also available from the Bank's Public Information Center [pic@worldbank.org]. See also Bank Information Center, *A Citizen's Guide to World Bank Environmental Assessment Procedures* (Washington, D.C.: BIC, 1992), and Cindy Buhl, *A Citizen's Guide to the Multilateral Development Banks and Indigenous Peoples* (Washington, D.C.: Bank Information Center, 1994) [bicusa@igc.apc.org].

66. Environmental Defense Fund and the Natural Resources Defense Council, *Power Failure: A Review of the World Bank's Implementation of its New Energy Policy* (Washington, D.C.: EDF, NRDC, 1994). For the World Bank energy policy paper, see *Energy Efficiency and Conservation in the Developing World: The World Bank's Role* (Washington, D.C.: World Bank, 1993).

67. World Wildlife Fund (Sweden), "A Megawatt Saved ... (ar en Megawatt Tjanad): Implementation of the World Bank's Energy Policy," unpublished draft (February 1996). See also Wysham, "The World Bank and the G-7."

68. See Office of Environmental and Scientific Affairs, *Tribal Peoples and Economic Development: A Five-Year Implementation Review of OMS 2.34 (1982–1986)*, unpublished (Washington, D.C.: World Bank, 1987).

69. See Jonathan Fox, "The World Bank and Social Capital: Contesting the Concept in Practice," *Journal of International Development* 9, no. 7 (1997). For a full list of the World Bank's rural development–related projects in Mexico, almost all of which eluded the application of operational directive 4.20, see Fox and Aranda, *Decentralization and Rural Development*, and the ongoing work of the Mexican NGO Trasparencia (www.laneta.trasparencia).

70. See Environment Department, World Bank, *Resettlement and Development* (Washington, D.C.: World Bank, 1996).

71. Quotes from Josette Murphy, *Gender Issues in World Bank Lending* (Washington, D.C.: World Bank Operations Evaluation Department, 1995), pp. 34, 3, 56, and 59. The official policy paper is *Enhancing Women's Participation in Economic Development* (Washington, D.C.: World Bank, 1994). See also Nancy Alexander, *Gender Justice and the World Bank* (Silver Spring, MD: Bread for the World Institute, September 1996); Myra Buvanic, Catherine Gwin, and Lisa Bates, *Investing in Women: Progress and Prospects for the World Bank*, policy essay no. 19 (Washington, D.C.: Overseas Development Council/International Center for Research on Women, 1996); Chris Chamberlain, *A Citizen's Guide to Gender and the World Bank* (Washington, D.C.: Bank Information Center, 1996); Nuket Kardam, *Bringing Women In: Women's Issues in International Development Programs* (Boulder, Colo.: Lynne Reinner, 1991); Shahra Razavi and Carol Miller, *Gender Mainstreaming: A Study of the Efforts by the UNDP, World Bank, and ILO to Institutionalize Gender Issues*, UNRISD Occasional Paper, 4 (Geneva: United Nations Research Institute on Social Development, 1995); Veena Siddharth, "Gendered Participation: NGOs and the World Bank," *Institute for Development Studies Bulletin* 26, no. 3, 1995; and Operations Evaluation Department, World Bank, *Population and the World Bank: Implications from Eight Case Studies* (Washington, D.C.: World Bank, 1992). Note also the ongoing monitoring by the Women's Eyes on the Bank NGO campaign and the External Gender Consultative Group. (See the Latter's "Recommendations to the World Bank," Washington, D.C., August 1996 [unpublished memo].)

72. World Bank, *Implementing the World Bank's Gender Policies*, progress report no. 1 (Washington, D.C.: World Bank, 1996), p. 7.

73. For example, one might wonder about the gender impact of a 1994 basic education project in Nicaragua that proposed to support "curriculum reform" to "promote family values" (*ibid.*, p. 12). On World Bank credit projects, see Lynn Bennett and Mike Goldberg, *Providing Enterprise Development and Financial Services to Women: A Decade of Bank Experience in Asia*, technical paper no. 236, Asia Technical Department (Washington, D.C.: World Bank , 1993).

74. From Buvanic, *Investing in Women*, pp. 19–20.

75. This section draws on John Gershman and Jonathan Fox, "Taking Aim at 'Poverty-Targeting,'" *Bankcheck Quarterly* 14 (May 1996).

76. World Bank, *From Vision to Action in the Rural Sector* (Washington, D.C.: World Bank, March 1996).

77. World Bank External Affairs staff, interview by Fox, Washington, D.C., February, 1996.

78. World Bank, *Poverty Reduction and the World Bank: Progress in Fiscal 1994* (Washington, D.C.: World Bank, 1995), annex 1, p. 27.

79. Although table 13.4 is limited to investment loans, the PTI category is also applied to structural adjustment loans. An adjustment loan qualifies if it meets one of the following criteria: (1) reforms social expenditures to better reach the poor; (2) removes distortions of particular harm to the poor; (3) contains safety nets or other targeted programs; (4) introduces poverty monitoring; or (5) develops a poverty policy. Assessing the poverty focus of adjustment lending is also problematic. According to *Poverty Reduction and the World Bank, Fiscal 1994* (Washington, D.C.: World Bank, 1994), and the World Bank Annual Report for 1994, seventeen of twenty-three adjustment operations in fiscal 1994 were poverty focused, compared with a third in fiscal 1993 and with just more than half in fiscal 1992. But the criteria for determining poverty focus are very weak. If an adjustment loan merely funds the *counting* of the poor (poverty monitoring), then the entire loan would count as a PTI loan. Meanwhile, the policies promoted by the loan could actually worsen poverty.

80. See Agriculture and Natural Resources and Environment Departments, World Bank, *Review of Implementation of the Forest Sector Policy* (Washington, D.C.: World Bank, 1994). For an earlier official evaluation of previous projects and their relationship with the 1978 policy paper, see Operations Evaluation Department, World Bank *Forestry: The World Bank's Experience* (Washington, D.C.: World Bank, 1991).

81. See Marcus Colchester's critique: "Towards Partnership? Community Participation in World Bank Forestry Projects. Comments on the World Bank's 'Forest Policy Implementation Review' (April draft)," World Rainforest Movement (May 1994). He stresses the lack of attention to indigenous rights in many forest projects and concludes:

What has struck me most forcefully about this document is its literature based and anecdotal approach. In the first place, no field visits were undertaken to check out even a handful of projects to see if the documentation available to the reviewers adequately reflects the reality on the ground. In the second place, the review appears to be very unsystematic. It reads as if the reviewers have trawled through the documents related to the Bank's recent forestry lending and then have picked out a project or two here and there to illustrate points that they wish to make.... It makes it very hard to gauge from the review whether commitments made in the 1991 Policy have been adhered to by none, some, many of all, the Bank's projects.

For an overview of the links between land tenure and forest issues, see Marcus Colchester and Larry Lohmann, eds., *The Struggle for Land and the Fate of the Forests* (Penang, Malaysia: World Rainforest Movement, Ecologist, Zed Books, 1993).

82. Michael Hansen, *The First Three Years: Implementation of the World Bank Pesticide Guidelines, 1985–1988* (Mt. Vernon, N.Y.: Consumer Policy Institute, Consumers Union, 1990).

83. See Tjaart W. Schillhorn van Veen, et al., *Integrated Pest Management: Strategy and Policy Options for Promoting Effective Implementation,* draft (Washington, D.C.: World Bank, Environmentally Sustainable Development Department, 1997). See also the critical NGO letter commenting on the "reformatted" pest policy (Pesticide Action Network et al., to Douglas Forno, Agriculture Division Chief, World Bank, 24 May 1996) and Consumer Policy Institute, Environmental Defense Fund and Pesticide Action Network, "Over 100 NGOs Worldwide Condemn World Bank for Cutting its Pesticide Policy," press release (Washington, D.C., 11 November 1996).

84. Land, Water, and Natural Habitats Division, Environment Department, World Bank, *The Impact of Environmental Assessment,* final draft, unpublished (Washington, D.C.: World Bank, August 1995).

85. See Environment Department, World Bank, *Mainstreaming the Environment,* p. 3.

86. High-impact or Category A projects *increased* their share of the nonadjustment portfolio from 11 percent in fiscal year 1991 to 24.5 percent in fiscal year 1994 (*ibid.,* p. 10). This increase may be due in part to better screening.

87. World Bank analyst, interview by Fox, Washington, D.C., April 1996.

88. For example, the December 1995 draft revised policy removed a clear requirement that the EA must be done before project appraisal, weakened analysis of "cumulative impacts," weakened the requirement to use independent experts in controversial cases, and reduced Bank responsibility for EAs in sector investments. On the other hand, the draft language on public consultation is strengthened from a Bank "expectation" to a "requirement." See letter from Center for International Environmental Law and other NGOs to Andrew Steer, director of Environment Department, World Bank, December 1995.

89. Operations Evaluation Department, World Bank, *An Overview of Monitoring and Evaluation in the World Bank,* report no. 13247 (Washington, D.C.: World Bank, June 1994), p. iii. See also Jonathan Fox, "Transparency for Accountability: Civil Society Monitoring and Multilateral Development Bank Anti-Poverty Projects," *Development in Practice* 7, no. 2 (1997).

90. Former senior environmental advisor Jane Pratt underscored the need for an institutionalized "Lorax" who would speak for the trees (citing Dr. Suess). Jane Pratt, interview with Fox, Washington, D.C., 31 October 1995.

91. For an overview of internal accountability problems, see Bruce Rich, *The World Bank: Institutional Problems and Possible Reforms* (Washington, D.C.:

Environmental Defense Fund, March 1995). For updates on changes in internal organization, see Nancy Alexander, "World Bank's Strategic Compact," *News and Notices for Bank Watchers*, no. 17, April 1997, and her "Chaotic Reorganization at the World Bank," *News and Notices for Bank Watchers*, no. 18, August 1997. Jane Pratt recalled that the original conception of the environmental assessment process, after the Bank's 1987 reorganization, did nominally give reviewers the responsibility of approving or rejecting projects. However, when the senior environmental official in the Latin American region began to attempt to veto problem projects, he was removed from operational authority over projects, which had a chilling effect on the whole environmental project review process.

92. Myrna Alexander, OPRDR, Operations Policy memo, on "Conversion of Remaining ODs," 15 March 1996, emphasis added.

93. For example, Susan George and Fabrizio Sabelli present a sophisticated explication of the dominant Bank ideology as a virtually religious dogma (*Faith and Credit* [Boulder, Colo.: Westview, 1994]. This approach helps to explain continuity, which is certainly the dominant trend, but it does not help to explain degrees of partial change. In other words, the assumption that the Bank is monolithic makes it difficult to explain advocacy victories. Claiming that the Bank was "forced" to make a concession is insufficient. External protest does not necessarily "force" concessions from large, authoritarian institutions; whether to respond with concessions or hardening depends on the impact of protest on the internal balance of forces. If protest strengthens an insider reform faction, then concessions are more likely.

94. Some sustainable development advocates have long argued that many destructive projects would lose their economic logic if the social and environmental costs were fully taken into account (e.g., loss of "natural capital," such as forests, or just compensation for involuntary resettlement) and if more socially and environmentally appropriate alternatives were considered. Insider environmentalists question the conventional criteria for valuing resources (see Ismail Serageldin and Andrew Steer, eds., *Valuing the Environmental: Proceedings of the First Annual Conference on Environmentally Sustainable Development* [Washington, D.C.: World Bank, 1994). One notable ecologist resigned, however, in frustration over the limited results so far (see Herman Daly, "Farewell Lecture to the World Bank," in John Cavanagh, Daphne Wysham, and Marcos Arruda, eds., *Beyond Bretton Woods* (London: Pluto, TNI, IPS, 1994).

Index